# INDUSTRIAL APPLICATIONS OF FUZZY TECHNOLOGY IN THE WORLD

## ADVANCES IN FUZZY SYSTEMS – APPLICATIONS AND THEORY



---

Vol. 1: Between Mind and Computer: Fuzzy Science and Engineering
(*Eds. P.-Z. Wang and K.-F. Loe*)

Vol. 2: Industrial Applications of Fuzzy Technology in the World
(*Eds. K. Hirota and M. Sugeno*)

Vol. 3: Comparative Approaches to Medical Reasoning
(*Eds. M. E. Cohen and D. L. Hudson*)

Vol. 4: Fuzzy Logic and Soft Computing
(*Eds. B. Bouchon-Meunier, R. R. Yager and L. A. Zadeh*)

*Forthcoming volumes:*

Vol. 5: Fuzzy Sets, Fuzzy Logic, Applications
(*G. Bojadziev and M. Bojadziev*)

Vol. 6: Fuzzy Sets, Fuzzy Logic, and Fuzzy Systems: Selected Papers by Lotfi A. Zadeh
(*Eds. G. J. Klir and B. Yuan*)

Vol. 7: Genetic Algorithms and Fuzzy Logic Systems: Soft Computing Perspectives
(*Eds. E. Sanchez, T. Shibata and L. A. Zadeh*)

Vol. 8: Foundations and Applications of Possibility Theory
(*Eds. G. de Cooman, D. Ruan and E. E. Kerre*)

Advances in Fuzzy Systems – Applications and Theory Vol. 2

# INDUSTRIAL APPLICATIONS OF FUZZY TECHNOLOGY IN THE WORLD


***Editors***

**Kaoru Hirota**

**Michio Sugeno**

*Tokyo Institute of Technology*



**World Scientific**

*Singapore • New Jersey • London • Hong Kong*

*Published by*

World Scientific Publishing Co. Pte. Ltd.

P O Box 128, Farrer Road, Singapore 9128

*USA office:* Suite 1B, 1060 Main Street, River Edge, NJ 07661

*UK office:* 57 Shelton Street, Covent Garden, London WC2H 9HE

**Library of Congress Cataloging-in-Publication Data**

Industrial applications of fuzzy technology in the world / editors,
Kaoru Hirota, Michio Sugeno.
p. cm. -- (Advances in fuzzy systems ; vol. 2)
Includes bibliographical references.
ISBN 9810223668
1. Automatic control. 2. Fuzzy systems. I. Hirota, K. (Kaoru)
II. Sugeno, Michio, 1940– . III. Series.
TJ225.I53 1995
629.8'3--dc20 95-35584
CIP

**British Library Cataloguing-in-Publication Data**

A catalogue record for this book is available from the British Library.



Printed in Singapore.

# PREFACE

It was L. A. Zadeh, of the University of California, who first proposed the concept of fuzzy sets in 1965. In the 1980s came practical applications, led by Japan, mainly in the field of process control. In the 1990s there has been a rush of applications to household electrical appliances, and the term "fuzzy" has become a high-tech fad in Japan. This decade has also seen North American and European nations becoming involved with the industrial applications of fuzzy technology. Very recently, the development of fuzzy theory itself has become more closely linked with other related fields such as neural networks and chaos.

Such trends in the industrial application of fuzzy technology are summarized in this volume. The topics include fundamentals of industrial fuzzy applications, recursive fuzzy inference, fuzzy chaos, and neuro-fuzzy systems. Various practical examples are covered, e.g. extra-corporeal circulation, cranes, speech recognition, automobiles, home electronics and space. The contributors are leading experts from various countries.

We hope the book will be a great help to those who have an interest in the applicational aspects of fuzzy theory, e.g. senior or graduate students and researchers in industry.

Finally, the editors would like to express their appreciation of the efforts of the contributors, and the support they have received from World Scientific Publishing Co., especially Mr Yew Kee Chiang for his help with the editorial process.

Kaoru HIROTA
Michio SUGENO
*Tokyo Institute of Technology*

# CONTENTS

# INDUSTRIAL FUZZY CONTROL REVIEW: A PERSPECTIVE FROM FEEDBACK AND MANUFACTURING


SATORU ISAKA

*Omron Advanced Systems, Inc.,*
*3945 Freedom Circle, Suite 700,*
*Santa Clara, CA 94054, USA*

VICTOR K. CHU

*Department of Mechanical Engineering,*
*University of California,*
*Berkeley, CA 94720, USA*


## 1. Introduction

The foundation of industrial control systems is feedback. Whether at the manufacturing or the process control, discrete events or continuous processes, feedback control allows automation of tasks. As stated by Otto Mayr[18] "Every animal is a self-regulating system owing its existence, its stability and most of its behavior to feedback controls." A feedback control system is a control system which tends to maintain a prescribed relationship of one system variable to another by comparing functions of these variables and using the difference as a means of control, according to the American Institute of Electrical Engineering.

The origin of feedback control goes way back to the third century BC in Alexandria. A Greek inventor named Ktesibios built a water clock, the earliest known construction of a device for feedback control.[18] The water clock measures the passage of time by means of a constant water flow into a tank. An indicator riding on the water tells the time as the water level rises (see Fig. 1). The engineering problem here is how to maintain the water flow rate at a constant. Ktesibios solved this problem by inventing a feedback device that resembles the modern automobile carburetor. A float valve inside a secondary vessel regulates the incoming water flow rate. When the supply water flow rate

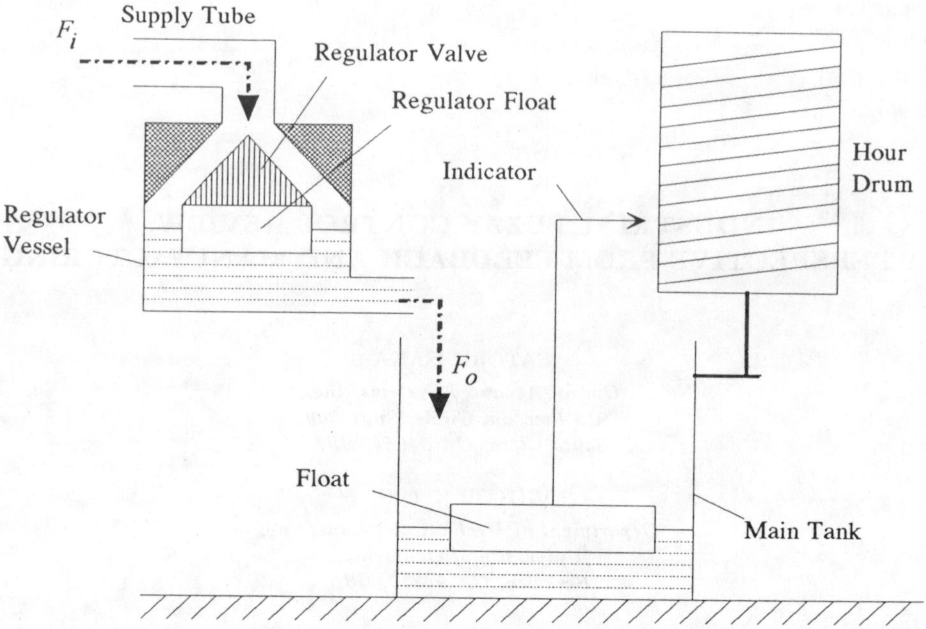


Fig. 1. This water clock was designd by a Greek inventor named Ktesibios in the third century BC. The time is marked by the indicator which points to the hour drum. The indicator rises when the water level of the main tank rises. The regulator on the left hand side functions as a simple feedback controller. It was designed to keep the outgoing flow rate, $F_o$, at a constant.

increases, the water level at the regulating vessel rises and so does the float. A valve attached to the float closes just enough to reduce the supply flow rate. When the water level is low, the float valve opens just enough to increase the supply flow rate (see Fig. 2).

Simple though it may seem, the feedback mechanism of Ktesibios displays a fundamental design principle in control systems. In particular, our interest lies in the control logic, for instance, *if the water level is low, then increase the valve opening.* We, as control engineers, attempt to build a feedback control system by exploiting control logic. The logic is not in the sense of formal logic, but it is rather an informal description law, that is based on the knowledge about the process characteristics as well as objectives. Control designers attempt to find an effective and feasible representation formula for the logic. The complexity depends somewhat upon applications. However, the point of interest here is the fact that the basis of control system design is the logic within the problem domain. Heuristic control system is a result of an explicit exploitation of this design philosophy in a representation formula by rules. In the last 10 years, a type of heuristic control — fuzzy control — has gained popularity in industry.

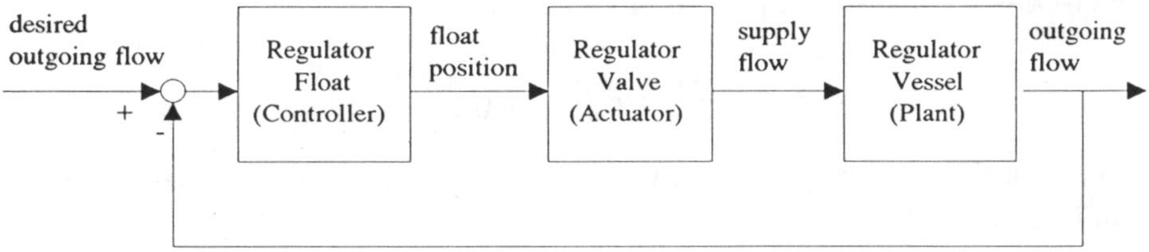


Fig. 2. Feedback control diagram for Ktesibios' regulator.

This chapter reviews the recent applications of fuzzy control in industry from the perspective of feedback and manufacturing. Our interest is to learn how we, as control engineers, utilize domain theories in control system design. In industrial applications, control must be viewed in the context of the overall objective. In a manufacturing environment, to produce a product is not just to keep a single variable at a given operation point, but to meet various conditions such as efficiency and safety. The challenges here are that most processes being controlled are not well-defined mathematically, and many processes involve the control of more than one variable.[21] We will see how explicit exploitation of domain theory benefit some difficult problems in industrial control systems. In the next section, we will begin our study by reviewing basic approaches in feedback design.

## 2. Conceptual Approaches in Feedback Control System Design

### 2.1. *Linear control approach*

Let us take a closer look at the Ktesibios' water clock. The engineering problem is to adjust the supply water flow rate $F_i(t)$ so that the outgoing water flow rate $F_o(t)$ remains constant. The mass balance equation gives the dynamics of the regulating vessel:

$$\frac{dV(t)}{dt} = F_i(t) - F_o(t) \tag{1}$$

where $V(t)$ is the volume of the water in the regulating vessel. The outgoing flow rate $F_o(t)$ depends upon the water volume $V(t)$. Assuming that the vessel has constant cross-sectional area, $S$, we can write

$$F_o(t) = k\sqrt{\frac{V(t)}{S}} \tag{2}$$

where $k$ is a constant. Equation (1) now becomes

$$\frac{dV(t)}{dt} = F_i(t) - k\sqrt{\frac{V(t)}{S}} \,. \tag{3}$$

Let us consider a steady-state situation where all quantities are constant. From Eq. (1),

$$0 = F_{io} - F_{oo} \tag{4}$$

where $F_{io}$ is the steady state value of $F_i(t)$ and $F_{oo}$ is the steady state value of $F_o(t)$. From Eq. (2), we obtain

$$F_{oo} = k\sqrt{\frac{V_o}{S}} \tag{5}$$

where $V_o$ is the steady state value of $V(t)$. Let us assume that only small deviations from steady-state conditions occur. We write

$$F_i(t) = F_{io} + u(t)\,, \tag{6}$$

$$V(t) = V_o + x(t)\,, \tag{7}$$

where we consider $u(t)$ an input variable and $x(t)$ a state variable. By assuming these quantities are small, we can linearize Eq. (3) and obtain

$$\frac{dx(t)}{dt} = u(t) - \frac{k}{2V_o}\sqrt{\frac{V_o}{S}}\,x(t)\,. \tag{8}$$

Substituting Eq. (5) into Eq. (8) gives

$$\frac{dx(t)}{dt} = -\frac{F_{oo}}{2V_o}\,x(t) + u(t)\,. \tag{9}$$

We define the output variable by

$$\begin{aligned} y(t) &= F_o(t) - F_{oo} \\ &= \frac{F_{oo}}{2V_o}\,x(t)\,. \end{aligned} \tag{10}$$

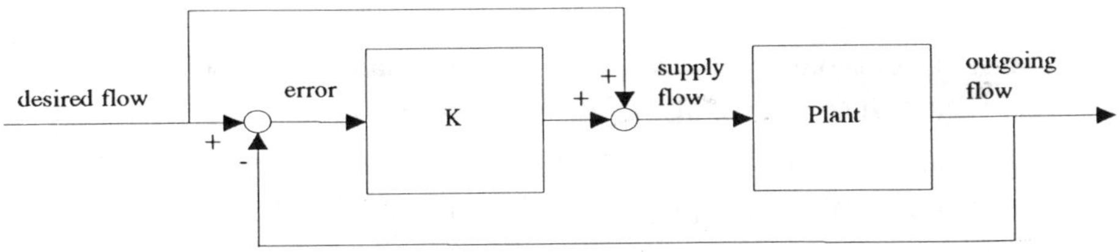


Fig. 3. Feedback control diagram for the water clock regulator by a proportional gain control law.

Equation (9) is the linearized state differential equation for the feedback system of Ktesibios. Equation (10) is the linearized output equation.

By using the state equation, we can construct a feedback controller to regulate the outgoing water flow rate at constant. We may employ a simple control strategy, proportional gain (Fig. 3). The strategy is that if the outgoing flow rate is too small, then adjust the supply water flow rate proportionally to the difference between the actual flow, $F_o(t)$, and the desired flow, $r$. The control law is now written as

$$\begin{aligned} u(t) &= K(r - F_o(t)) \\ &= K(F_{oo} - F_o(t)) \\ &= -K\,y(t)\,. \end{aligned} \tag{11}$$

The value for $K$ depends on the specification of desired performance. In case of the linear single-input–single-output system with asymptotically stable property, choosing $K$ is not a difficult task. However, for a multivariable system, the choice of proportional gains is a complex task, and it is typically approached by a trial-and-error method.[15] This is the design method based on the linear control theory.

## 2.2. *Fuzzy control approach*

Now, let us consider an alternative approach, expert system and fuzzy logic. Suppose that we cannot construct a state differential equation such as Eq. (9) because of various reasons. Perhaps, we do not know the mass balance equation. However, we do know how to regulate the water flow. Let us exploit the knowledge to construct a feedback control law. Suppose we have the following control rules:

> If the outgoing flow rate $F_o(t)$ is larger than the desired rate $r$, then reduce the supply water flow rate $F_i(t)$.
>
> If the outgoing flow rate $F_o(t)$ is smaller than the desired rate $r$, then increase the supply water flow rate $F_i(t)$.
>
> If the outgoing flow rate $F_o(t)$ is equal to the desired rate $r$, then maintain the current supply water flow rate $F_i(t)$.

These control rules can be implemented as a heuristic control, but we need to quantify the matter of degree in control variables. This is where fuzzy logic is applied. Fuzzy logic is concerned with handling assertions like "flow rate is high" whose truth is a matter of degree. We can construct a set of fuzzy sets to characterize the degree of difference between the outgoing flow and the desired flow. By associating the fuzzy sets based on the control logic, we obtain a fuzzy controller (Fig. 4).

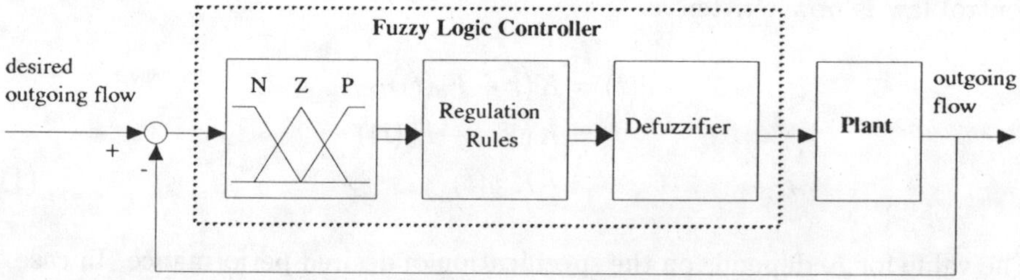


Fig. 4. Feedback control diagram for the water clock regulator by a fuzzy logic control law. A fuzzy logic controller is in essence a nonlinear controller. In the diagram, it is composed of three blocks — fuzzification, rule inference, and defuzzification.

Although building basic rules and membership functions may be relatively easy, tuning the rules and membership functions is not a simple task. This is the Achilles' heel of a fuzzy system in practice.[14] Trial-and-error method is still a common way. There have been, however, many reports on tuning methodologies recently. Many of them employ parameter estimation or numerical optimization techniques, such as neural networks and genetic algorithms. This is the design approach based on expert knowledge and fuzzy logic.

## 2.3. *Classic approach*

Let us reconsider how Ktesibios built his controller. It is safe to assume that Ktesibios did not use the state differential equation of (9) to construct his mechanical feedback controller. However, he knew the relationships among the water volume, the supply flow, and the outgoing flow. He specifically knew that the constant water volume in the vessel yields a constant outgoing flow. He also knew that the constant supply flow yields a constant outgoing flow. He was aware that the supply flow is subject to disturbance, thus a regulating device was necessary. His knowledge of the water flow led to the construction of the water volume measurement device (regulator float) and the supply flow valve. He established a feedback loop by attaching the float and the valve, yielding an automatic regulation device. Using his knowledge and ingenuity, it must have been relatively easy to build a prototype of the controller. Nevertheless, tuning the device must have been a tough task. He probably spent a lot of time shaving and carving the regulating valve and the float until he could achieve the desired control performance. Of course, that must have been the way of control engineering in the 3rd BC.

## 2.4. *Knowledge and control*

We have illustrated feedback controller design methods by linear control theory and fuzzy logic. Although there has been significant progress in control systems theory and practice in the last 2000 years (or more specifically in the last 50 years), there is a fundamental commonality among the approaches in feedback control. Building a feedback controller requires knowledge about the plant being controlled. Where and how the knowledge is used and represented is the differentiation factor in design approaches. How much of the knowledge is available is the deciding factor in the choice of approach.

The consequential issue to the design approach is performance. Fine tuning of the controller was and still is difficult. Just Ktesibios had spent long hours figuring out the perfect shape and weight of the float valve, we, as today's control engineers, spend long hours figuring out the perfect values of the control gain or the membership functions. A reason for the difficulty is that the system behavior in actual operating environment may be unforeseeably different from the state equations or other models including fuzzy models. A possible advantage of exploiting domain theory such as fuzzy control is that such deviation in nature may already be captured by the domain experts. The extent to how well such knowledge can be captured and represented is an open issue. This argument can only be justified in the actual operation environment.

Applicability of fuzzy logic was made known to us by early explorers in the field. Particularly, Mamdani's steam engine control[16] and Hitachi's subway control[26] are significant contributions to the progress in fuzzy logic control. Applications in home appliance products, initiated by Matsushita's washing machine, are significant contributions to today's popularity of the control design approach. These early applications, in general, concerned direct implementation of fuzzy logic to control systems. Most of them have directly applied their known control logic to the control system in place of the existing methods, particularly PID-type. More recent applications of fuzzy logic have flavors of diversity in their use and focus on benefits. Successful applications of fuzzy logic have the tendency to take a hybrid structure with linear controllers. This trend is substantiated by the recent reports of successful applications.

In the subsequent sections, we will begin our review on the latest applications of fuzzy control in industry. We chose six published reports: Overhead-traveling cranes, industrial temperature controller, mineral mine processing, weld pool, chemical baths, and steam turbines. Most of them have been installed and tested in real-time plants, and their operations have been successful. In each of them, we attempt to see what goals the designers seek to achieve and what they benefit from fuzzy logic.

## 3. Industrial Applications

### 3.1. *Crane control*[10]

We begin with a relatively straight forward application of fuzzy logic. This application seeks fuzzy logic to be a stand-alone nonlinear controller. The domain of the application is the industrial overhead-traveling crane. In the crane system, the objective is to transport a load from point A to point B with minimum swing (see Fig. 5). However, delays and friction may cause residual swing. To prevent the swing while simultaneously improving the positioning, the designer chose fuzzy logic to incorporate expert's control logic (see Fig. 6).

The inputs to the fuzzy controller are the acceleration and the distance to the destination. The output is the trolley velocity. There are 16 rules. A sample rule goes like this: *If the acceleration is small and the distance is medium, then the trolley velocity is small negative.*

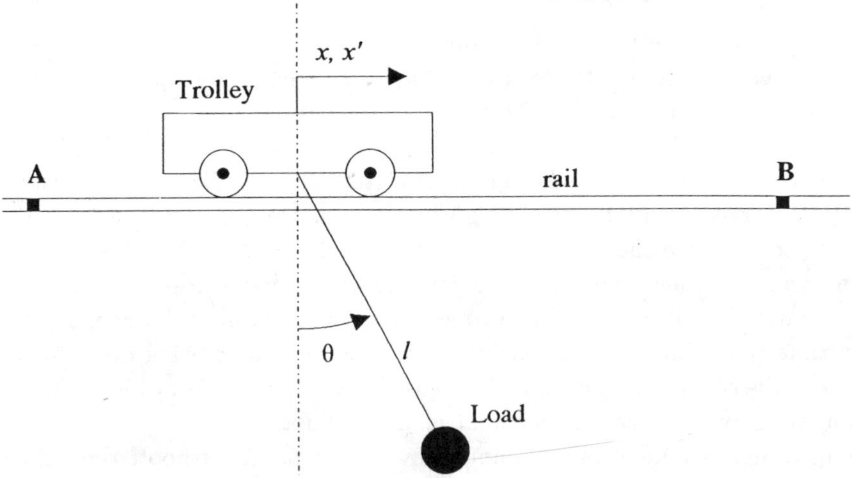


Fig. 5. Diagram for an industrial overhead-traveling crane. The objective is to transport the load from point A to point B with minimum swing.

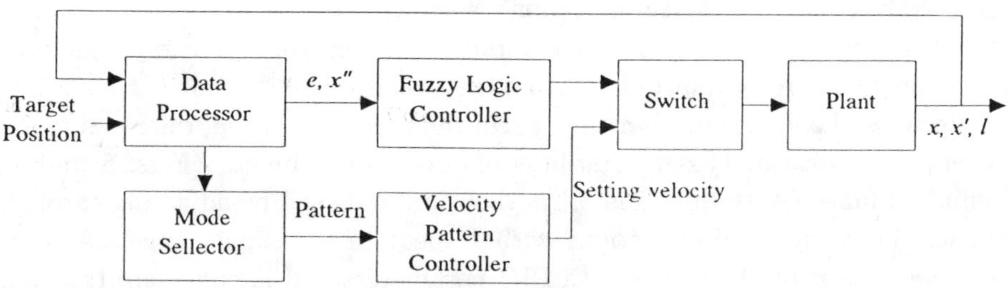


Fig. 6. Control diagram for the crane. Note that fuzzy logic controller (FLC) is used in parallel with a velocity pattern controller to control the system.

The result was validated in simulation and in real-time operation. The performance of the fuzzy controller was as good as the expert's manual operation. The benefit of fuzzy logic in this application is that two separate objectives, swing prevention and trolley positioning, were achieved by a small number of control rules.

### 3.2. *Industrial temperature control*[8,23]

This application uses fuzzy logic for a specific purpose: Disturbance compensation. The overall controller takes a hybrid architecture: A fuzzy logic unit and a PID control unit. The domain of application is industrial temperature control.

This industrial temperature regulator (model code: E5AF) employs a two-degree-of-freedom PID method. The control method uses a feed forward operator to enhance the system's response to a desired value and allows simultaneous adjustments on target tracking and disturbance response. The E5AF automatically tunes PID control parameters to minimize disturbance recovery time (i.e., time to come back to a target value after being disturbed). To achieve the objective, the controller sensitivity is set always at high. Such high gain, however, may cause overshoot or undershoot.

In order to reduce the amount of overshoot or undershoot, the gain may be decreased. The low gain, however, slows the disturbance recovery. In order to solve the problem, fuzzy logic is utilized to reduce the amount of overshoot or undershoot in case of disturbance, while the PID controller sensitivity is set such that the disturbance recovery time is minimized. The controller structure takes a hybrid form of a PID and a fuzzy logic units in parallel (see Fig. 7). When no disturbance is observed or when the process output is close to the target value, the PID control unit alone is sufficient. Under this circumstance, an output signal from the fuzzy logic unit is set to zero. The fuzzy logic unit generates non-zero output only when a disturbance is observed.

Because of constraints on memory size and execution speed, rules and membership functions are fixed regardless of process dynamics. (Figs. 8 and 9) Tuning of fuzzy control parameters is thus accomplished by adjusting through scaling the membership functions with respect to the input range. A limit cycle method is used to tune both PID parameters and membership function scale values.

In this application, there exists an unique control logic which was implemented in the control structure by the designers. It involves a trade-off in the hybrid control structure during the step-disturbance compensation by the fuzzy logic unit. As the fuzzy logic unit attempts to compensate the overshoot or undershoot, the integral area is reduced by this action. Because the integral time for the PID unit remains unchanged, it takes longer for the integral action to kick in. In other words, the lack of compensation on the integral action of the PID control unit causes the disturbance recovery time to increase. In order

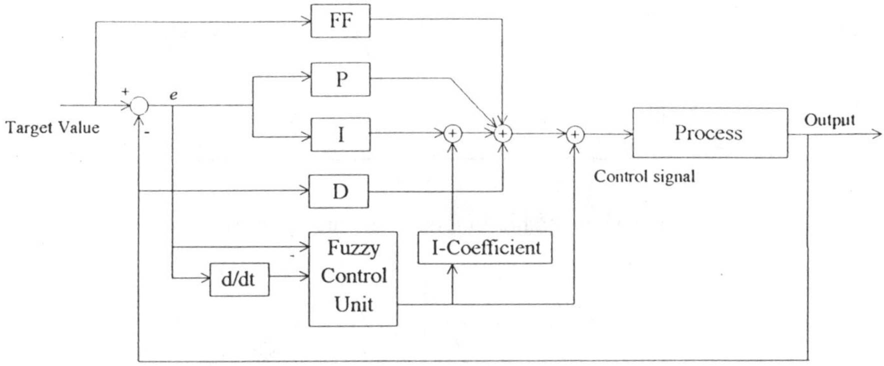


Fig. 7. Structure of the temperature controller.

| *de* \ *e* | NB | NS | ZR | PS | PB |
|---|---|---|---|---|---|
| PB | PB | PB | PM | PB | PB |
| PS | PS | ZR | ZR | PM | PB |
| ZR | NS | ZR | ZR | ZR | PS |
| NS | NB | NM | ZR | ZR | NS |
| NB | NB | NB | NM | NB | NB |

Fig. 8. Fuzzy Association Map for the temperature controller, where *e* = error, *de* = change in error, P = positive, N = negative, ZR = zero, B = big, M = medium, and S = small.

to compensate the overshoot amount and the recovery time simultaneously, it is necessary to adjust the integral action of the controller. This was accomplished by multiplying a coefficient to the fuzzy control output and adding it to the PID's integral action. The result is shown in Fig. 10.

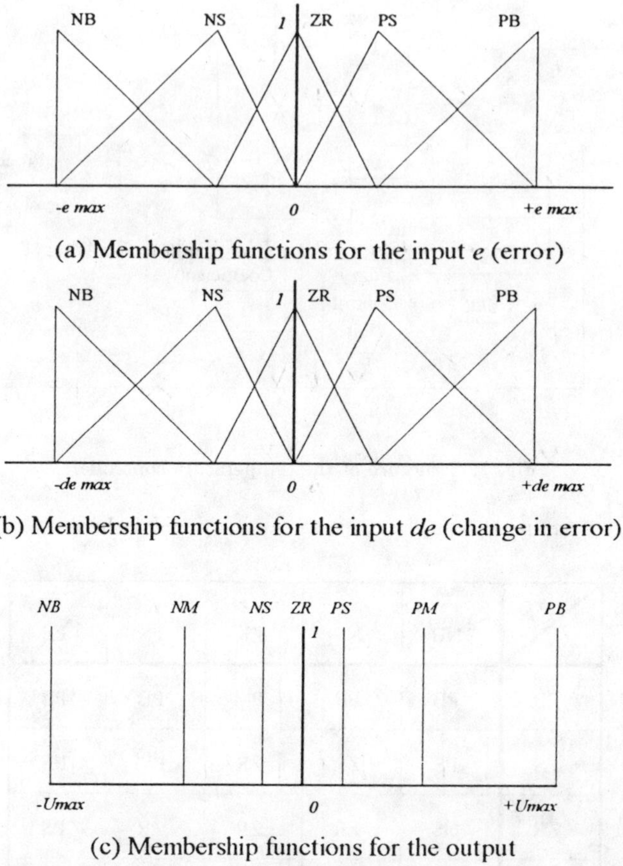


Fig. 9. Membership functions of the temperature controller, where NB = negative big, NM = negative medium, NS = negative small, ZR = zero, PS = positive small, PM = positive medium, and PB = positive big.

The merits of the proposed controller are: (1) Control performance under disturbance response is improved over PID control alone. (2) Fuzzy control parameters can be tuned automatically according to the process dynamics; therefore, users do not have to generate fuzzy rules or membership functions. (3) All algorithms are implemented in software; thus, it is easy to implement into an existing temperature control unit. This control system is available commercially for various industrial applications.

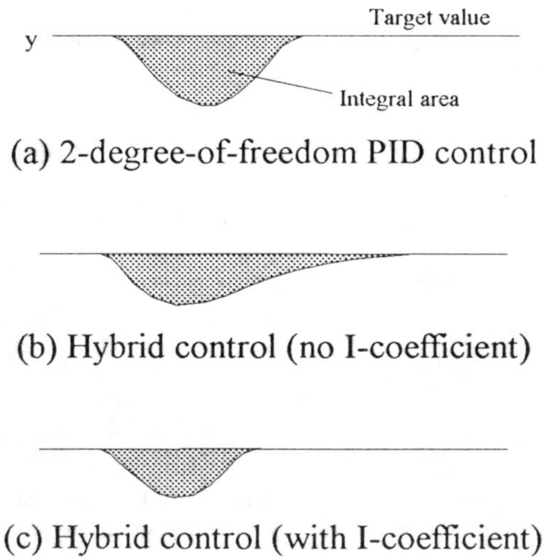


Fig. 10. The PID controller alone achieves a quick recovery but causes overshoot (a), while the fuzzy logic unit reduces the overshoot amount but increases the recovery time (b). An addition of I-coefficient reduces the recovery time (c).

### 3.3. *Iron ore processing*[3]

This application involves a supervisory control system for iron ore processing. The overall complexity exceeds a typical feedback controller; however, the designers successfully implemented and installed the fuzzy logic system at the operation site.

Mineral processing unit operations such as grinding, floatation, and dewatering, are difficult to control because of long time lags, nonlinearity, lack of sensor equipment, variable responses to loads, and variable input characteristics (see Fig. 11). Wabush Mines and Camdale Technologies have jointly developed and installed a fuzzy logic expert system to improve spiral iron recovery and mill productivity (see Fig. 12). The system is a supervisory controller to achieve the following objectives:

1. To monitor continuously the mill power draw and to regulate the mill feed rate to prevent damage to the mill motor while maintaining the feed rate at the maximum rate determined by the process supervisor.

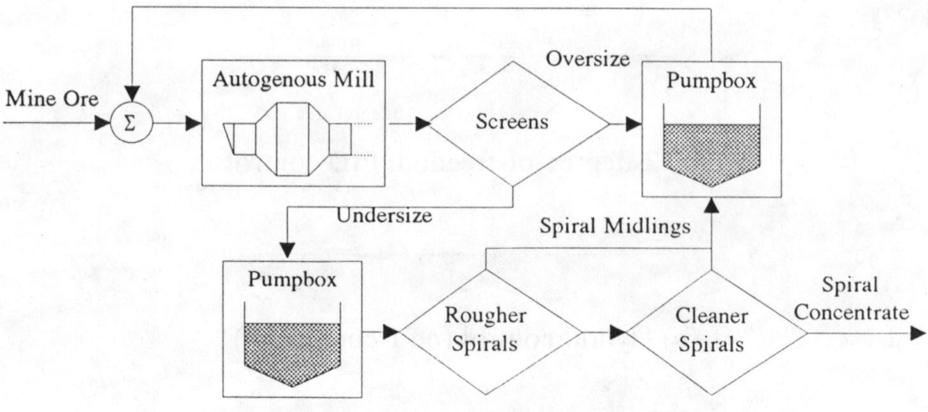


Fig. 11. Flow diagram for an iron ore process line. Note that oversized mine ore and spiral midlings are sent to a pumpbox, which will be recirculated to the autogenous mill.

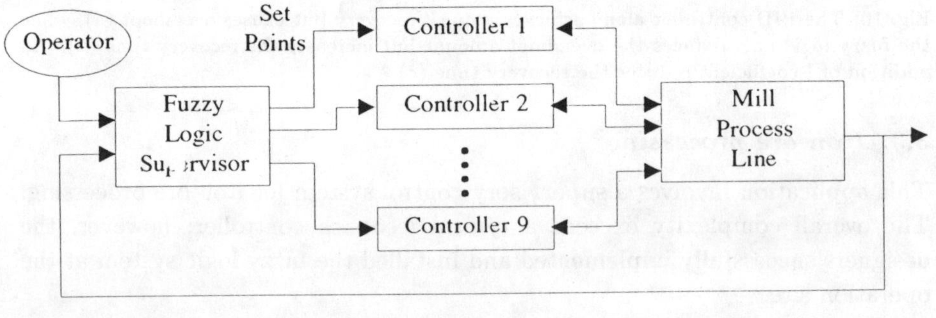


Fig. 12. Supervisory control diagram for the iron ore processing line. Nine PID controllers are used to control different part of the process line. Fuzzy logic is used as a supervisory controller which provides set points for those PID controllers.

2. To improve control over the spiral density and feed rate.
3. To improve control over both the rougher and cleaner wash water.
4. To reduce the frequency with which the pumps are plugged.
5. To structure the knowledge base so that modifications can be made easily and quickly without the need for rewriting the entire system.

The supervisory system consists of 110 rules, 35 fuzzy sets for recognition of process states, 19 fuzzy sets for control actions, 25 fuzzy sets to identify trends, and 24 variables monitored for time variation. The sample control logic goes like this: *If mill power draw is high and trending upward fast, and if recirculating density is not too high and not trending upward, and if recirculating sump level is too high and trending upward, then reduce mill feed water by a small amount.*

The resulting system was able to meet the goals with improved quality, improved production performance, and reduced down-time. It has been in operation for four years, and plans are in place to install the system on all other mill lines at the site.

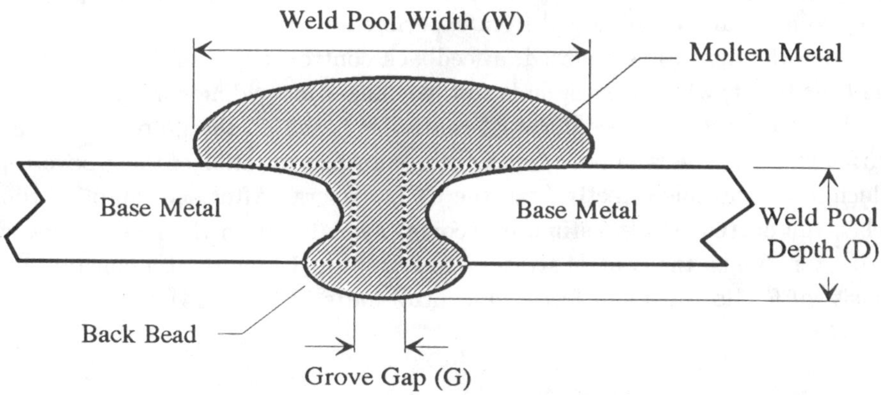


Fig. 13. Conceptual figure of the weld pool. The weld pool depth ($D$) is defined as the distance from the top surface of the base metal to the bottom of the back bead. This distance is important for the quality of welding; however, it is usually difficult or impossible to measure during the welding process. An artificial neural network can be used to relate the weld pool width ($W$) and the grove gap width ($G$) to the depth.

### 3.4. *Weld pool depth control*[25]

This application exhibits a unique control logic by the designers to solve problems in two areas: state estimation and control. The domain of this application is industrial welding. In order to achieve high quality welding, the weld pool depth is referred to be constant. During welding, however, the weld pool depth changes by the variation of the groove gap (see Fig. 13). Feedback control is needed to maintain constant depth by adjusting the welding current. However, because of the structural limitations under the operation environment, direct measurement of the weld pool depth during welding is not feasible.

To solve the measurement problem, a neural network is used to estimate the depth from measurable information. The designers utilized CCD cameras to observe the surface shape of the weld pool and the state of the groove gap. They observed that the weld pool width, which is measurable by the camera, and the weld pool depth are correlated with respect to the welding current. The correlation depends on the state of the groove gap. Although the welding phenomena can be described by partial differential equations, it is difficult to construct the state equations for the depth. Therefore, the designers chose a neural network to estimate the depth. The network structure is a feedforward sigmoidal type, with one hidden layer. It was trained by back propagation using numerical data obtained from experiments.

Once the depth is estimated, a feedback controller can adjust the state of the heat input, which corresponds to the changes in welding current. It is also possible to construct a feedforward controller based on the groove gap, since during welding, the weld pool depth changes by the variation of the groove gap, which is measurable directly from the CCD camera. After one sampling time delay, the current depth estimate becomes available from the neural network estimator. Thus, the control strategy to maintain the depth at a constant can consist of feedforward and feedback control units (see Fig. 14).

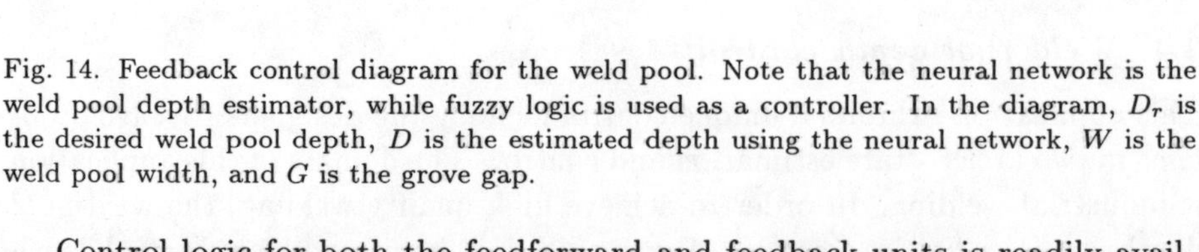


Fig. 14. Feedback control diagram for the weld pool. Note that the neural network is the weld pool depth estimator, while fuzzy logic is used as a controller. In the diagram, $D_r$ is the desired weld pool depth, $D$ is the estimated depth using the neural network, $W$ is the weld pool width, and $G$ is the grove gap.

Control logic for both the feedforward and feedback units is readily available from the welding operators. Therefore, the designers chose fuzzy logic to construct the controller. The feedforward control logic goes like this: *When the*

| $\Delta u_0$ \ $e$ / $\Delta e$ | N | Z | P |
|---|---|---|---|
| P | Z | N | N |
| Z | P | Z | N |
| N | P | P | Z |

(a)

| $\Delta u_1$ \ $\Delta G$ | N | Z | P |
|---|---|---|---|
| | P | Z | N |

(b)

Fig. 15. Fuzzy rule tables for the weld pool control, where P is positive, Z is zero, and N is negative. The feedback control signal, $\Delta u_0$, is inferred from the deviation $e$ and its variation $\Delta e$. The feedforward control signal, $\Delta u_1$, is inferred from the variation of the groove gap width, $\Delta G$, only. The final control signal is a combination of these two control signals and the change in previous control signal.

*groove gap becomes wide, the weld pool depth becomes deep; thus, the heat input must be reduced. When the groove gap becomes narrow, the depth becomes shallow; thus, the heat input must be increased.* The feedback control logic is described in Fig. 15. There is a particular control logic that indicates nonlinearity in the process. When the depth is shallow, it does not immediately become deep even if the heat input becomes large. However, when the depth is deep, it changes sharply even if the heat input change is small. Therefore, the gain must be larger when the deviation of the depth is positive. This can be accomplished by adjusting the membership functions (see Fig. 16). The control output, $\Delta u$, is described in the equation:

$$\Delta u[n] = \Delta u_0[n] + \Delta u_1[n] - 0.4\,\Delta u[n-1] \tag{12}$$

where $u_0$ is the feedback unit control signal and $u_1$ is the feedforward unit signal. The resultant fuzzy controller was tested in real-time operation, and showed improved performance over constant heat input.

### 3.5. *Chemical bath control*[2]

Many industrial manufacturing processes use chemical baths for cleaning, etching, and plating metal surfaces. In order to maintain a proper balance of chemi-

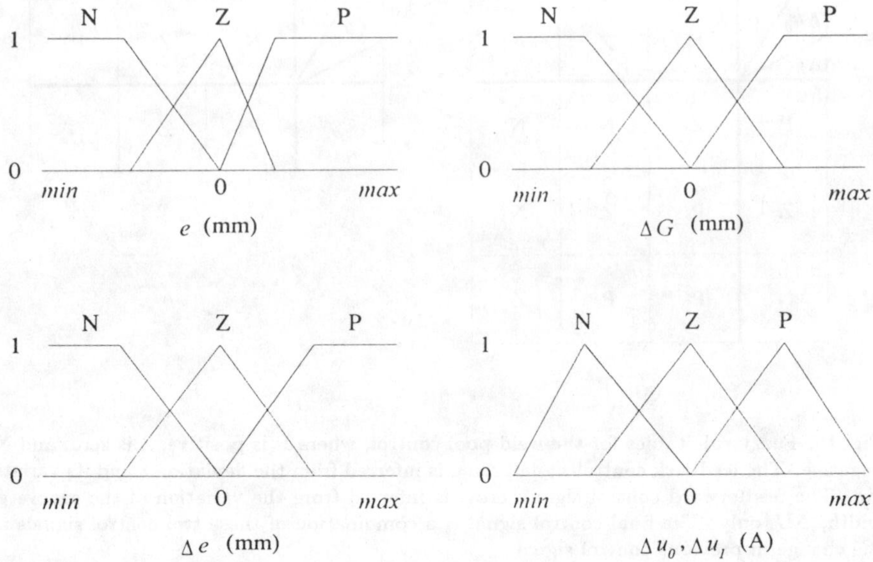


Fig. 16. Membership functions for the weld pool control, where P stands for positive, Z for zero, and N for negative.

cal components in the bath, replenisher products must be added appropriately. The conventional approach is to use a PLC (programmable logic controller). It cycles the replenisher pumps on and off at preset intervals to maintain the pH level within a range. There are plenty of engineering problems in this process.

First, the replenisher is a chemical mix which specifically matches the depletion rates of the individual components. It is not only inflexible but also costly to utilize custom replenisher mix. Second, the pH level is sensed by a conductivity probe. It is subject to sludge build-up and corrosion which may change the sensor characteristics. Third, the pH process is inherently nonlinear; making control by a linear controller difficult. Forth, periodic manual titration and manual injection of replenisher products may be necessary to assure the pH level. These factors add cost to the manufacturing process.

To improve the overall performance, the sensing devise was upgraded to an on-line analyzer which measures the levels of free acid, total acid, and individual chemical components. This increases the number of observation signals. Second, control outputs are directed to individual chemical tanks instead of a nominal replenisher mixture. This increases the number of control outputs. Because of the increased number of inputs and outputs, the designers found fuzzy logic to be the appropriate strategy (see Fig. 17). The resultant system eliminated the need for manual titration because of the continuous sensing. Maintenance costs and manual operating costs were significantly reduced. The results were validated in real-time operations.

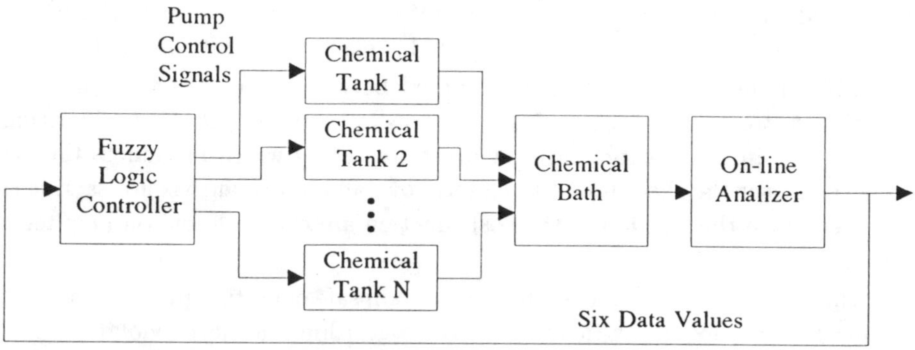


Fig. 17. Control diagram for a chemical bath. An on-line analizer measures the levels of accelerator, free acid, total acid, and individual chemical components. These data values are used by the fuzzy logic controller to control the pump of individual chemical tank.

### 3.6. *Steam turbines control*[17]

This application provides a unique coordinated control scheme by applying fuzzy logic to an optimal controller. The domain of this application is steam turbines control during load cycling in power plant operation. The control objective is to meet power demand at minimal cost with available generation equipment. An effective control strategy must assess the cost of operation at different modes.

An optimal controller was designed to minimize turbine stress and load tracking error by coordinated control of both the turbine governor valve and boiler pressure during a cycling operation. According to the designers, there are four key ideas underlying the design of this controller. They are described in the following excerpts.[17]

"Future load demand from projected load profiles, which are accurate and readily available information at power plants, is considered when deciding on the current control action. Optimal future performance is therefore not jeopardized by focusing only on short term demand, and the possibility that current modes of operation will result in future constraint violation is greatly reduced."

"Fuzzy logic is utilized to encode operator experience and intuition in establishing priority among conflicting performance objectives. For example, load tracking can be traded off against stress minimization or turbine valve throttling against varying boiler pressure. This intuition provides the long term guidance for overall cycling cost optimization, and is transferred to the model predictive controller (MPC) through weights in a cost function."

"MPC is used to optimize the trade off between good load following and equipment stress minimization during cycling operations. The MPC algorithm finds the optimal valve/pressure set point trajectory which minimizes the cycling costs over the short term. Long term overall performance is not sacrificed however, since the weights in the cost function are chosen based on long term objectives."

"Hard system constraints such as maximum safe operating pressure or valve closure rate are explicitly included in the algorithm and never violated. Soft constraints such as desired pressure at the end of a load change are also included. These soft constraints and the priority given to satisfying them is determined by the fuzzy logic."

A sample fuzzy logic rule goes like this: *If initial loading rate is slow, and if initial load change is small, and if constant load duration is short, and if subsequent loading rate is fast, and if subsequent load change is large, then pressure variation weight is medium and valve change weight is medium and temperature variation weight is high.* The details of the controller are in the literature.[17]

The resulting controller was compared against a standard optimal controller in simulation. By utilizing fuzzy logic to predict the extended future load, the optimization algorithm became computationally compact without sacrificing long term performance (see Fig. 18).

## 4. Applications Summary

### 4.1. *Controller types and merits*

The cited applications are the most successful ones reported in recent literature. Their objectives are clearly defined, and the results are validated mostly in real-time operations. We have observed that fuzzy control can be realized in various

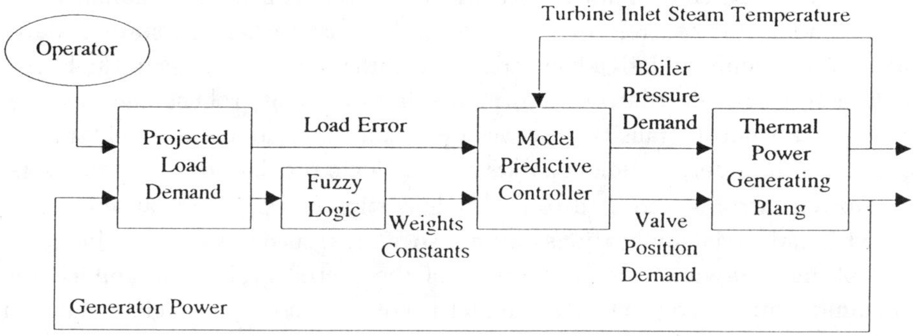


Fig. 18. Steam turbines control diagram. Note that fuzzy logic uses the projected load demand to provide performance weight constants for the model prdictive controller (MPC). MPC uses these constants to optimize the short term system performance.

forms depending upon the objective and the designer's choice. In the case of the crane control, weld pool, and chemical bath, the fuzzy systems are realized as stand-alone controllers. In the case of temperature control, the fuzzy system is a sub-system within a hybrid architecture with a PID controller. Its design objective is disturbance compensation. In the case of iron ore processing, the fuzzy system is a supervisory controller whose objective is not only to control the mill feed rate but also to monitor and supervise the overall operation. In the case of steam turbines, the fuzzy system is a weight adjuster for an optimal controller.

We have also observed that fuzzy control is used to achieve various design objectives. Most often it is used to compensate nonlinearity in the process. This is due to the universal estimation capability of fuzzy systems. By partitioning the input space through membership functions, it is possible to adjust the control surface locally. It is not possible to do so by a linear PID controller alone. When a PID control parameter is changed, the entire control surface is affected. If the process response is nonlinear, the linear controller must keep the operation point within the linearized region. Therefore, comparison between fuzzy logic and PID-type linear controller requires careful consideration in the process linearity.

Because of the ability to adjust the control surface locally, combining a fuzzy control unit with a linear controller often improves the overall performance. This hybrid approach can be considered as a form of variable structure control. A linear control law is used when the state is near the nominal value, while a nonlinear control law (fuzzy control) is used when the state deviates far from the nominal design condition. Advantages of fuzzy logic in the hybrid architecture include control performance improvement, robust and easy design, and a smooth transition between controllers without control chattering. Because of its interpolation capability, fuzzy logic provides compact rule sets.

Fuzzy control is also used to reduce operation costs. This is an important aspect in industrial applications, but not so often argued in theory. In industry, control must be viewed in the context of the overall goal of the application. In manufacturing, to produce a product is not just to keep a single variable at a given operation point. Control designers must consider conditions such as efficiency, safety, reliability, and economy. The application reports on chemical bath and steam turbines indicate the advantage of fuzzy control in this regard.

### 4.2. *Theoretical issues*

Despite the reported success of fuzzy control applications, many theoretical issues remain open for further research. The most basic issue is on the foundation of fuzzy logic. It is still a strongly debated topic. The recent publication of *IEEE Expert* (August 1994) contains informative discussions on the formalism of fuzzy logic. The details of which are beyond the scope of this writing. However, an important point from the debate deserves mentioning here. Opponents of fuzzy logic agree that critiques of fuzzy logic have no impact on fuzzy control theory; the areas of fuzzy logic that get criticized are simply not employed in the control arena.[20]

An area of fuzzy control that needs further study is stability and robustness. Although good efforts have been made, for instance,[22] fuzzy control theory has a lot of room to fill. Because of the lack of theoretical foundation, compared to linear control theory in particular, performance in fuzzy control must depend on designer's skill. Construction of a robust rule-base is a necessary condition for acceptable performance. Since the majority of control applications require only a small number of rules (less than 100), control rules are relatively manageable. It is no longer simple when the number of rules or the inference complexity increases. An application such as iron ore processing, where supervisory control rules are involved, is a good example. Processes in real-world applications are highly complex in nature. For this reason, we need theoretical and practical tools. Nonetheless, this is an issue that is not unique to fuzzy control.

### 4.3. *Fuzzy system tuning*

It is relatively easy to construct initial rules and membership functions. However, tuning the controller to obtain desired performance is not simple. In theory, a number of automatic tuning techniques exist. In practice, they are not yet utilized fully. In fact, none of the reviewed systems in this chapter utilized any automatic tuning technique. They were designed and tuned manually. Perhaps, it is a matter of time to see more applications generated with automatic tuning techniques. In the next section, we list some available tuning methods for fuzzy systems.

## 5. Automatic Tuning Approaches for Fuzzy Systems

Tuning fuzzy systems involves modification and generation of fuzzy sets and rules. Modifications of fuzzy sets can be accomplished by changing the shape and location of the membership functions, while generation of rules involves finding appropriate causal relations between inputs and outputs. Typically, systematic synthesis and tuning methodologies require an input–output data set, or a simulation model to describe target system behavior. Such reference models can be obtained either in the form of input–output data or a simulation model, not necessarily an analytical model. We will briefly review each one of the following tuning methods. It should be noted here that this is not an exhaustive list of existing techniques.

1. Fuzzy to neural net transformation by manifold simulation,
2. Fuzzy to neural net transformation by structural conversion,
3. Neural-network driven fuzzy inference,
4. Rule generation by differential competitive learning,
5. Membership function (MF) parameter adjustment by a descent method,
6. MF parameter adjustment by optimization methods,
7. Rule generation method, and
8. Rule and MF generation method.

### 5.1. *Fuzzy to neural net transformation by manifold simulation*[9]

The basic principle of this approach is to transform a given fuzzy system into a neural network whose mapping characteristics are similar to the ones of the fuzzy system. A fuzzy system and a neural network are both function estimators. A feedforward network with an arbitrary squashing function can approximate any continuous function to any desired accuracy, provided that many hidden units are sufficiently available. This representation theorem implies that it is possible to find a neural system which approximates a fuzzy system, provided that such a fuzzy system is a continuous function. When such a transformation is performed, a resultant neural system can be tuned by applying any recursive parameter estimation algorithm or an optimization algorithm.

In order to transform a given fuzzy system into a neural system, parameters of an arbitrary neural system must be adjusted so that a manifold of the neural system closely matches the one for the fuzzy system. This can be achieved by numerical optimization algorithms to minimize an energy function, based on a geometrical comparison of dynamic characteristics between the two systems.

### 5.2. *Fuzzy to neural net transformation by structural conversion*[6,12]

This approach transforms a given fuzzy system into a pre-wired network structure. Membership functions and fuzzy logic inference scheme are functionally represented by simple networks separately. The networks are then connected properly in series to yield a roughly-equivalent feedforward neural network. A more accurate neural network is then obtained by training the initial network.

A fuzzy system can be converted to a 7-layer feedforward neural network. The first layer is an input layer. The second and third layers generate membership functions of antecedent parts. The fourth layer determines the true value of each rule via a bounded product operation, and the node number is equal to that of the fuzzy rules. Each node of the fifth layer corresponds to a fuzzy set of outputs. The sixth layer implements the membership functions of consequent parts. The last layer functions as a defuzzification scheme using the center of gravity method. Membership functions of an antecedent part are realized by a combination of sigmoidal functions. The resultant network which is roughly equivalent to a fuzzy model can be trained to increase the accuracy.

### 5.3. *Neural network driven fuzzy logic inference*[5]

Neural network-driven fuzzy reasoning uses a set of neural networks to simulate the function of fuzzy reasoning. Each network performs a partial task of fuzzy inference such as calculation of the membership grade and representation of fuzzy rules. All the networks are trained using a back propagation algorithm.

Training data must be divided into different groups to train corresponding fuzzy rules.

### 5.4. *Differential competitive learning*[13]

In this approach, fuzzy rules are extracted from training quantization vectors. The quantization vectors are defined on the input–output product space ($X \times Y$); $X$ is in $R^n$ and $Y$ is in $R^p$. The product space is clustered into $K$ subspaces, while each cluster represents a fuzzy rule. If there are $r$ fuzzy sets in $X$, and $s$ fuzzy sets in $Y$, then there will be totally $r \times s$ possible fuzzy rules (or clusters) in the product space. These clusters are stored as column vectors of a weight matrix $M$. Then the differential competitive learning (DCL) adjusts the weight matrix $M$ according to the density distribution of the training data. Fuzzy rules are then extracted from the final weight matrix.

### 5.5. *Fuzzy MF adjustment by a descent method*[19]

This approach adjusts membership functions by updating a set of parameters, $\mathbf{z}$, in a direction which an objective function, $E(\mathbf{z})$, decreases. The parameter-updated algorithm is expressed as

$$z_i(t+1) = z_i(t) - K\frac{\partial E(\mathbf{z})}{\partial z_i} \qquad (i = 1, \dots, p) \tag{13}$$

where $t$ is an iteration number and $K$ is a learning constant. The objective function $E(\mathbf{z})$ is typically a quadratic function.

### 5.6. *Fuzzy MF adjustment by optimization methods*[7,11]

This approach adjusts membership function parameters by optimization methods. Any multidimensional numerical optimization algorithm can be applied to this method. Examples of such algorithms include downhill simplex, simulated annealing, and genetic algorithms. In essence, optimization methods attempt to search for good parameter values which minimize a given objective function. A typical objective function is a sum of squared errors.

### 5.7. *Fuzzy rule generation*[24]

This approach generates fuzzy rules based on a set of input–output data and a set of pre-defined membership functions. No initial rule set needs to be provided. The algorithm begins by defining a set of fuzzy sets for input and output variables to divide the input–output space into fuzzy regions. The second step is to apply numerical data to the fuzzy regions, and determine the maximum matching degree between the data set and the fuzzy regions. A new rule is obtained by combining the fuzzy sets with maximum matching degree. When identical rules are generated, some rules must be discarded. In order to do so, a weight value is assigned to each rule. A weight value is computed by multiplying the matching degrees for each input and output variable. Repeat these steps until all the input–output data are applied.

### 5.8. *Fuzzy rule and MF generation*[1]

This approach generates both rules and membership functions. Parameter adjustment and rule generation are performed by monitoring an inference error and its differential change. A descent method updates the parameters. If the parameter adjustment cannot attain a desired accuracy, new membership functions and consequently new fuzzy rules are generated in the region of the highest inference error. This is a local adjustment of the control surface.

## Conclusions

Manufacturing is the foundation of economic activities. Mass production using numerical control, machine tools, flexible manufacturing, and lately, computer-integrated manufacturing are the technologies and tools that have advanced the global economy. However, socio-economic changes have caused a number of problems to industrial nations, such as changes in human factors, changes in market requirements, and preservation of natural resources and the environment.[5]

To meet the needs, control must be viewed in the context of the overall goal of the application. While keeping a single variable at a given operation point is important, control designers must consider global conditions such as efficiency, safety, reliability, and human and environmental factors.

We view fuzzy control as one of many approaches available to design a control system. From the perspective maintained in this writing, fuzzy logic provide a means to incorporate heuristic control laws in a control structure. Exploitation of domain theory and the explicit use of decision logic allow the effective use of multiple sensory data. We have observed such evidence from the application reports in this chapter. We have also observed that more theoretical tools are desired to improve the efficiency and reliability of fuzzy control. All things considered, the practical merits of the fuzzy control are appealing as a technique for global system design.

## References

1. S. Araki, H. Nomura, I. Hayashi, and N. Wakami, "A self-generating method of fuzzy inference rules," in *Proc. IFES'91* (1991), pp. 1047–1058.
2. E. Dohmann, "Fuzzy logic control of a chemical bath," in *Proc. IEEE International Conference on Fuzzy Systems* (1994), pp. 958–961.
3. M. Hall and A. Harris, "Fuzzy logic expert system for iron ore processing," in *Proc. IEEE Industry Applications Conference* (1993), pp. 2190–2199.
4. H. Hayashi, "A preview of the 21st century," *IEEE Spectrum* **30**, 9 (1993), pp. 82–84.
5. I. Hayashi, H. Nomura, and N. Wakami, "Artificial neural network driven fuzzy control and its application to the learning of inverted pendulum system," in *Proc. 3rd International Fuzzy Systems Associations [IFSA] World Congress* (1989).
6. H. Ichihashi, " Iterative fuzzy modeling and a hierarchical network," in *Proc. 4th IFSA Congress* (1991).
7. S. Isaka, A. V. Sebald, and A. Karimi, "On the design and performance evaluation of adaptive fuzzy controllers," in *Proc. the 27th Conference on Decision and Control* (1988).
8. S. Isaka, "Fuzzy temperature controller and its applications," in *Proc. Applications of Fuzzy Logic Technology* (SPIE – The International Society for Optical Engineering, 1993), pp. 59–65.
9. S. Isaka, "On neural approximation of fuzzy systems," in *Proc. IJCNN'92* (1992).
10. O. Itoh, H. Migita, H. Itoh, and Y. Irie, "Application of fuzzy control to automatic crane operation," in *Proc. International Conference on Industrial Electronics, Control, and Instrumentation* (1993), pp. 161–164 .
11. C. L. Karr, L. M. Freeman, and D. L. Meredith, "Improved fuzzy process control of spacecraft rendezvous using a genetic algorithm," in *Proc. SPIE Conference on Intelligent Control and Adaptive Systems* (1989), pp. 274–288.
12. A. Kawamura, N. Watanabe, Y. Owada, R. Masuoka, and K. Asakawa, "Neurofuzzy system — an overview and a prototype system," *Micro-computer and Workstation* (1991), pp. 1–8.
13. B. Kosko, *Neural Networks and Fuzzy Systems* (Prentice Hall, 1991).
14. B. Kosko and S. Isaka, "Fuzzy logic," *Scientific American* **269**, 1 (1993), pp. 76–81.
15. H. Kwakernaak and R. Sivan, *Linear Optimal Control Systems* (John Wiley & Sons, 1972).
16. E. H. Mamdani, "Application of fuzzy algorithms for control of simple dynamic plant," in *Proc. IEE* (1974), pp. 1585–1588 .
17. K. Marcelle, K. Chiang, P. Houpt, and P. Bonissone, "Optimal load cycling of large steam turbines," in *Proc. IEEE International Conference on Fuzzy Systems* (1994), pp. 667–672.

18. O. Mayr, "The origins of feedback control," *Scientific American* **223** (1970), pp. 110–118.
19. H. Nomura, I. Hayashi, and N. Wakami, "A learning model of fuzzy inference rules by descent method," in *Proc. IEEE Int'l Conf. on Fuzzy Systems Conference* (1992), pp. 203–210.
20. F. J. Pelletier, "Fuzzy logic: A misplaced appeal," *IEEE Expert* (1994), pp. 29–31.
21. M. G. Rodd and G. J. Suski, "Computer control systems: An introduction," in eds. L. Boullart, A. Krijgsman, and R. A. Vingerhoeds, *Application of Artificial Intelligence in Process Control* (Pergamon Press, 1992), pp. 205–222.
22. K. Tanaka and M. Sugeno, "Stability analysis and design of fuzzy control systems," *Fuzzy Sets and Systems* **45**, 2 (1992), pp. 135–156.
23. T. Ueda, T. Tsukabe, and Y. Nagami, "Temperature control by hybrid controller of 2-degree-of-freedom pid and fuzzy," *OMRON Technics* **30**, 4 (1990), pp. 53–57.
24. L. X. Wang and J. M. Mendel, "Generating fuzzy rules from numerical data, with applications," *USC-SIPI Report #169* (1991).
25. S. Yamane, N. Kitahara, K. Ohshima, and M. Yamamoto, "Neural network and fuzzy control of weld pool with welding robot," in *Proc. IEEE Industry Applications Conference* (1993), pp. 2175–2180.
26. S. Yasunobu and S. Miyamoto, "Automatic train operation system by predictive fuzzy control," in ed. M. Sugeno, *Industrial Applications of Fuzzy Control* (North-Holland, 1985), pp. 1–18.

# FUZZY LOGIC CONTROL IN FINNISH INDUSTRY

HEIKKI N. KOIVO
*Department of Electrical Engineering, Tampere University of Technology, PO Box 692, FIN 33101 Tampere, Finland*

## 1. Introduction

Finnish industry has become more and more interested in applying fuzzy logic in recent years. A number of unrelated research and development projects have been carried out. In 1994 a major national research program on *Learning Systems* was started. This has generated enthusiastic response on the part of the Finnish industry to participate in order to add extra value to their products or processes. Many of the projects in the program are fuzzy logic related. The purpose of this chapter is to summarize some of the current industrial applications in the Finnish industry.

Two different applications are discussed: Recovery boiler bed control and active magnetic bearing control.

## 2. Fuzzy Control in Recovery Boiler

### 2.1. *Introduction*

In the preparation of pulp for paper manufacturing, the fibres of tree or other vegetation are separated from each other either mechanically or chemically. In chemical pulp making, acidic cooking liquor is used in a digester. The more chemical treatment is applied the better the separation of fibres, but the more tree substance appears in cooking liquor instead of in pulp. Both the dissolved tree substance and the used chemicals pose tough environmental and economical challenges for recovery and chemical regeneration. Recovery boilers are designed for this purpose.

The dual task for the recovery boiler is to recover the chemicals in cooking liquor and also the heat in it. Thus a recovery boiler acts also as a boiler, in which the dissolved organic matter is burned. The new, stricter environmental requirements mean that there is more inorganic material mixed with the used cooking liquor, black liquor. This eventually implies also more corrosive liquor and lower heat content. The new requirements also imply that the automation system has to be improved and optimized for better control action to be taken at the right time.

## 2.2. *Process description*

Figure 1 shows the operation principle of a recovery boiler. The black liquor from the evaporator plant is pumped into the mixing tank, in which a fresh portion of natriumsulfate and fly ash separated from flue gases are mixed with black liquor.

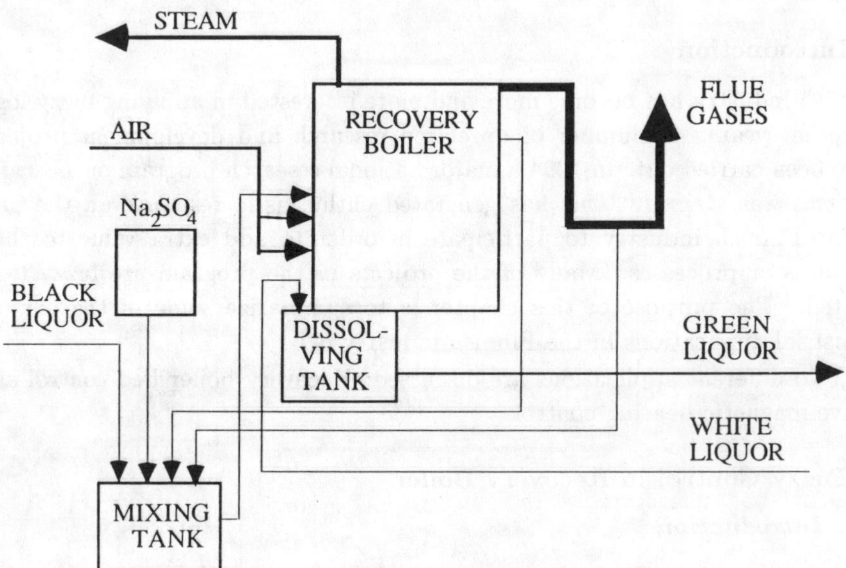


**Fig. 1. Burning of liquor in the recovery boiler.**

The black liquor is then sprayed into the recovery boiler. Air needed in combustion is added. A charbed is formed at the bottom, the state of which must be carefully controlled for efficient combustion and melting. During combustion the water remaining in the liquor is evaporated and the organic substance burns. The inorganic substance in solid melts in high temperature and runs in molten form from the bottom into dissolving tank. The quality of the molten substance is usually determined by the reduction degree and sulfur ratio. Modern recovery boilers reach reduction degrees of 90–95 %. The molten substance is dissolved into water or white liquor coming from lime kiln. This results in green liquor, which will be processed into fresh cooking liquor in a causticizer. The hot exhaust gases formed during combustion are led into the heat recovery part of the boiler, where the heat in exhaust gases is recovered in the form of superheated steam. As a steam generator the recovery boiler acts like conventional boilers satisfying the energy needs of chemical pulp making.

### 2.3. *Control of recovery boiler*

Control of the recovery process is to manage the simultaneous effects of several combustion variables. An exact model of combustion process does not exist and prediction of future behavior is practically impossible. Continuous measurements of many process variables do not exist or are imperfect and dirty operating conditions impose additional restrictions.

Control strategies for a recovery boiler can be subdivided into furnace control and sootblowing.[1] The most important control objectives are to achieve effective control of combustion over the whole load range, continuous management of quality variations and load changes, low emissions and high reduction, optimum temperature profile, and optimization of sootblowing.

### 2.4. *Bed control using fuzzy logic*[2]

Fuzzy logic has been applied to bed control in several recovery boilers by Valmet Automation Inc. This includes both black liquor droplet size and vertical temperature profile optimization. The main idea in control strategy is to recognize the disturbances in combustion process, to determine the cause of the problem, and then adjusting the process variables towards the right conditions. This results in better recovery of heat and chemicals. User interface is considered to be an integral part of the overall solution. It supports the determination and verification of the fuzzy logic rule base. The basic principle of the system is shown in Fig. 2.

The measurements in Fig. 2 are first filtered and then fuzzified. Three membership functions for each measurement are used: small, normal and big. In addition to this, fuzzy logic is used to form classes small *or* normal, normal *or* big, and small *or* big. The *or* operation is carried out taking maximum

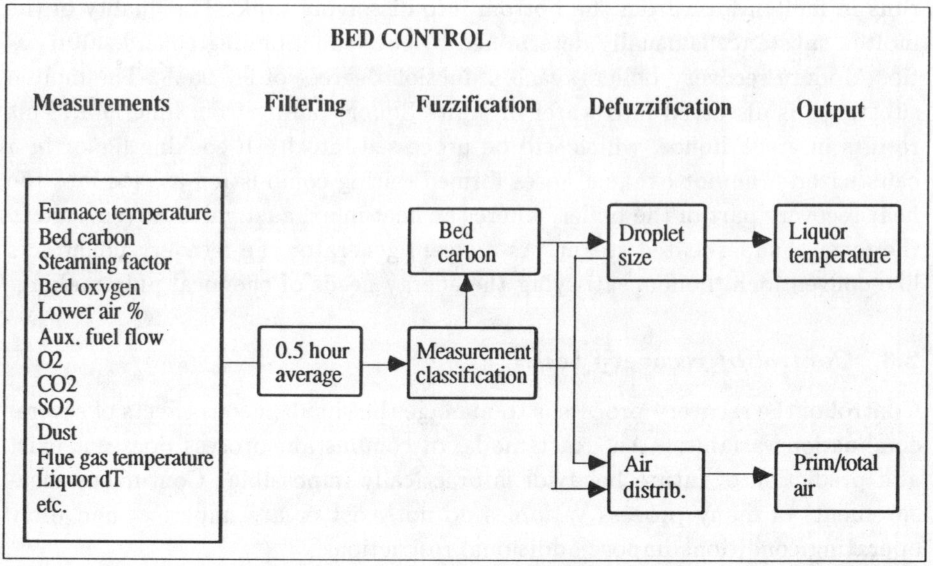


Fig. 2. Principle of bed control.

of the corresponding membership functions, e.g. max [membership function of small, membership function of normal]).

Fuzzy rules are of the form: *If A and B then C.* Each rule has a weighting factor connected to the value. The overall result is achieved by combining all the rules using weighted average. The output is formed so that each result of the reasoning affects the final output value. The final output will be between small and big, not small and big.

In the first phase of operation the process status and disturbances are identified. This is followed by classification of the output values, which form the necessary data for control signal calculations. The main objective is to keep optimum combustion conditions. If there are deviations from the optimum, the fuzzy logic recognizes them and gradually adjusts the air distribution and liquor droplet size to correct values.

The system is configured into a distributed automation system.

## 2.5. *User interface and tuning*

In everyday use the user interface becomes the most important issue. If the user makes changes in the process, he must be able to see their effect.

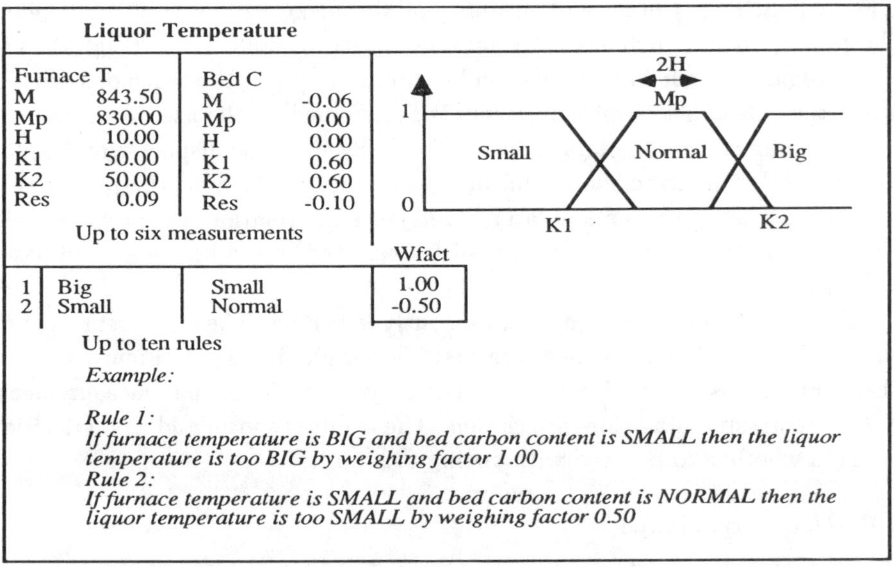


**Fig. 3.** Bed control tuning display.

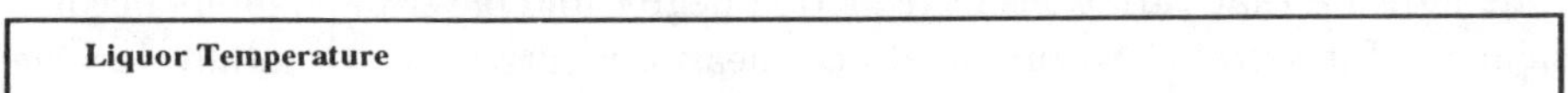


**Fig. 4.** Bed control follow-up display.

This will help in tuning and retuning of the fuzzy logic system. Therefore the tuning display includes both parameter settings and trends (Fig. 3). The measurements needed for tuning can be chosen simply at the touch of a button, while the actual list of measurements is determined by the user.

In tuning the parameters are set and displayed on one display, which shows how the rules describe the actual process situation. At first, the preliminary measurements are chosen and later, when more information becomes available, the most suitable measurements are picked and the final rules are formed based on these.

The two displays are used simultaneously in tuning. The measurements and the rules associated with them are first chosen like in Fig. 3. Figure 4 shows the result of this choice. Based on this display a new selection of measurements and the corresponding rules are chosen. The result is shown and again decision is made whether to proceed with tuning or stop it.

### 2.6. *User experience*

A conventional control solution aims at compensating disturbances, whereas the fuzzy control tries to optimize the process. The control result is improved because the interactions between variables can better be taken care of. Results indicate that variations in reduction degree and other parameters become smaller. Inaccurate measurements or measurements that are totally lacking can be replaced by combining other measurements and calculated values such as bed carbon consumption.

The user interface makes the control functions and logic easy to comprehend for the operator. The different displays allow the operator to concentrate in the process follow-up and changes occurring in it.

Figure 5 shows a real-life example of one of the bed control functions. Target values exist for the bed and furnace final temperatures. The purpose of the control is to achieve these by deducing from all available data the signal to the air controller. The control variables are the size of black liquor droplets and the air distribution control. It can be observed that when the target state is reached the control signals become zero, i.e., no further corrections are needed.

Significant improvement has been achieved in process operation. For instance, the typical average reduction degree in the same plant as in Fig. 5, with conventional control, was 84.8 % with a standard deviation of 2.5 %; and with fuzzy logic the average was raised to 87.5 % with a standard deviation of 2.3 %.

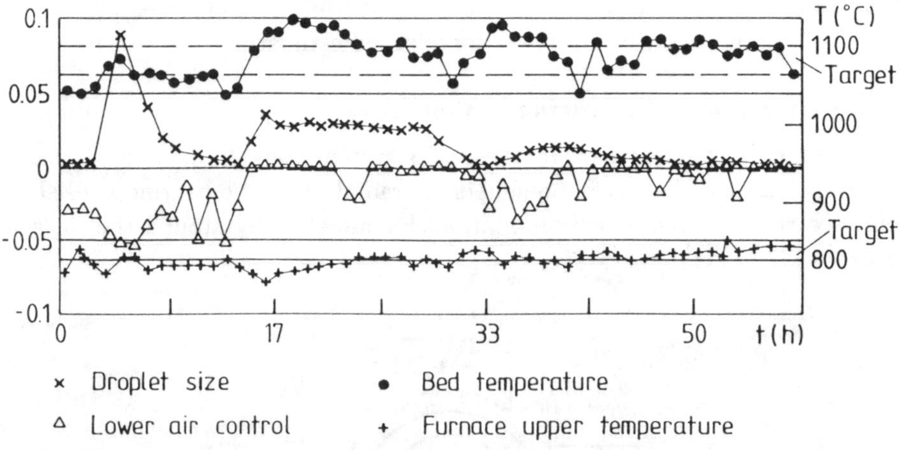


Fig. 5. Process data from a mill over a period of 56 hours. Temperature scales are on the right hand side and the control situation and balance on the left hand side.

## 3. Fuzzy Logic in Active Magnetic Bearing Control

### 3.1. *Introduction*

Active magnetic bearings (AMB) have proved to be an attractive technology in applications like turbo molecular pumps, flywheels for energy storage, reaction wheels for attitude control of artificial satellites, the main shafts of electric motors of tool machines and turbo compressors. In such applications the motor speeds are very high ranging from 10.000 to 200.000 1/min. Traditional bearings will not last or their life span is quite short. Therefore non-contact bearings like gas, liquid or magnetic bearings are used.

Magnetic bearings have the following advantages: small losses, good damping properties, very reliable, maintenance free with long life span, high rotating speeds, good operation even in dirty conditions. The disadvantages include: external power source, relatively expensive, lamination of radial bearing, tailored tuning of controllers. Non-linearities in the AMB make their control a very demanding task. A complete magnetic bearing system consists of a controller with position sensor, current amplifier and magnetic bearing itself.

A fuzzy derivative gain adjusting method for AMB control has been developed by High Speed Tech Ltd with VTT Automation. The method is based on the error and the error derivative signals that are used as the basis of the fuzzy inference. The approach provides a powerful method for elimination of the vibration noise effects and rotational speed estimation.

### 3.2. *Active magnetic bearing system*

The electro-mechanical part of the AMB system consists of a rotor that is hovering between the electric magnets. A radial magnetic bearing consists of four electric magnets, which are stationed symmetrically around the rotor as in Fig. 6.

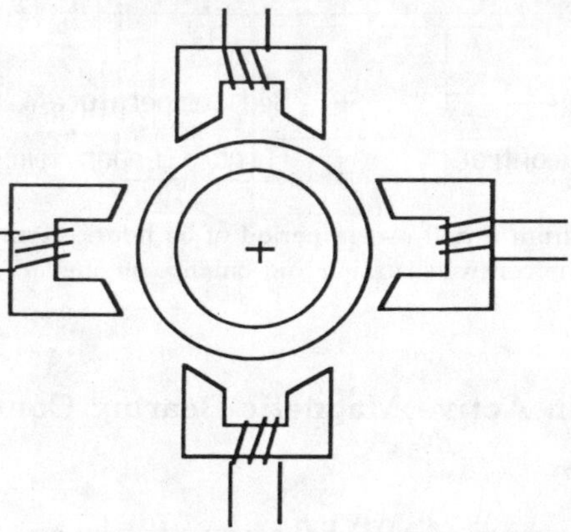

**Fig. 6. Principle of a radial magnetic bearing.**

The rotor is manufactured out of laminated iron to decrease the eddy currents. The lamination needed in radial bearings is not able to stand as high radial speeds as the rest of the rotor and the place of the bearing must be made fairly shallow. This means that the rotor becomes elastic and the critical speeds decrease. On the other hand, active magnetic bearings can efficiently decrease the bending oscillation, in which case the rotor can be driven over the critical rotational speeds to safer speeds.

The rotor position is measured and the electric current of magnets is controlled according to a control algorithm. Mathematically, the electro-mechanical part of the AMB system can be described using force equation. If the force equation is linearized around the operating point this results in

$$m\ddot{\delta} - K_b\delta = K_i i + F_e \tag{1}$$

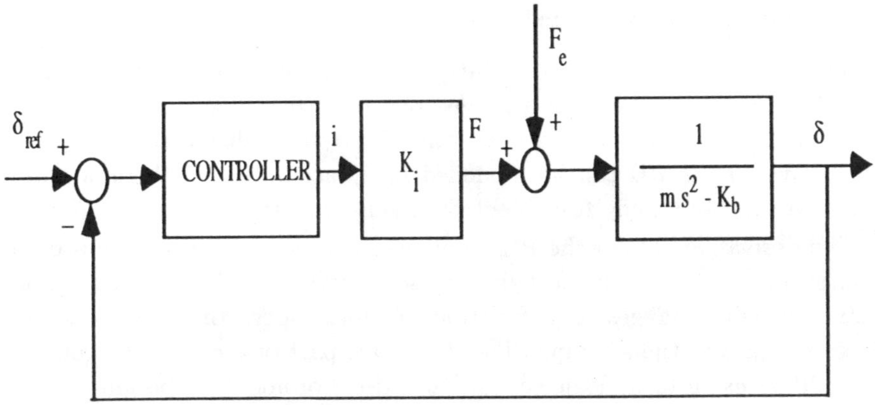


Fig. 7. Block diagram of AMB system.

where $m$ is the lumped mass of the rotor, $\delta$ is the air gap between the shaft and the electric magnet, $F_e$ is the external disturbance force, $K_i$ is the current gain, and $K_b$ is the position stiffness. The linear system in (1) can be displayed as a block diagram of Fig. 7.

It is clear that the open loop system is unstable having two unstable poles on the imaginary axis.

### 3.3. *Control of the active magnetic bearing system*

The most common primary control algorithm for an AMB is a PID controller. With a PID controller both the demands of the dynamic properties and the steady-state accuracy of the AMB system can be fulfilled. However, conventional PID controllers are insufficient to control the AMB in all situations. The unbalanced rotating shaft fluctuates sinusoidally according to the rotating speed of the shaft and generates noise in the air gap measurement. Even a small eccentricity in the shaft can cause problems in the stability of the system because of the derivative term of the controller. The PID-controlled system has high stiffness when the rotating frequency $\omega$ exceeds the natural frequency $\omega_n$. Usually some kind of compensator has to be added to meet the dynamic demands especially at high frequencies. Unfortunately, the compensating circuit can become very complicated and hard to tune.

### 3.4. *Fuzzy control structures*

There are several potential possibilities to apply fuzzy logic in the control of an AMB. The first possibility is the use of a basic fuzzy logic controller (FLC) which can be considered as a conventional way of utilizing fuzzy logic in a control system. AMB can be modelled fairly accurately so there is not too much advantage in using fuzzy control in this manner.

The derivative gain of the PID controller can be a source of vibration due to noise in the displacement signal caused by the unbalanced shaft. Rotor balance usually changes as a function of motor speed or because warming causes changes in the air gaps. The derivative part of the PID controller can be modified using fuzzy logic so that the effects of noise can be minimized as shown in Fig. 8.[3,4]

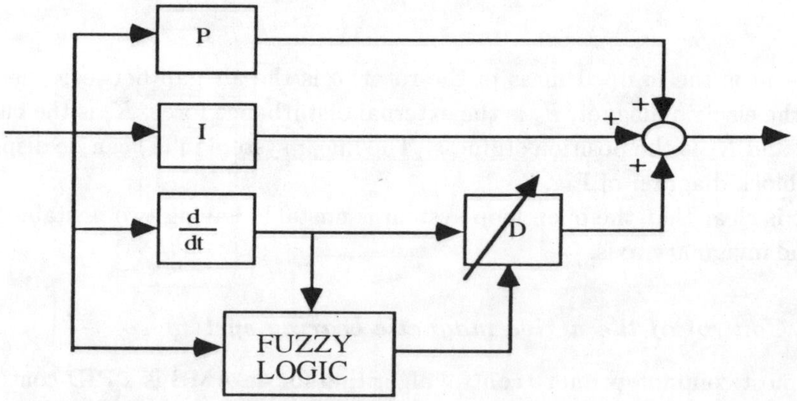


Fig. 8. Fuzzy control of the D part in the AMB control.

The problem with a PID controller is a bathtub-shaped magnitude response. The high gain at low frequencies is desired because stiffness improves the response during the start-up and at low frequencies noise is not a problem. The increased gain at high frequencies makes the system sensitive to the noise. This implies that the high gain at low frequencies should be maintained, but the gain at high frequencies should be decreased.

One way to decrease the gain at high frequencies is to change the derivative gain as a function of the rotating speed. But this requires the speed to be measured and this would require additional sensors which is undesirable. Since the position measurement is available, its change can also be computed and used to set up fuzzy rules. The rules are determined from the engineering knowledge of the system. When the speed of the motor is high the system is operating at high frequencies and the gain of the derivative part must be reduced. When the error in displacement is big and the absolute value of speed is also big, the system is vibrating a lot and therefore the derivative gain should be reduced significantly. The basic rule table can be summarized as:

Rule 1: If *error* is *small* and *change in error* is *small* then *gain* is *big*
Rule 2: If *error* is *small* and *change in error* is *big* then *gain* is *moderate*
Rule 3: If *error* is *big* and *change in error* is *small* then *gain* is *big*
Rule 4: If *error* is *big* and *change in error* is *big* then *gain* is *small*

For inference min-max was applied and for defuzzification center of gravity method. The membership functions are displayed in Fig. 9.

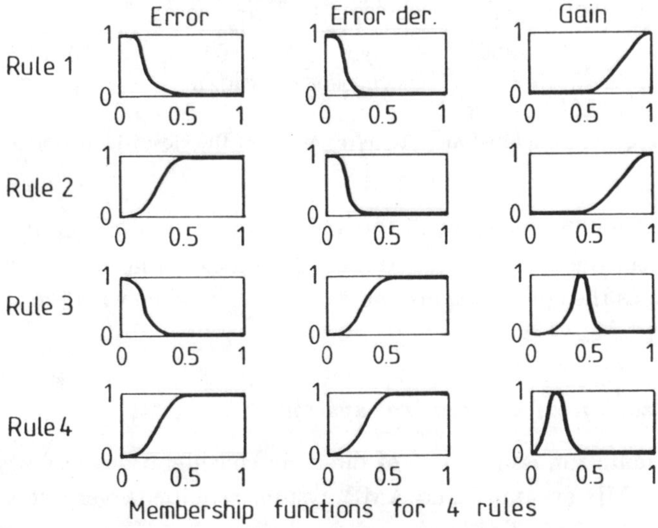


Fig. 9. The membership functions used in the rule base.

Figure 10 shows the amplitude responses of the closed-loop system simulations of both conventional PID controller and the proposed PID controller with fuzzy derivative gain adaptation. The conventional PID control system has a small amplitude at low and high frequencies and an amplitude peak at the natural frequency. The fuzzy derivative gain adjusting controller has a small

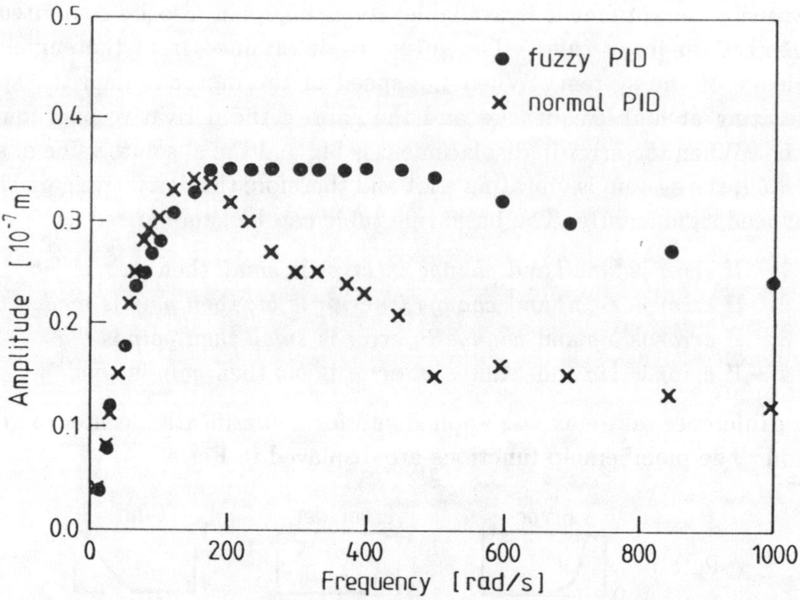


Fig. 10. Simulated amplitude responses of the closed-loop system.

amplitude at low frequencies, but at high frequencies the amplitude is bigger than in the conventional PID controller. This means that the fuzzy realization better satisfies the specifications and is also able to cope with the audible noise problem.

### 3.5. *Realization of the control system*

The implementation issue is one of the main considerations of fuzzy logic control of the AMB system. The AMB system requires a very short sampling time, when the controller is implemented digitally. Usually a sampling frequency of 10 to 15 kHz is appropriate. When the sampling frequency is so high, strong demands are posed for the hardware. The timing can be critical, if any additional calculations are needed. A common situation is that the position measurement has to be digitally filtered to obtain a proper and feasible signal for the control algorithm. The filter algorithms together with the fuzzy inference, even if the min-max calculations require only little computation, generate difficult timing problems. Another problem arises with the derivative action of the controller. A smooth signal is required, which again means filtering posing additional demands for the hardware.

Figure 11 shows the experimental hardware setup. The signal processor board used for development work is a SPIRIT30 board made by Sonitech Inc. The board includes Texas Instruments TMS320C30-33 digital signal processor with 64 k words (32 bits) of zero wait-state memory, peripheral interface, two serial channels, interface for memory board and a 16-bit PC bus interface. The communication between the DSP board and the host PC is provided via I/O and DMA interface and the maximum transfer rate is 3.3 Mbytes. The signal processor board is replaced with a stand alone DSP board which exploits TMS320C31 digital signal processor. The board includes SCSI and CAN interface.

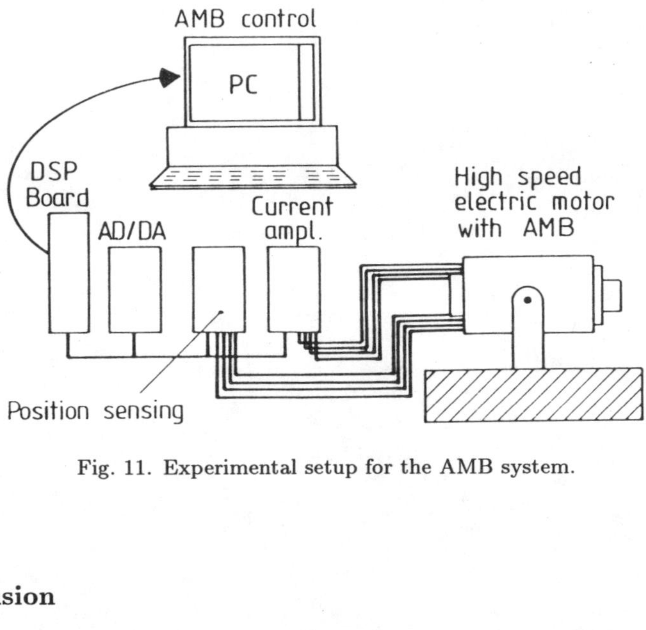


Fig. 11. Experimental setup for the AMB system.

## 4. Conclusion

In this chapter some fuzzy control applications in Finland have been reviewed. Boiler bed control is a process automation application, where time is not a critical factor, but user interface is of utmost importance. In the second application, active magnetic bearing, time considerations are essential in control implementatation. In both cases the use of fuzzy logic has proven to be a useful technique with which extra value can be added to the system. In the near future there will be significantly more industrial fuzzy logic applications in Finland.

## References

1. K. G. Sundström, *Proc. on Pulp Technology and Energy. New Available Techniques and Current Trends. 24th EUCEPA Conference* (Stockholm) Sweden, p. 401.
2. J. Mikkonen, J. Koskinen and J. Hanhinen, *Paperi ja Puu — Paper and Timber* **76** (1994).
3. H. Koskinen, O. Lindgren and O. Karasti, *Fourth Int. Symp. on Magnetic Bearings*, Zürich, Switzerland (August, 1994).
4. H. Koskinen, *EUFIT'93–The First European Congress on Fuzzy and Intelligent Technologies*, Aachen, September 7–10, 1993.

# RECURSIVE FUZZY REASONING AND ITS APPLICATION TO AN AUTO-TUNING CONTROLLER

KOHEI NOMOTO
*Electro-Optics and Microwave Systems Laboratory,*
*Mitsubishi Electric Corporation,*
*5-1-1 Ofuna, Kamakura, Kanagawa, 247, Japan*

## 1. Introduction

Fuzzy theory is now popular in the industrial world. Following success of early works,[1,2] many efforts have been made to apply the fuzzy theory to industrial control. This application is called "fuzzy control" and it has been a leading area among many applications of the fuzzy theory.

In the fuzzy control, fuzzy reasoning takes the place of a conventional controller. The fuzzy control is an intelligent control because it is able to perform skilled operator's technique. The fuzzy control is hence considered as a promising method to control complex systems.

In spite of the worth of the fuzzy control, there are not many actual control systems using it even now, and are especially few in process control. Most controllers used in the process control are conventional ones. They are PID (Proportional Integral and Derivative) controllers. The PID controllers have been used widely in many areas of industry. Engineers and operators have been using the PID controllers for many years, and they have the know-how of making the best use of them in their own plants and rely on them very much. The PID controllers work well in many kinds of plants if their parameters are proper. It depends on experience of skilled operators to tune the parameters properly. Recently, the number of the controllers is increasing, but on the other hand the number of skilled operators is decreasing. An auto-tuning method of the PID controllers is necessary to the plant.

Many of auto-tuning controllers have been proposed.[3,4] Most of the auto-tuning controllers assume a linear model of the control systems and tune the parameters of the PID controllers according to it. These auto-tuning controllers are, however, not applicable to practical plants, because the assumption does not hold. Most of controlled systems are nonlinear; besides, actual PID controllers are also nonlinear. A different auto-tuning method is necessary for industrial plants.

This chapter presents a new auto-tuning PID controller. In this auto-tuning PID controller, the fuzzy reasoning works as a tuner instead of a controller. The fuzzy reasoning tuner simulates skilled operator's manual tuning, but the fuzzy reasoning is not a conventional one. The conventional fuzzy reasoning is not useful for the auto-tuning because of the following reasons: (1) Signs of the control condition appear only occasionally at signals in the control system, and consequently, inputs of the fuzzy reasoning are not always available. (2) The proper parameters of the PID controller are almost constant, and hence the outputs of the fuzzy reasoning are not variable.

A fuzzy reasoning method that is able to reason constant outputs using scarce inputs is necessary. It is called "recursive fuzzy reasoning".[5]

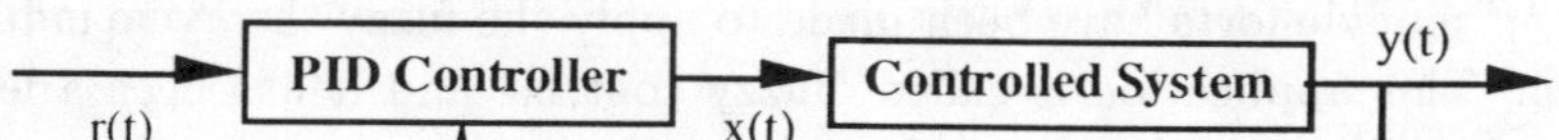


Fig. 1. Typical control systems.

## 2. Fuzzy Auto-Tuning PID Controller

Most control systems use PID controllers as shown in Fig. 1. In this figure, $r(t)$ is reference input, $x(t)$ is manipulated variable and $y(t)$ is the controlled variable. The PID controllers account for 84 percent of practical controllers and 90 percent including improved ones according to an investigation in Japan in 1990.[6] Tuning the parameters properly, the PID controller is applicable to various kinds of processes. The parameters of the PID controller are proportional gain, integral time and derivative time.

In fuzzy control, fuzzy reasoning is used as a controller instead of PID controller as shown in Fig. 2. Fuzzy reasoning is able to perform more sophisticated control than PID controller. A problem in it is that designing it is

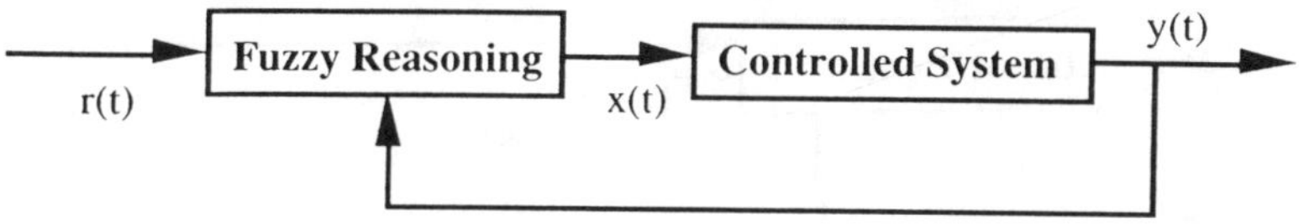


Fig. 2. Conventional fuzzy control systems.

very difficult because it has much more parameters than the PID controller. No general design theory of the fuzzy control has been established yet, and consequently, many parameters must be designed by trial and error. It is a heavy burden for its users.

The tuning of the PID controller is not easy indeed, but skilled operators in plants actually tune the three control parameters. The operators watch signs appearing at $r(t)$, $x(t)$ or $y(t)$ occasionally and use tuning rules according to their experience as shown in Fig. 3.

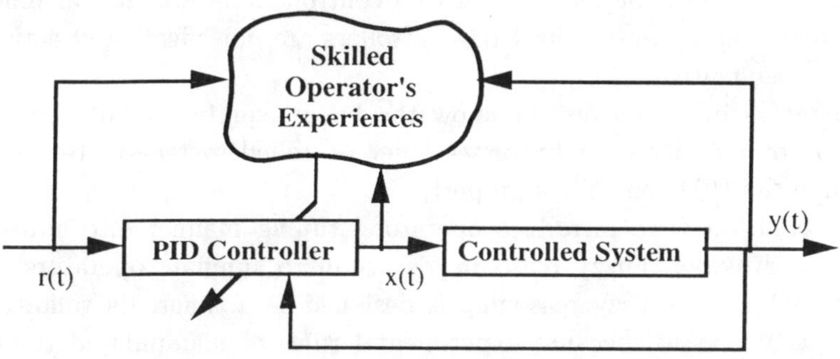


Fig. 3. PID control tuned by skilled operator.

Auto-tuning PID controllers have been developed in order to tune the parameters in the place of skilled operators who are fewer as compared with the number of controllers. The algorithm of conventional PID auto-tuning consists of two steps as shown in Fig. 4. The first step is the identification of controlled system. Linearity is usually assumed in this step and the controlled system is modeled as a linear equation. The second step is PID tuning according to the linear model. The performance of the conventional PID auto-tuning is

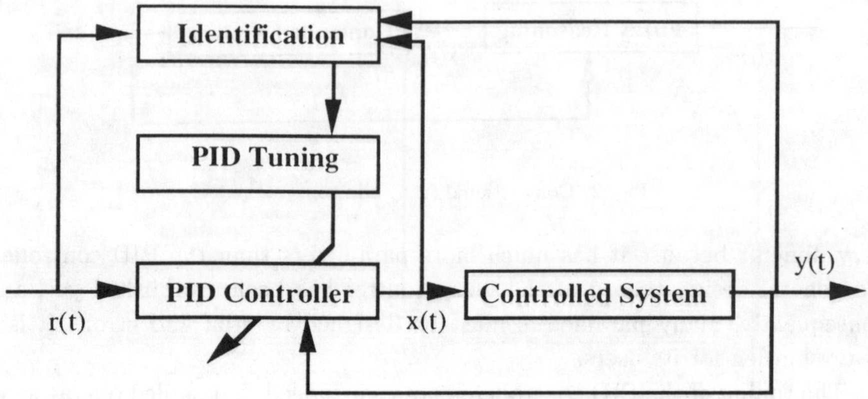


Fig. 4. Conventional auto-tuning PID controller.

determined by the accuracy of the identification. Unfortunately, most processes are nonlinear; besides, actual PID controllers have nonlinear functions. The conventional auto-tuning PID controllers are not effective in actual processes consequently.

Operators in plants do not know the linear equation models and hence they do not identify their processes (the controlled systems). Nevertheless, they tune the PID controllers properly.

The solution is to introduce operators' tuning manner into auto-tuning algorithm directly. Fuzzy reasoning is useful to simulate operators' tuning manner. Once the fuzzy reasoning is designed as a tuner, its validity holds in various processes, because experimental rules of manipulated tuning are invariant with the processes.

The fuzzy auto-tuning PID controller (1) controls processes with PID control algorithm which is predominant in industry, and (2) tunes its parameters automatically through use of fuzzy reasoning which simulates skilled operators' experimental manner. The fuzzy reasoning used in the fuzzy auto-tuning PID controller is not an usual one. It is "recursive fuzzy reasoning," whose inputs are not always available and whose outputs are constant or almost constant. Figure 5 illustrates the fuzzy auto-tuning PID controller using the recursive fuzzy reasoning.

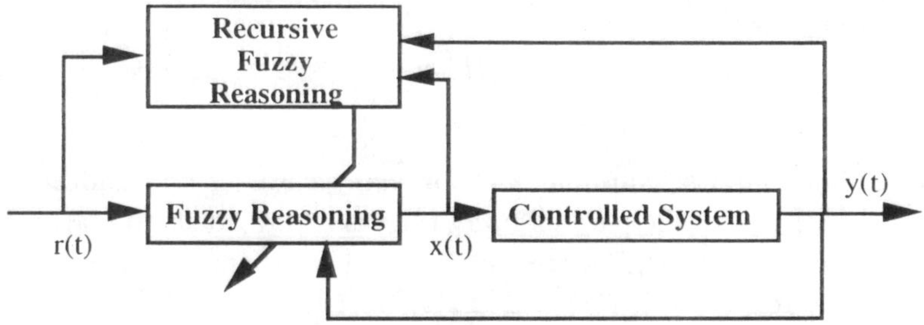


Fig. 5. Fuzzy auto-tuning PID controller using recursive fuzzy reasoning.

## 3. Usual Fuzzy Reasoning Used in Conventional Fuzzy Control

Though usual fuzzy reasoning used in conventional fuzzy control is not useful in PID auto-tuning, it is described here as a contrast of the recursive fuzzy reasoning, which is used in PID auto-tuning.

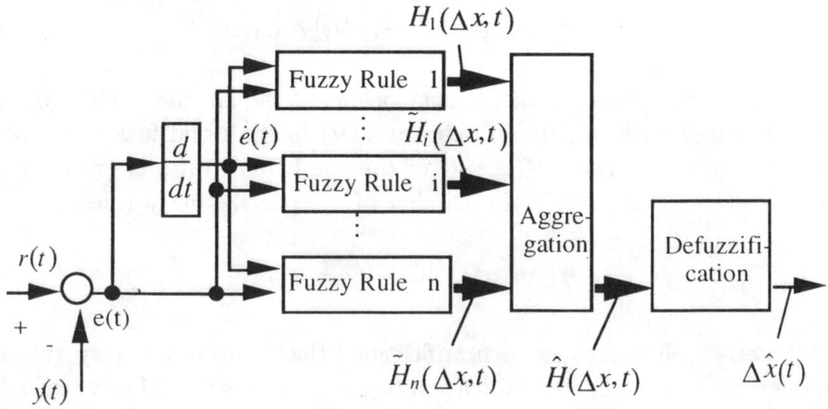


Fig. 6. Usual fuzzy reasoning used in conventional fuzzy control. (Thin lines denote numerical values and thick lines denote fuzzy quantities.)

Figure 6 illustrates the usual fuzzy reasoning used in conventional fuzzy control, which appeared in Fig. 2. Error signal $e(t)$ between the reference input $r(t)$ and the controlled variable $y(t)$ and its derivative $\dot{e}(t)$ are inputs of parallel fuzzy rules.

$$e(t) = y(t) - r(t)\,. \tag{1}$$

The parallel fuzzy rules describe proper values $\Delta x(t)$ to increase or decrease the present manipulated variable $x(t)$ for many cases of $e(t)$ and $\dot{e}(t)$. For example, the $i$th one of the parallel fuzzy rules is described as "If $e(t)$ is $F_i$ and $\dot{e}(t)$ is $G_i$, then $\Delta x(t)$ is $H_i$," where, $F_i$, $G_i$, and $H_i$ are membership functions which define states of $e(t)$, $\dot{e}(t)$ and $\Delta x(t)$, respectively. The states are, for example, "positive large", "positive small," "nearly zero," "negative small" and "negative large." The degrees at which the inputs satisfy the corresponding states are evaluated with the membership functions, $F_i[e(t)]$ and $G_i[\dot{e}(t)]$. The proper value of $\Delta x(t)$ to increase or decrease the present manipulated variable $x(t)$ in the $i$th fuzzy rule is also the membership function $H_i(\Delta x)$.

The level at which each rule holds true is evaluated by the if-part, and the then-part is suppressed by the level. In the usual fuzzy reasoning, "and" in the if-part and "then" in the fuzzy rule are calculated with minimum operation ($\wedge$). The conclusion $\tilde{H}_i(\Delta x,\, t)$ of the $i$th fuzzy rule at time $t$ is then calculated by the following equation:

$$\tilde{H}_i(\Delta x,\, t) = F_i[e(t)] \wedge G_i[\dot{e}(t)] \wedge H_i(\Delta x)\,. \tag{2}$$

The next step is aggregation of the conclusions of the all rules. The aggregation is calculated with maximum operation ($\vee$) in the usual fuzzy reasoning. The synthesized conclusion $\hat{H}(\Delta x,\, t)$ of the total fuzzy rules is given by the following equation, where $n$ is the number of the parallel fuzzy rules:

$$\hat{H}(\Delta x,\, t) = \bigvee_{i=1}^{n} \tilde{H}_i(\Delta x,\, t)\,. \tag{3}$$

The last step is referred to as "defuzzification" that transforms fuzzy quantity into numerical value. The defuzzification technique in the usual fuzzy reasoning is to calculate the center of gravity of the synthesized conclusion $\hat{H}(\Delta x,\, t)$. The value obtained in this step is the reasoned proper value, $\Delta\hat{x}(t)$ to increase or decrease the present manipulated variable $x(t)$.

$$\Delta\hat{x}(t) = \frac{\int \Delta x \cdot \hat{H}(\Delta x,\, t) d\Delta x}{\int \hat{H}(\Delta x,\, t) d\Delta x}\,. \tag{4}$$

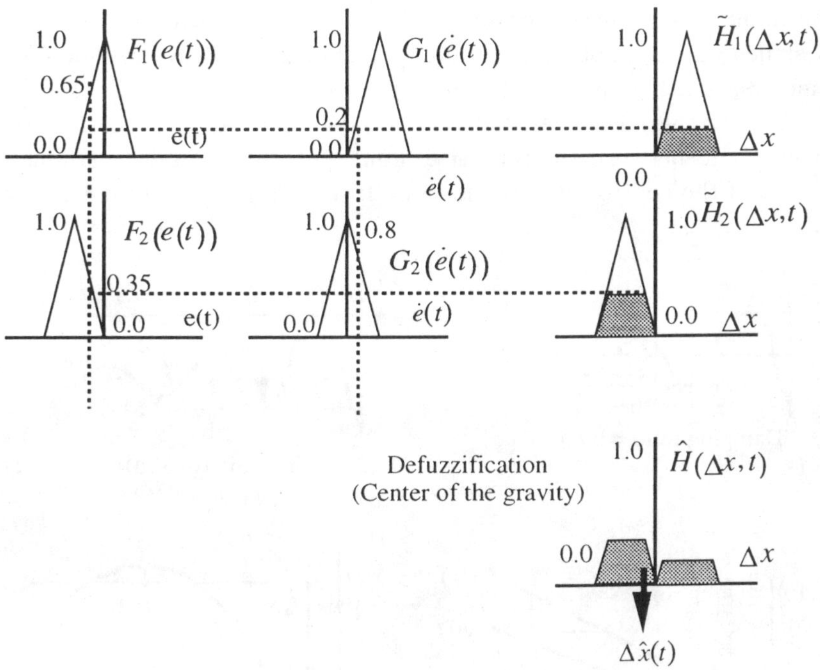


Fig. 7. The summary of the calculations of the usual fuzzy reasoning.

The present value of the manipulated variable is then determined as:

$$x(t) = x(t-1) + \Delta\hat{x}(t)\,. \tag{5}$$

Figure 7 illustrates the summary of the calculations of the usual fuzzy reasoning.

## 4. Fuzzy Reasoning for PID Auto-Tuning

### 4.1. *Inputs and outputs of the reasoning*

As outlined in the previous section, the inputs of the usual fuzzy reasoning in conventional fuzzy control are error signal $e(t)$ and its derivative $\dot{e}(t)$ and its output is the change of the manipulated variable $\Delta\hat{x}(t)$. We can observe the inputs all the time and we must always update the output.

Our purpose is to develop a fuzzy reasoning that tunes the PID parameters in the manner of skilled operators. The inputs of this fuzzy reasoning are what skilled operators observe to tune the parameters. They are characteristic parameters, which appear at the reference input, manipulated variable and controlled variable. For example, they are damping ratio of the error signal, follow-up ratio between the reference input and the controlled variable, and overshoot of the controlled variable caused by step input. Figure 8 illustrates them.

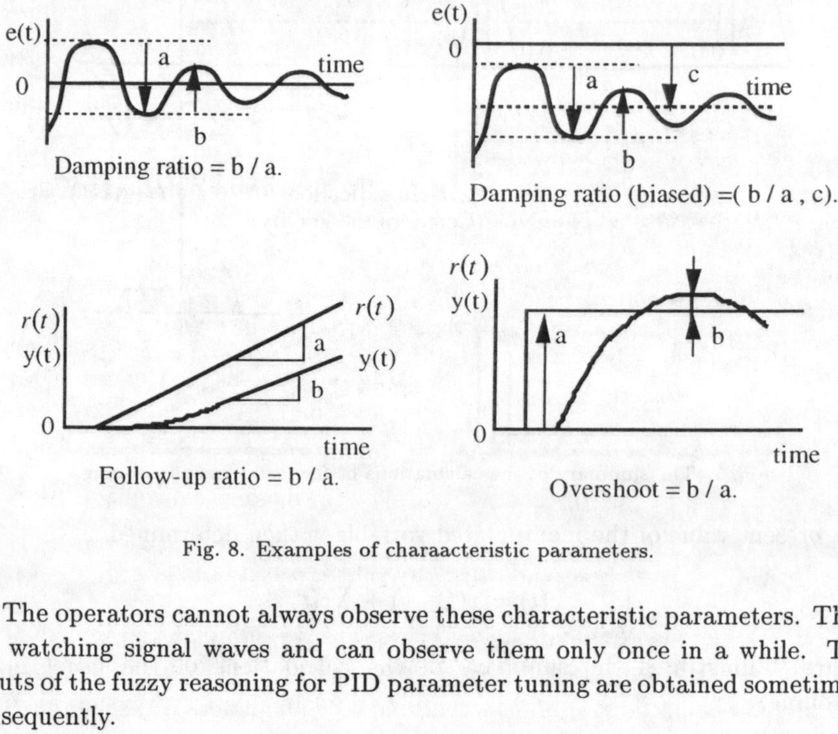


Fig. 8. Examples of charaacteristic parameters.

The operators cannot always observe these characteristic parameters. They are watching signal waves and can observe them only once in a while. The inputs of the fuzzy reasoning for PID parameter tuning are obtained sometimes, consequently.

The outputs of the fuzzy reasoning for PID parameter tuning are of course PID parameters, which are proportional gain, integral time and derivative time. These outputs are constant or they change very slowly even if they do.

Skilled operators watch the characteristic parameters, which appear once in a while, accumulate fragmentary knowledge of the control system and finally tune the PID controllers. We must develop a fuzzy reasoning that tunes the PID controllers in the same way as the skilled operators' one.

### 4.2. *Evaluation in the reasoning*

The basic operation of PID controller is:

$$\begin{aligned}\Delta x(t) &= x(t) - x(t-1) \\ &= K_c \left[ \{e(t) - e(t-1)\} + \frac{T_s}{T_I} e(t) + \frac{T_D}{T_s} \{e(t) - 2e(t-1) + e(t-2)\} \right] ,\end{aligned} \tag{6}$$

where $\Delta x(t)$ is the change of the manipulated variable, $T_s$ is sampling time, $K_c$ is proportional gain, $T_I$ is integral time and $T_D$ is derivative time.

The actual PID controller, however, is modified from the basic one. The actual PID controller has, for example, saturation function, gap function and anti-reset windup function. These functions are nonlinear, and hence actual PID controller does not have linearity. Besides, most controlled systems are nonlinear. The PID auto-tuning method using linear models is not successful because of these nonlinearities.

These nonlinearities bring about difficulty in operators' PID tuning, too. Even skilled operators cannot evaluate how the PID parameters are agreeable, in many cases. Judging from the stability and speed of response, they evaluate how inadequate the PID parameters are instead. They compare degrees of inadequacy among other PID parameters and tune them gradually. Since stability and speed of response are not compatible with each other, the optimum values of the PID parameters are the balanced points between them.

A fuzzy reasoning in PID auto-tuning also evaluates the degrees of inadequacy.

## 5. The Recursive Fuzzy Reasoning

The recursive fuzzy reasoning has been developed to tune the PID parameters. Its inputs are characteristic parameters which are obtained once in a while, and its outputs are parameters which are constant or change very slowly. It evaluates the degree of inadequacy, while usual fuzzy reasoning evaluates satisfaction of the reasoned value.

Figure 9 illustrates the recursive fuzzy reasoning used in PID auto-tuning, which appeared in Fig. 5.

Figure 10 illustrates the summary of the calculations of the recursive fuzzy reasoning.

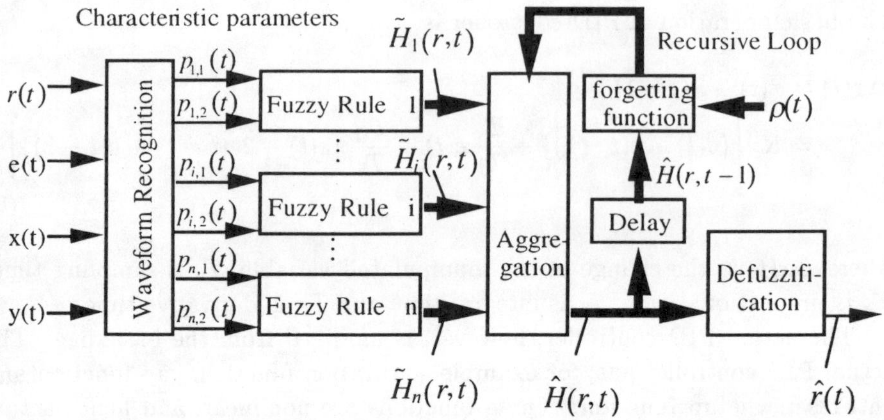


Fig. 9. Recursive fuzzy reasoning used in PID auto-tuning. (Thin lines denote numerical values and thick lines denote fuzzy quantities.)

## 5.1. *Characteristic parameters and conclusion of each rule*

The first block recognizes signal waves and calculates characteristic parameters, which are shown in Fig. 8.

The parallel fuzzy rules describe the degree of inadequacy of the control performance determined by the present PID parameters. For example, a rule states "If damping ratio of the error signal is small and following ratio between the reference input and the controlled variable is small, then the present proportional gain is too small". To generalize the proposition of the parallel fuzzy rules, we describe the $i$th rule as "If $p_{i,1}(t)$ is $F_{i,1}$ and $p_{i,2}(t)$ is $F_{i,2}$, then $r(t)$ is $H_i$," where $p_{i,1}(t)$ and $p_{i,2}(t)$ are the characteristic parameters. The number of the characteristic parameters in each rule is arbitrary, while there are two parameters, error signal and its derivative, in each rule of the usual fuzzy reasoning. $F_{i,1}$ and $F_{i,2}$ are membership functions which define states of the characteristic parameters and the number of membership functions is the same as that of the characteristic parameters in each rule. The degrees at which the characteristic parameters satisfy the corresponding states are evaluated with the membership functions, $F_{i,1}[p_{i,1}(t)]$ and $F_{i,2}[p_{i,2}(t)]$, when $p_{i,1}(t)$ and $p_{i,2}(t)$ are observed. On the other hand, when $p_{i,1}(t)$ and $p_{i,2}(t)$ are not observed, we define $F_{i,1}[p_{i,1}(t)]$ and $F_{i,2}[p_{i,2}(t)]$ as zero.

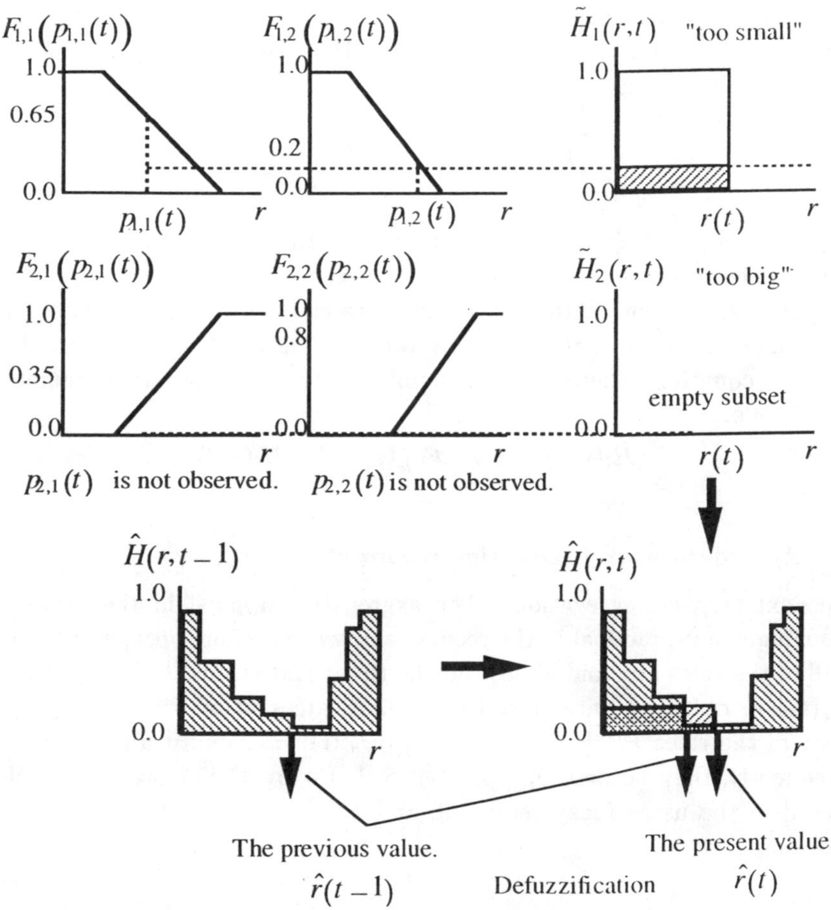


Fig. 10. The summary of the calculations of the recursive fuzzy reasoning. (Generally, $p_{1,1}(t)$, $p_{1,2}(t)$, $p_{2,1}(t)$ and $p_{2,2}(t)$ are different characteristic parameters.)

$r(t)$ is the present value of one of the PID parameters which we tune. $H_i$ is a membership function which define inadequacy of the PID parameter. There are two kinds of the inadequacy: One is "too large" and another is "too small." If the present value of a PID parameter $r(t)$ is "too large," larger values than $r(t)$ are, of course, "too large." Hence we define membership function of "too large" as:

$$H_i(r,\, t) = \begin{cases} 0; & r < r(t) \\ 1; & r(t) \le r \end{cases} . \tag{7}$$

Similarly, if the present value of a PID parameter $r(t)$ is "too small," smaller values than $r(t)$ are, of course, "too small." Hence we define membership function of "too small" as:

$$H_i(r, t) = \begin{cases} 1; & r \le r(t) \\ 0; & r(t) < r \end{cases} . \tag{8}$$

The level at which each rule holds true is evaluated by the if-part and the then-part is suppressed by the level. In the recursive fuzzy reasoning, "and" in the if-part and "then" in the fuzzy rule are calculated with minimum operation ($\wedge$). The conclusion of the $i$th fuzzy rule at time $t$ is then calculated by the following equation, where $m_i$ is the number of the characteristic parameters in the $i$th rule:

$$\tilde{H}_i(r, t) = \left[ \bigwedge_{j_i=1}^{m_i} F_{i,j_i}(p_{i,j_i}) \right] \wedge H_i(r, t) . \tag{9}$$

## 5.2. *Aggregation and recursive reasoning*

The next step is aggregation. The aggregation applied in the usual fuzzy reasoning is not practical in the recursive fuzzy reasoning, because conclusions of all of the rules are void almost all the time. Since characteristic parameters $p_{i,j_i}(t)$ are obtained once in a while, the degrees at which they satisfy the states in the rules $F_{i,j_i}[p_{i,j_i}(t)]$ are $F_{i,j_i}[p_{i,j_i}(t)] = 0$ almost all the time and, infrequently, they become $F_{i,j_i}[p_{i,j_i}(t)] > 0$. Figure 11 illustrates examples of $F_i[e(t)]$ in the usual fuzzy reasoning and $F_{i,j_i}[p_{i,j_i}(t)]$ in the recursive fuzzy reasoning.

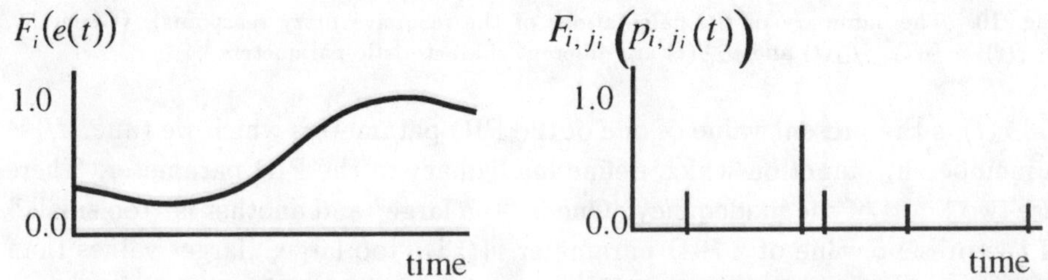


(a) The usual fuzzy reasoning. (b) The recursive fuzzy reasoning.

Fig. 11. Examples of degrees at which input parameters satisfy the states in the rules.

The conclusions of each rule are given as a fuzzy subset. If at least one characteristic parameter in a rule is not observed, the conclusion of the rule is an empty subset, which is void. If conclusions of all the rules are empty subsets, the synthesized conclusion obtained by the aggregation applied in the usual fuzzy reasoning is also an empty subset. In this case, the synthesized conclusion is meaningless and defuzzification is not feasible.

Fortunately, we are able to compensate this lack of new information of the characteristic parameters with past information of the synthesized conclusion, because the outputs are constant or change very slowly. Namely, we are able to construct the present synthesized conclusion $\hat{H}(r, t)$ with not only the rules' conclusions $\tilde{H}_i(r, t)$, but also past synthesized conclusion $\hat{H}(r, t-1)$. This idea suggests a recursive scheme.

Delay in the recursive loop in Fig. 9 brings about time shift. There is a forgetting function in the recursive loop, when the output changes slowly. The forgetting function is to multiply the fuzzy set of the past synthesized conclusion $\hat{H}(r, t-1)$ by a scalar $\rho(t)$; $0 < \rho(t) \leq 1$. The equation of the present synthesized conclusion $\hat{H}(r, t)$ is, as a result:

$$\hat{H}(r, t) = \left[\bigvee_{i=1}^{n} \tilde{H}_i(r, t)\right] \vee [\rho(t) \cdot \hat{H}(r, t-1)]\,. \tag{10}$$

## 5.3. *Defuzzification technique and reasoned value*

The last step is defuzzification. The center of gravity is calculated for defuzzification in the usual fuzzy reasoning, because the fuzzy subset of the synthesized conclusion $\hat{H}(\Delta x, t)$ means satisfactory values of $\Delta x(t)$. On the other hand, the fuzzy subset of the synthesized conclusion in the recursive fuzzy reasoning $\hat{H}(r, t)$ means inadequate values of $r$, and hence we must develop alternative defuzzification technique.

Defuzzification is to choose a value of $r$ from the fuzzy subset $\hat{H}(r, t)$ in the recursive fuzzy reasoning. As shown in Fig. 10, we choose a value of $r$ at which the degree of inadequacy $\hat{H}(r, t)$ is the smallest. Since such a value of $r$ is not unique, we define the defuzzificated value $\hat{r}(t)$ as the nearest value to the previous value $\hat{r}(t-1)$ of all values of $r$ at which $\hat{H}(r, t)$ is the smallest.

We express the defuzzification in the recursive fuzzy reasoning as $f(\cdot)$. The value obtained by the defuzzification is the output of the recursive fuzzy reasoning. This output is the proper value of a PID parameter.

$$\hat{r}(t) = f(\hat{H}(r, t))\,. \tag{11}$$

## 6. Application to the PI Auto-Tuning Controller

Examples of computer simulation are presented to illustrate the use of the recursive fuzzy reasoning as a means of auto-tuning. In this simulation, control mode is PI control, which is most common in practical applications. The tuning mode is on-line tuning, in which parameters of a controller in operation are tuned. Desirable quality of control varies according to its applications. We suppose the desirable quality of control to be one that does not make overshoot, which is required in many applications such as in chemical plants.

In the first example, the controlled system has second-order lag and dead-time. It is given by:

$$G(s) = \frac{1}{T^2s^2 + 2\zeta Ts + 1} \cdot e^{-Ls} \qquad (T = 10,\ L = 10,\ \zeta = 0.75)\,. \tag{12}$$

This system is difficult to control, since this second order system is oscillatory itself, and besides, time lag is large. Figure 12 shows a simulation result of this system.

In this figure, time is normalized by sampling interval and amplitudes of the reference input and the controlled variable are normalized by step magnitude of the reference input.

At the beginning of the control, (see Figs. 12(a1) and (a2)), the proportional gain and the reciprocal of the integral time were too large and hence the control was very oscillatory. The tuning operation decreased these values to suppress the oscillation. When the reference input showed a stepped change at about time 500, control was then too weak. In such a case, we seldom observed the characteristic parameters and consequently, changes of the control parameters were rare.

From time 5000 through 6000, (see Figs. 12 (b1) and (b2)), control parameters were more proper than at the beginning. The tuning operation increased the proportional gain at time 5150, because the speed of response was slow. The tuning operation, however, put it back again at time 5250, because the controlled variable became oscillatory. Since stability and speed of response are not compatible with each other in such a difficult system, we must search the balanced points and permit the inadequacy of them to a certain extent.

From time 7000 through 8000, (see Figs. 12 (c1) and (c2)), control performance was much better than before. The speed of response was the fastest to the extent that the controlled variable does not overshoot. Though the proportional gain changed between 1.00 and 0.79, the control performance was

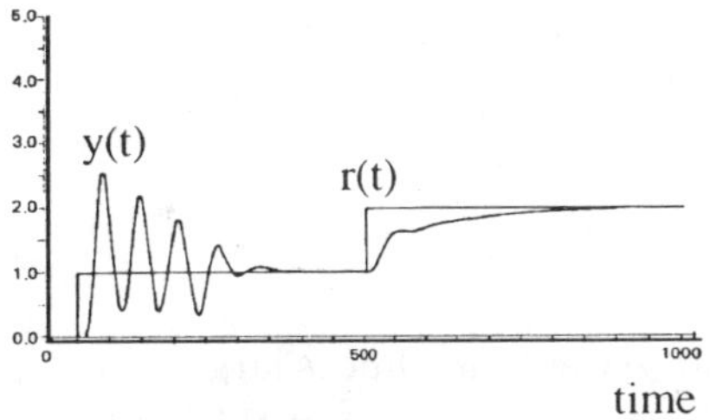


(a1) Reference input and controlled variable from 0 through 1000.

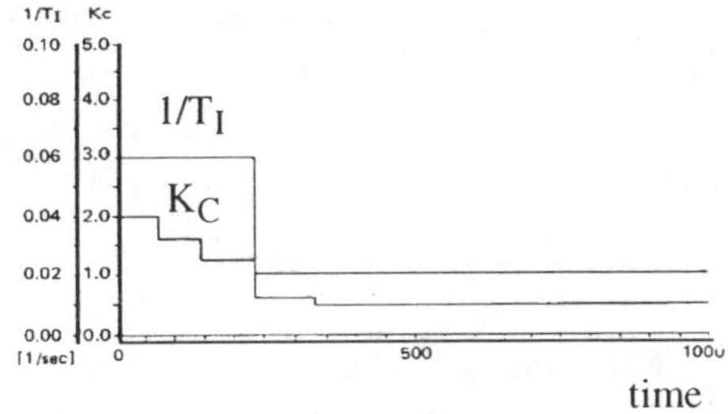


(a2) Proportional gain and the reciprocal of integral time from 0 through 1000.

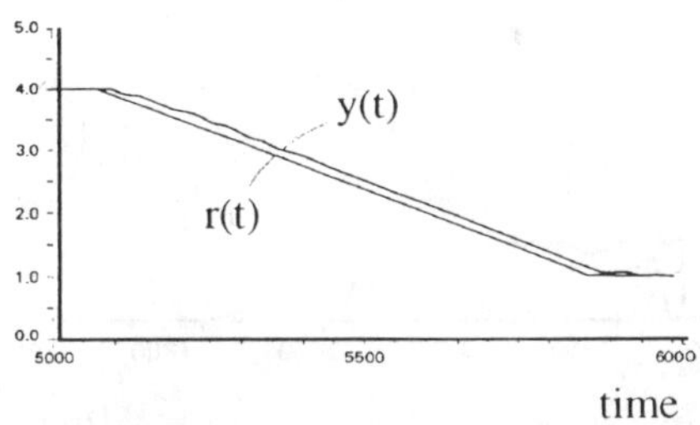


(b1) Reference input and controlled variable from 5000 through 6000.

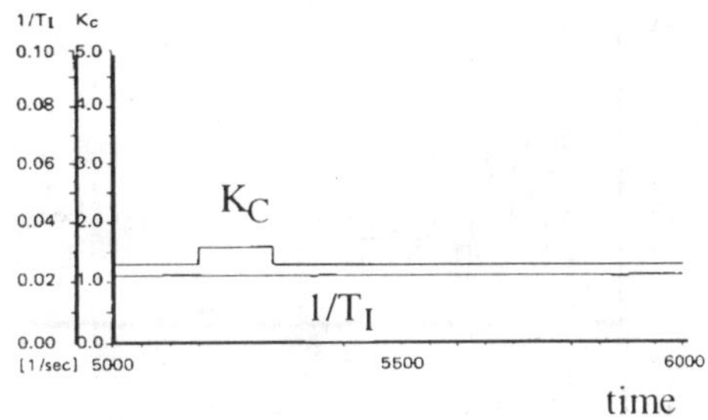


(b2) Proportional gain and the reciprocal of integral time from 5000 through 6000.

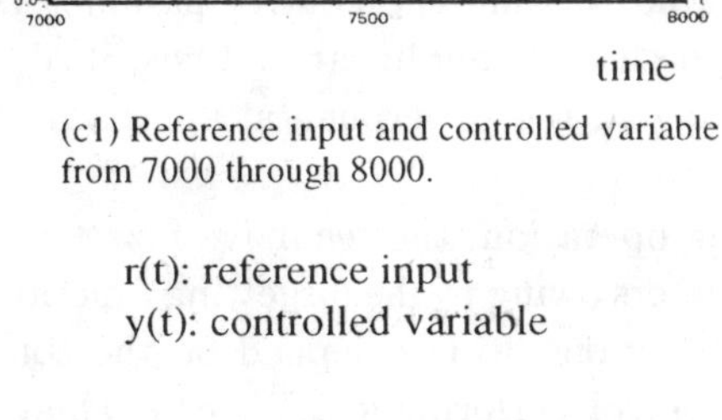


(c1) Reference input and controlled variable from 7000 through 8000.

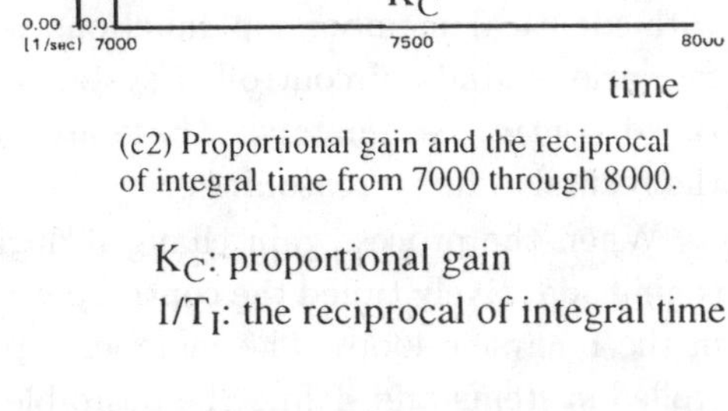


(c2) Proportional gain and the reciprocal of integral time from 7000 through 8000.

r(t): reference input
y(t): controlled variable

$K_C$: proportional gain
$1/T_I$: the reciprocal of integral time

**Fig. 12. A simulation result of second-order lag and dead-time.**

almost the same. This change is due to digitization of the proportional gain. It is expected that the proportional gain will converge in the process of tuning.

In another example, the controlled system has first-order lag and dead-time. It is given by

$$G(s) = \frac{1}{Ts+1} \cdot e^{-Ls} \quad (T = 10,\ L = 20)\,. \tag{13}$$

This system is also difficult to control, since the dead-time is large compared with the time constant. Figure 13 shows a simulation result of this system.

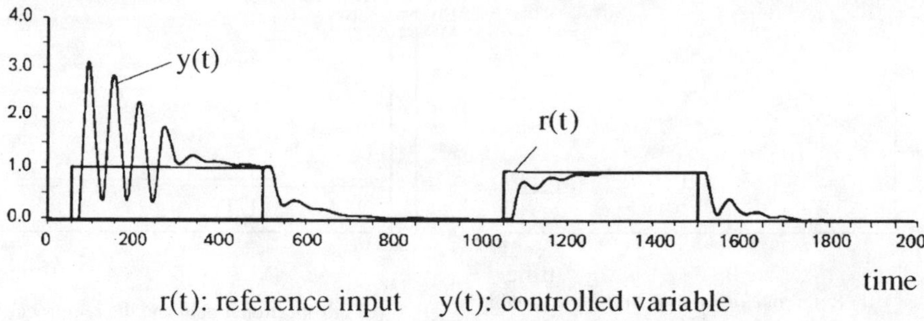


Fig. 13. A simulation result of first-order lag and dead-time.

In this figure, time and amplitude are also normalized.

The membership functions were identical with the first example, however, the desirable control performance was established in this case too.

We have simulated the auto-tuning by use of the recursive fuzzy reasoning in many cases. The results have confirmed that the recursive fuzzy reasoning with identical membership functions is able to tune the control parameters for various kinds of controlled systems. In cases of nonlinear systems, it also tuned control parameters. These are owing to the experimental approach of the recursive fuzzy reasoning.

When the process gain changed during operation, the recursive fuzzy reasoning adaptively tuned the control parameters owing to the forgetting function in the recursive loop. The membership functions do not depend on the controlled systems and define the desirable control performance. We can achieve another desirable control performance such as 25 percent overshoot, if we change the membership functions.

## 7. Conclusions

The fuzzy auto-tuning controller presented in this chapter controls processes with PID algorithm and tunes the PID control parameters with fuzzy reasoning. The fuzzy reasoning is the recursive fuzzy reasoning that observes characteristic parameters and optimizes the PID control parameters. The recursive fuzzy reasoning is also applicable to many other applications whose inputs are seldom observed and whose outputs are almost constant.

## References


1. S. Yasunobu and S. Miyamoto, "Automatic train operation system by predictive fuzzy control", in ed. M. Sugeno, *Industrial Applications of Fuzzy Control* (North Holland, Amsterdam, 1985), pp. 1–18.
2. O. Yagishita, O. Itoh, and M. Sugeno, "Application of fuzzy reasoning to the water purification process", ed. M. Sugeno, *Industrial Applications of Fuzzy Control* (North Holland, Amsterdam, 1985), pp. 19–39.
3. A. B. Corripio and P. M. Tompkins, "Industrial application of a self-tuning feedback control algorithm", *ISA Transactions*, **20**, 2 (1981), pp. 3–10.
4. H. S. Hoopes, W. M. Hawk, Jr. and R. C. Lewis, "A self-tuning controller", *ISA Transactions* **22**, 3 (1983), pp. 49–58.
5. K. Nomoto and M. Kondo, "The recursive fuzzy reasoning and its application to an auto-tuning controller", in *Proc. International Workshop on Fuzzy System Applications* (1988), pp. 81–82.
6. K. Furuta and T. Tomita, *Journal of the Society of Instrument and Control Engineers*, **29**, 10 (1990), pp. 953–958 (in Japanese).

# A PRACTICAL APPLICATION OF FUZZY THEORY TO AN AUTO-REGULATION SYSTEM FOR EXTRA-CORPOREAL CIRCULATION(ECC)

TOSHIKAZU TOBI*

*Technical Research Laboratory, DAI-DAN Co., Ltd.*
*390 Kitanagai, Miyoshi-Machi, Iruma-Gun, Saitama 354, Japan*

## 1. Introduction

ECC is an extremely vital technique used during operative cardiac surgery. Its automatization was first attempted during the latter half of the 1950s.[1–3] However, at that time there were many factors which hindered the development of an auto-regulation system. Recently, remarkable developments in ECC techniques have been attained through improvements such as increased performance and decreased apparatus size. In addition, sufficient processing speeds and performance are available at a relatively low cost because of the epoch-making progress in computers over the past decade.

Under these circumstances, we previously attempted the development of a microcomputer-based auto-regulation system for ECC.[4] However, with respect to range specifications, our static regulation method lacked smoothness. This time we have attempted to develop a preliminary experimental model of an auto-regulation system for ECC using fuzzy theory[5] to account for the insufficiencies of the static regulation method.

In this paper we describe the control theory of the artificial heart–lung circuit, the operating environment of the regulatory apparatus, and the performance evaluation results of the system.

## 2. Purpose and Scope of Development

The system is comprised of many regulatory elements including hemodynamic regulation, a regulatory circuit for cardioplegic solution, and withdrawal

*Phone: +81-492-58-1511 Fax: +81-492-58-7601 E-mail: tobi@lab.daidan.co.jp

venting. There is also a regulator for blood/moisture balance which combines in-line sucking, bleeding, and urination. In this study, we have focused on hemodynamic regulation, and our goal is to contribute to the establishment of a safe and reliable means for ECC operation.

## 3. Fuzzy Inference and Theory

The basic concept of fuzzy inference involves the handling of a language variable in terms of a basic term and its value. A language variable is a word or phrase from natural or synthetic language that is assigned a specific value.

In the example "*blood pressure is low*", *blood pressure* is the language variable, and *low* is the basic term. In another example "*and if it is not high*", *and* is the conjunction, *high* is the antonym of *low*, and *not* is the modifying element. In other words, a language variable name is formed by a basic term, an antonym, a set of modifiers, and a set of conjunctions. Furthermore, the value of the basic term or language variable is expressed as a possibility distribution.[6] In fuzzy set theory, the set that is characterized by a possibility distribution function, or *membership function*, is called a fuzzy set. In other words, a fuzzy set is defined as the following[7,8]:

The fuzzy set $A$ is considered over the domain $U = \{u\}$, where the membership function $\mu_A$ is characterized by $\mu_A : U \to [0,1]$. Here the value of $\mu_A(u)$ expresses the level of *belongingness* of $u$ (over the domain $U$) with respect to set $A$. In the finite set $U$ composed of $n$ elements, fuzzy set $A$ is expressed by Eq. (1):

$$A = \mu_A(u_1)/u_1 + \mu_A(u_2)/u_2 + \cdots + \mu_A(u_n)/u_n\,. \tag{1}$$

In fuzzy inference, using the computation according to the fuzzy inference mechanism, a conclusion is reached based on real facts and knowledge from fuzzy sets.

Therefore, a basic fuzzy system is comprised of a fuzzy inference mechanism and knowledge base. In general, applications of fuzzy inference method to control, fuzzy production rules such as in Eq. (2) are the most commonly used.[9] Furthermore, fuzzy production rules are represented as *if $x$ is $A$ then $y$ is $B$* statements. In this *if-then* rule structure, the antecedent portion *if $x$ is $A$* and the consequent portion *then $y$ is $B$* are said to be a statement of fuzziness.

$$\begin{array}{ll} \textit{knowledge} & \textit{if } x \textit{ is } A \textit{ then } y \textit{ is } B \\ \textit{fact} & x \textit{ is } A' \\ \hline \textit{conclusion} & y \textit{ is } B' \end{array} \tag{2}$$

This method where inference proceeds from cause to result is called "*forword fuzzy reasoning*".[10] Furthermore, the aforementioned fuzzy set points to the $A(B)$ portion. When this operation is used in control, an expert's intuition and experience is encoded in the form of *if-then* knowledge, lending the characteristic of information legibility. This reduces the control method's decision making process into conditional statements, and it can be furthermore concluded that this method provides definite conclusions. This indicates the elimination of the discrepancies attained by a certain control process. Methods to determine the adjustment amount during control can be broken into two large groups, the MAX–MIN composition/center-of-gravity method and the MAX–MIN composition/maximum-height method.

### 3.1. *MAX–MIN composition/center-of-gravity method*

The MAX–MIN composition/center-of-gravity method is often applied in the case where the resulting inference is continuous. The result is calculated as the center of gravity of the membership function obtained from composition. The center-of-gravity method is outlined in Fig. 1. Here, the $i$th control rule among the $N$ control rules is expressed as *if* $x_1$ *is* $A_i$, $x_2$ *is* $B_i$ *then* $u_1$ *is* $C_i$. A control rule of this form means that when input $x_1$ is $A_i$ and input $x_2$ is $B_i$, then the output $u_1$ is $C_i$. $A_i$, $B_i$, and $C_i$ are fuzzy sets that express an ambiguous condition.

Furthermore, functions $\mu_{A_i}$, $\mu_{B_i}$, and $\mu_{C_i}$ are the membership functions that characterize the fuzzy sets $A_i$, $B_i$, and $C_i$, respectively. Here, the input value is set to a fixed value. In the first part of the control rule $x_1$ *is* $A_i$, the grade of the input decision value of $x_1 = x_{1j}$ is the value of the membership function $\mu_{A_i}(x_{1j})$.

Likewise, the input decision value of $x_2 = x_{2j}$ of $x_2$ *is* $B_i$ is the value of the membership function $\mu_{B_i}(x_{2j})$. Furthermore, the total result from the antecedent of the $i$th control rule is the grade $w_i$, which can be determined by Eq. (3). In this case, Min implies a calculation operation where the minimum value of $\mu_{A_i}(x_{1j})$ and $\mu_{B_i}(x_{2j})$ is chosen. Next, the membership function $\mu_{C_i}$ that represents the fuzzy set $C_i$ of the latter half of the control rule $u_1$ *is* $C_i$ is multiplied by $w_i$ to obtain the function $\mu_{C_i^*}$. However, in a general digital

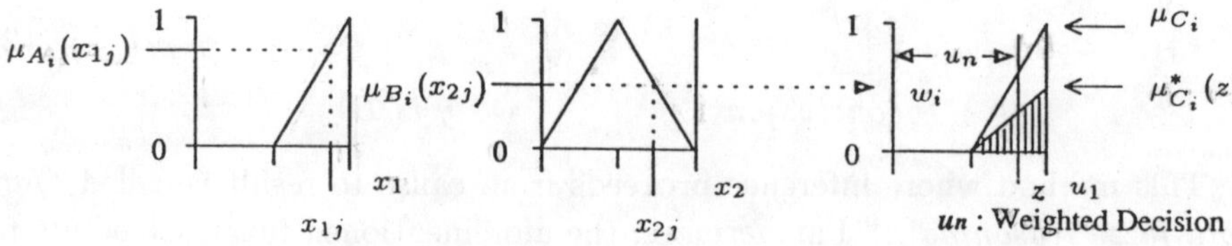


Fig. 1. MAX–MIN composition/center-of-gravity method.

processing system, the set variable $z$ is limited to being a finite number. It follows that the determination of the weight of the $i$th control rule is shown in the function in Eq. (4).

$$w_i = \text{Min}\{\mu_{A_i}(x_{1j}),\ \mu_{B_i}(x_{2j})\} \tag{3}$$

$$\mu_{C_i^*}(z) = w_i \times \mu_{C_i}(z) \tag{4}$$

Furthermore, the membership function $\mu_{C_i^*}$ that relates to the output of the $i$th control rule is calculated in a manner where all $N$ control rules are considered. In addition, the maximum value $\mu(z)$ is selected as in Eq. (5).

$$\mu(z) = \text{Max}_{i=1}^{n} \mu_{C_i^*}(z) \tag{5}$$

In other words, Max implies the computation process of taking the decision weights of the $N$ control rules and calculating the maximum value. The quantization operation is determining the quantitative center of gravity $u_n$ of the area of function $\mu$. The method for computing the center of gravity is expressed in Eq. (6).

$$u_n = \sum_{z=0}^{z} z \times \mu(z) / \sum_{z=0}^{z} \mu(z) \tag{6}$$

### 3.2. *MAX–MIN composition/maximum-height method*

The MAX–MIN composition/maximum-height method is often applied in the case where the inference result is discrete. As a result, out of the various values from the MAX–MIN composition, the rule with the highest value is chosen. In some cases, this result is modified or converted. This maximum-height method is outlined in Fig. 2. Here, the $i$th control rule among the $N$ control rules is expressed as *if* $x_1$ *is* $A_i$, $x_2$ *is* $B_i$ *then* $u_1$ *is* $C_i$. When the input values are fixed, the function $\mu$ that determines the weighting of the rules and attains the maximum value can be determined from Eqs. (3)–(5). The inference result uses the grade from the function that corresponds to the rule with the highest MAX–MIN composition value.

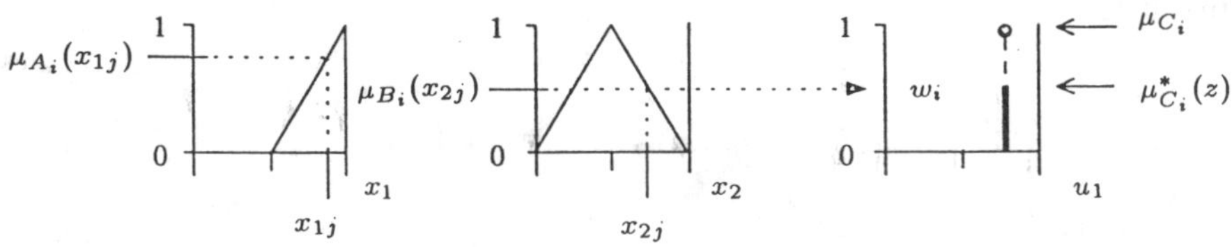


Fig. 2. MAX–MIN composition/maximum-height method.

## 4. Auto-Regulation System for ECC

Since 1939, ECC has undergone remarkable progress. In fact, manual ECC has recently been performed virtually safely. However, a large portion of the operation relies upon high-level expertise combined with a perfusionist's (ECC technician) intuition. Furthermore, the control system is comprised of several regulatory components including hemodynamic regulation, a regulatory circuit for cardioplegic solution, each type of withdrawal venting, as well as blood and water balance involved in in-line sucking, blood withdrawal, and urination. Therefore, even now it is difficult for a single perfusionist to be responsible for the entire operation of ECC. The most important aspect of ECC operation is maintaining the targeted blood volume and ensuring that oxygen is supplied to the arteries. However, there is a difference between hemodynamic regulation in ECC by an artificial heart–lung apparatus and hemodynamic regulation in natural internal circulation. Furthermore, to realize progress from a physiological perspective, for example extending ECC time and expanding surgical applications, radical mechanical improvements are necessary. It follows that even when using the most recent manual artificial heart–lung apparatus, it is probably still difficult to maintain sufficient support of many bodily phenomena throughout all stages of ECC. It is these conditions that make an attempt at automating ECC regulation significant.

### 4.1. *Composition of circuit and fundamentals of control*

A representative system diagram is shown in Fig. 3. The heart–lung circuit is comprised of the blood withdrawal line, the line occluder (occluder), the venous reservoir (reservoir), the blood flow pump (pump), the artificial lung, and the blood infusion line. In addition, the regulatory apparatus consists of a microcomputer, a fuzzy inference processing board, data input, and a controller.

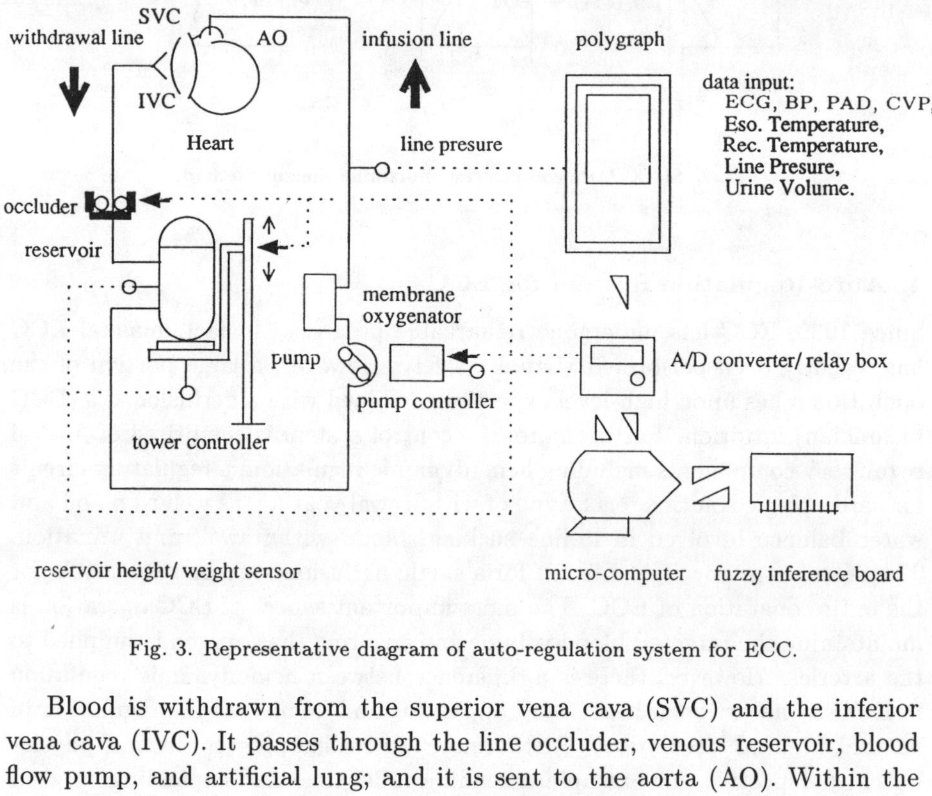


Fig. 3. Representative diagram of auto-regulation system for ECC.

Blood is withdrawn from the superior vena cava (SVC) and the inferior vena cava (IVC). It passes through the line occluder, venous reservoir, blood flow pump, and artificial lung; and it is sent to the aorta (AO). Within the body, the blood flows from the aorta to the arteries and then to the capillaries before returning to the greater vein. Furthermore, in the artificial heart–lung, oxygen gas bubbles flow through the blood, and the removal of oxygen and $CO_2$ from the venous blood is executed. At both the start and end of ECC, blood is sent out from the blood flow pump and the heart. However, when it reaches the greater portion of the heart–lung bypass, the blood that passes through the heart is isolated; and blood circulation within the body is maintained by the blood flow pump. In addition to ensuring the necessary circulatory blood volume, it is necessary to stabilize the systemic arterial pressure and diastolic pulmonary arterial pressure (venous pressure) in order to maintain the organs in a favorable state. Therefore, the basic theory of regulation is to control blood withdrawal and infusion such that the target regulatory values for the systemic arterial pressure and diastolic arterial pressure are maintained. The line occluder, venous reservoir, and blood flow pump are used for this purpose.

In order to regulate the withdrawal blood volume, an electromagnetic valve-like, four-stage clamp-type line occluder is used to provide a phased stop mechanism for changing the effective cross-sectional area of the withdrawal line. In addition, blood withdrawal from the body due to gravitation is regulated by the rising and falling behavior of the venous reservoir; and the amount of blood sent to the body is regulated by the output of the blood flow pump.

#### 4.1.1. *Microcomputer section*

The microcomputer used was a NEC PC-9801 $RX^{TM}$. In addition to the base system, the following system extentions were used: an extended RAM board, an A/D(D/A) conversion board with built-in digital i/o, and a relay box for operation signal conversion to the pump flow controller and line occluder.

#### 4.1.2. *Fuzzy inference section*

The fuzzy inference section is encapsulated in a single fuzzy inference processing board inserted into the i/o expansion slot of the microcomputer. A diagram of the board is shown in Fig. 4. The fuzzy inference board has a set of standard membership functions and has the functions for basic inference. The set of standard functions are convenient for modeling general membership functions for fuzzy sets. In addition, the basic inference functions operate as a single operator of the computation module. That basic processing occurs in the form of 4 inputs in parallel with one fuzzy production rule MAX–MIN composition value as the output.

#### 4.1.2.1. *Base membership functions*

Figure 5 shows the base membership function shapes used to model the system behavior as a fuzzy process. Functions $Z$, $P$, $S$, as well as the impulse function $SF$ are defined by Eqs. (7)–(10).

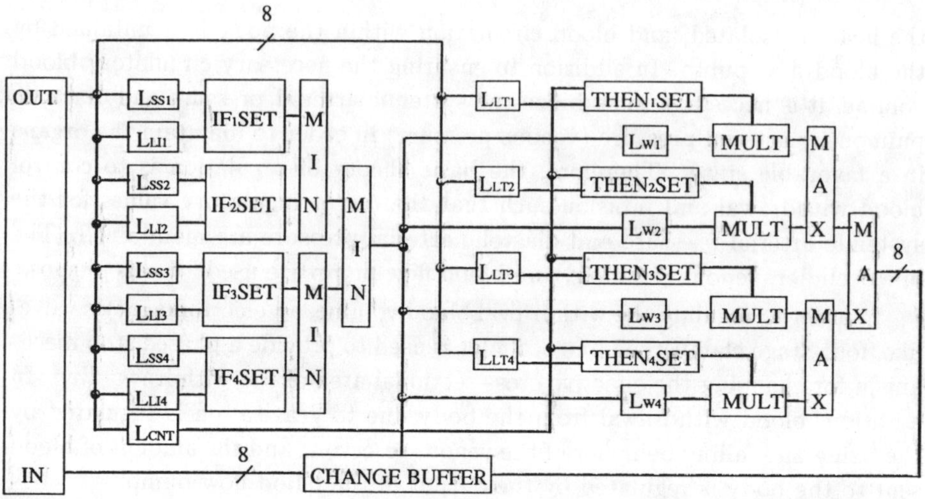


**Fig. 4. Outline of the fuzzy inference board.**

$$Z(u;a,b,c)=\begin{cases}1 & \text{if } u\leqslant a\,;\\ 1-(u-a)/(c-a) & \text{if } a\leqslant u\leqslant b\,;\\ 1-(u-a)/(c-a) & \text{if } b\leqslant u\leqslant c\,;\\ 0 & \text{if } c\leqslant u\,;\end{cases} \tag{7}$$

$$\text{where } b=(a+c)/2\,.$$

$$P(u;a,b,c,d,e,f)=\begin{cases}0 & \text{if } u\leqslant a\,;\\ (u-a)/(c-a) & \text{if } a\leqslant u\leqslant b\,;\\ (u-a)/(c-a) & \text{if } b\leqslant u\leqslant c\,;\\ 1 & \text{if } c\leqslant u\leqslant d\,;\\ 1-(u-d)/(f-d) & \text{if } d\leqslant u\leqslant e\,;\\ 1-(u-d)/(f-d) & \text{if } e\leqslant u\leqslant f\,;\\ 0 & \text{if } f\leqslant u\,;\end{cases} \tag{8}$$

$$\text{where } b=(a+c)/2,\ e=(d+f)/2\,.$$

$$S(u;d,e,f)=\begin{cases}0 & \text{if } u\leqslant d\,;\\ (u-d)/(f-d) & \text{if } d\leqslant u\leqslant e\,;\\ (u-d)/(f-d) & \text{if } e\leqslant u\leqslant f\,;\\ 1 & \text{if } f\leqslant u\,;\end{cases} \tag{9}$$

$$\text{where } e=(d+f)/2\,.$$

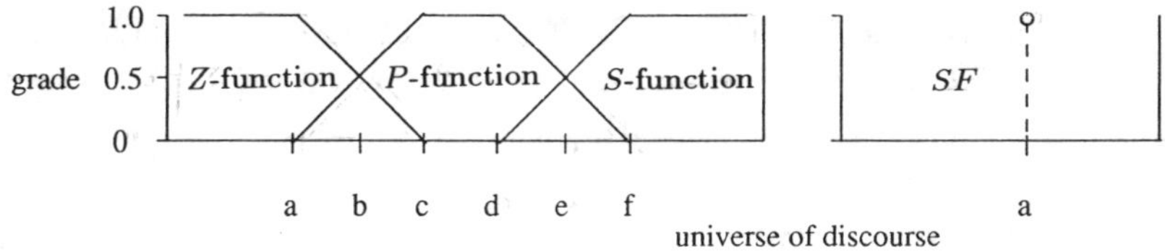


Fig. 5. Ordinary membership functions.

$$SF(u;a) = \begin{cases} 1 & \text{if } u = a\,; \\ 0 & \text{if } u \neq a\,. \end{cases} \tag{10}$$

In Table 1, an example of membership functions characterizing fuzzy sets as triangle and trapezoids is shown. In each function diagram, the horizontal axis represents the entire finite set and the vertical axis is the grade. The complete set and the grades of the standard functions are internally expressed as 256 discrete values.

Table 1. One example of definition of membership functions.

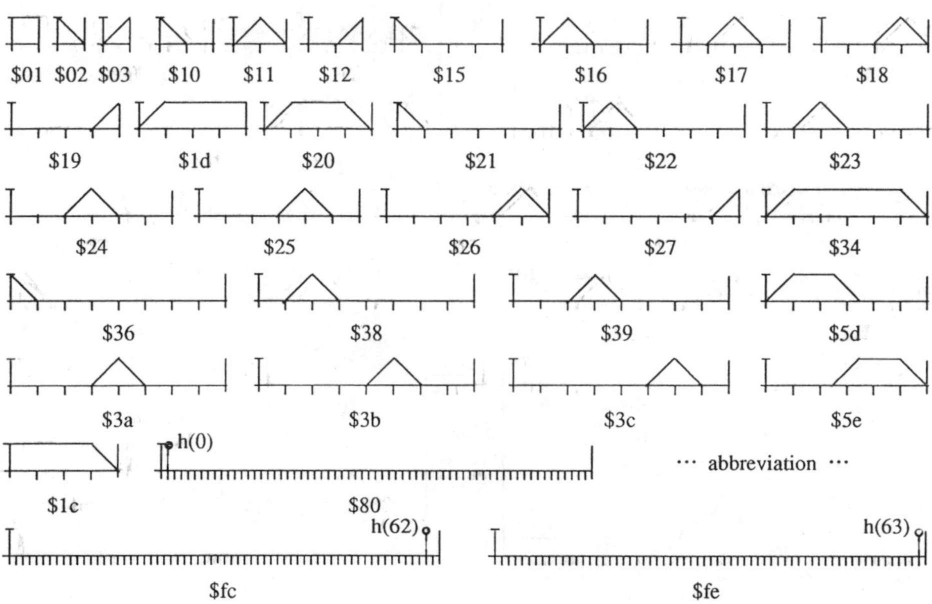

These are stored as $IF_1SET \sim IF_4SET$ and $THEN_1SET \sim THEN_4SET$ of the memory circuit. The processor selects from the set of 256 standard functions using the hexadecimal notation system of \$00–\$ff.

4.1.2.2. *Basic inference function*

The basic inference functionality for the MAX–MIN calculation involves many circuits including the fuzzy logic product, fuzzy logic sum, and algebraic product. With the objective of a simple circuit design, using the structure of a logic integrated circuit (IC) combined with a memory IC is a reasonable solution. Figures 6 and 7 illustrate examples of fuzzy inference ICs. Although there is no functional difference between the two ICs, the memory-based method is superior with respect to design simplicity. In contrast, the logic-based IC is difficult to compose from standard logic ICs. Therefore, on our fuzzy inference board, the fuzzy logic product circuit MIN, fuzzy logic sum circuit MAX, and the algebraic product circuit MULT all use the memory IC design.

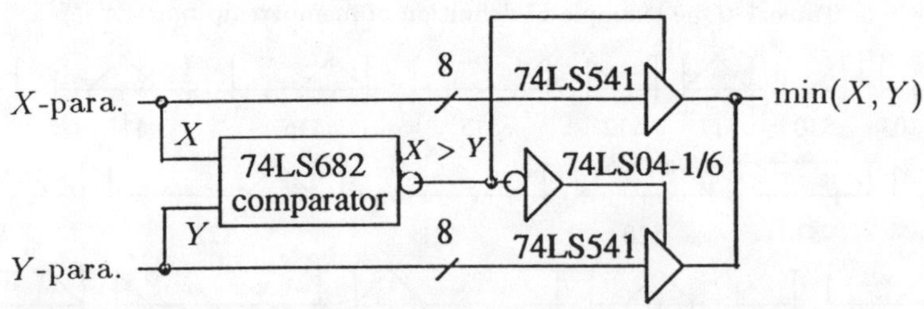


Fig. 6. Fuzzy inference circuit (logic IC).

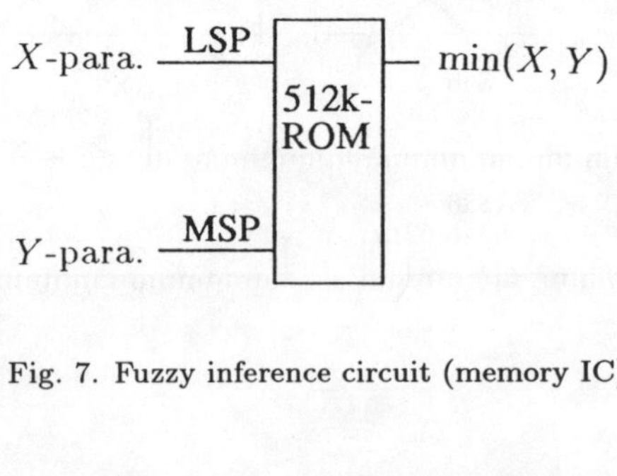


Fig. 7. Fuzzy inference circuit (memory IC).

4.1.2.3. *Basic processing*

Commands to the fuzzy inference board's inference engine are routed to the data latch circuit of the processor. The standard functions are selected by holding the numeric labels converted to bit representation at the outputs of $L_{LI1} \sim L_{LI4}$ and $L_{LT1} \sim L_{LT4}$. The MAX–MIN calculation is composed of the fuzzy logic product, algebraic product, and fuzzy logic sum.

The $i$th control rule of the set of 4 production rules with the 4-input/1-output structure is expressed as *if* $x_1$ *is* $A_i$, $x_2$ *is* $B_i$, $x_3$ *is* $C_i$, $x_4$ *is* $D_i$ *then* $u_1$ *is* $E_i$. First, the grade of the antecedent of each control rule is obtained by the following operation.

Membership function $\mu_{A_i}$ expressing the contents of the antecedent of fuzzy set $A_i$ is selected alone from among the group of $IF_1SET$ standard functions by the driving signal from $L_{LI1}$. In a similar manner, membership functions $\mu_{B_i}$, $\mu_{C_i}$, as well as $\mu_{D_i}$ are each selected from the $IF_2SET \sim IF_4SET$ of standard functions. The value of each selected membership function $\mu_{A_i} \sim \mu_{D_i}$ specifies the (essential) elements of the entire set by the outputs $L_{SS1} \sim L_{SS4}$ of the data latch and are output by the $IF_1SET \sim IF_4SET$. At the same time, the minimum value is calculated by the 2-stage fuzzy logic product circuit MIN and is presented to the $L_{W1} \sim L_{W4}$ inputs of the data latch circuit. Finally, the calculation of the grade of the $i$th control rule drives $L_{Wi}$ and with the passing of the value at the input end to the output side, the process is completed. It follows that the grade of the 4 control rules is computed by simply repeating this operation. In other words, the grade of each control rule is not obtained by parallel processing but by serial processing.

Next, the membership functions that express the inference composition are obtained by the following operation. Membership function $\mu_{E_i}$ expressing the contents of fuzzy set $E_i$ of the $i$th control rule is selected from among the group of $THEN_1SET$ by the driving signal of $L_{LTi}$. This operation is executed in the same manner for the 4 control rules. The value of each membership function $\mu_{E_1} \sim \mu_{E_4}$ specifies the essential elements of the entire set by the output of data latch circuit $L_{CNT}$ and is output from $THEN_1SET \sim THEN_4SET$. At the same time, those values are passed along with the cached control rule grade to the algebraic product circuit MULT. The value expressed at the output of MULT is the value of the function expressing the weighting inference of the control rules. Finally, the maximum value of the weighting inference for the 4 control rules is calculated by the 2-stage fuzzy logic sum circuit MAX, and is expressed by the input unit IN of the i/o control unit.

Ultimately, based on the membership functions expressing the composition of the inference defined by the 4-rule unit and obtained by the above process, it is necessary to execute the composition process by methods such as center of gravity and maximum height. Furthermore, the bus switching circuit CHANGE BUFFER is used to output the grade of the antecedents from the unit of 4 inputs.

### 4.1.3. *Data input section*

The following temperature and pressure points are recorded: systemic arterial pressure, diastolic pulmonary arterial pressure, central venous pressure, line pressure, withdrawal blood temperature, supply blood temperature, rectal temperature, esophagus temperature, and minor temperatures. The pressure data is displayed and recorded by the polygraph and is sent to the data input unit after the data is adjusted. The venous reservoir height is measured by using the change in resistance of a rotary potentiometer attached to the power mast that rotates relative to fluctuations within the reservoir. The blood volume inside the venous reservoir is estimated by calculating the weight of the total venous reservoir apparatus including inlet and withdrawal lines. Finally, the pump flow volume is captured by placing a rotary encoder sensor along the pump flow axis where the pulses generated by the device are counted with respect to unit time and are converted to flow volume.

### 4.1.4. *Operation section*

The operation section includes the line occluder, pump flow controller and the venous reservoir fluctuation power mast. Based on operation of the electromagnetic valve initiated by a pattern of digital signals from the processor, the line occluder drives the direct insertion of the withdrawal line, which in turn makes it possible for the cross-sectional area inside the tube to be divided into 4 states. This staged regulation controls the blood withdrawal from the body, and the regulation of the pump flow controller controls the blood supply to the body. Finally, the regulation of the venous reservoir fluctuation powered mast causes a change in the level difference between the body and venous reservoir and controls the blood flow volume inside the venous reservoir.

### 4.2. *Regulation principle and method of control*

The main structural elements of the regulation system are functionally classified into measurement, control, and regulation. With respect to the software development environment, Japanese MS-DOS (v3.3B)$^{\text{TM}}$ was used as the operating system, Microsoft C (v6.0A)$^{\text{TM}}$ as the programming language, and MS-Windows (v3.0)$^{\text{TM}}$ as the window system. Directly after the program starts, the 3 choices of patient data input, printout, and data sampling are displayed on a menu.

#### 4.2.1. *Patient data input*

The patient's characteristic data before surgery are input sequentially in a simple dialog format. The input items include the patient's name, ID number, age, sex, birthdate, height, weight, blood type, clinical diagnosis, surgical method, technician, perfusionist, Hct and Hb before surgery, the composition and volume of the artificial heart–lung filling fluid, type of artificial lung, pump tube size, and type and amount of antibiotic. Furthermore, the general volume of the heart–lung filler fluid, the dilution percentage before ECC, the body surface area, and the optimal pump flow volume are calculated from the height, weight, and other data.

#### 4.2.2. *Printout*

The ECC progress table can be written out in any established output format. Furthermore, a patient's data table or surgery-in-progress trendgram can be accessed by inputting the patient's name or relevant time, and it can be subsequently printed out using the processor's copy function.

#### 4.2.3. *Data sampling*

The display is composed of the trendgraph, heart–lung circuit diagram, and the display window for each type of data. In the trendgraph display window, 5 options including *sampling* and *execution* are provided. When *sampling* is chosen, the data for a specific period is collected and displayed in a trendgraph. When *execution* is selected, the 2 choices of *starting* and *weaning* appear. Fundamental operation and regulation are illustrated in Fig. 8. Furthermore, with respect to the basic operation and its relevance, the selection of *starting* becomes SW1 and the selection of *weaning* becomes SW2.

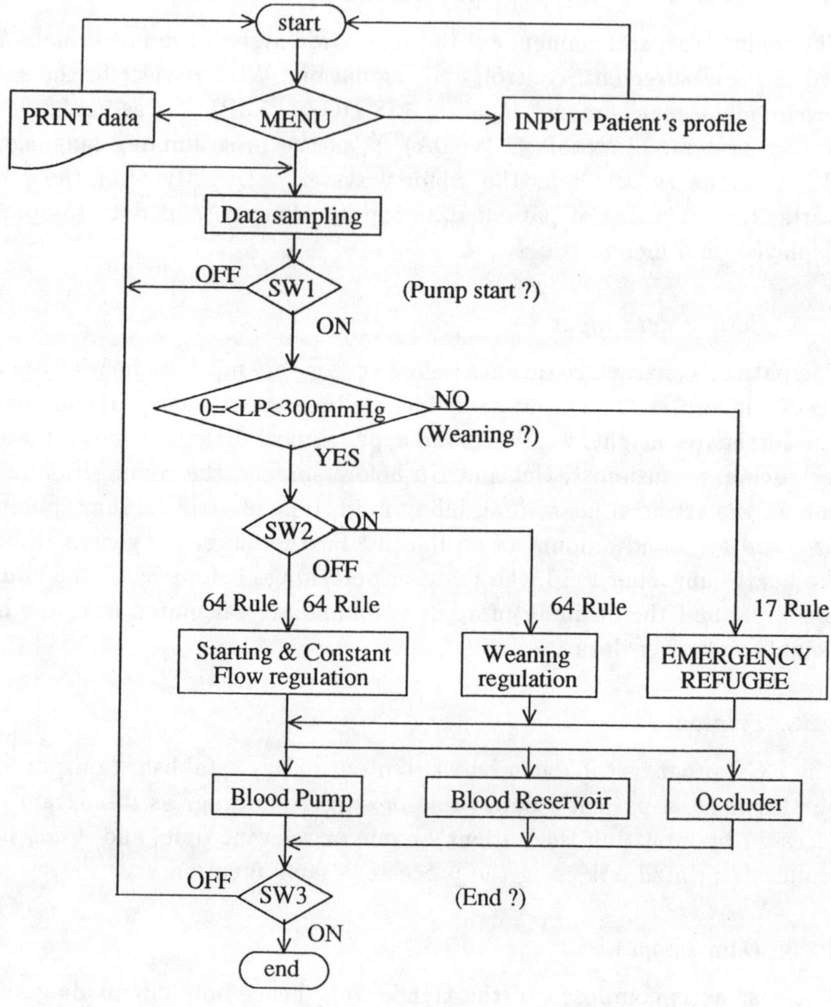


Fig. 8. Fundamental operation and regulation.

The flowchart for hemodynamic regulation during surgery, as well as before and after, is shown in Fig. 9.

Pump start regulation (Starting) commences with SW1-ON. The pump start regulator monitors the systemic arterial pressure, diastolic pulmonary

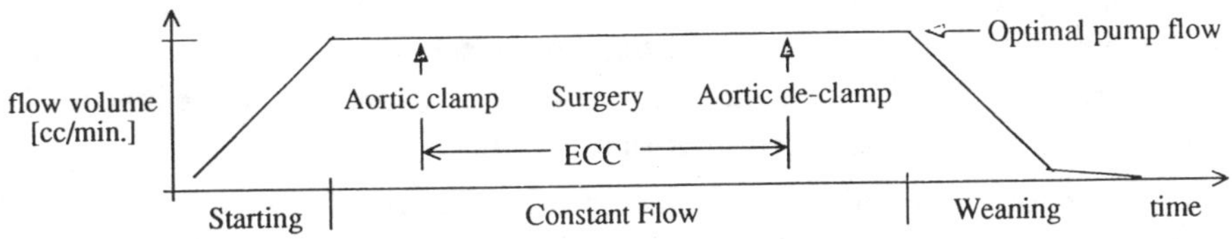


Fig. 9. Flowchart for hemodynamic regulation.

arterial pressure, and withdrawal blood volume. At the same time it obeys fixed regulatory conditions and slowly raises the frequency of the blood flow pump. The pump volume soon reaches over 90% of the optimal flow volume, and the pump switches automatically to normal regulation (Constant Flow). The systemic arterial pressure and withdrawal blood volume are still monitored, and the pump flow volume is maintained at a level close to the optimal pump flow volume. Pump weaning (Weaning) commences with SW2-ON. The pump weaning regulator monitors the systemic arterial pressure, diastolic pulmonary arterial pressure, and withdrawal blood volume. At the same time it obeys fixed regulatory conditions and slowly decreases the frequency of the blood flow pump. While systemic arterial pressure and diastolic pulmonary arterial pressure are maintained in a favorable state, *weaning* is completed if the blood flow pump stops. A safety feature is also included during ECC such that the system switches to a secure operating regulatory mode (EMERGENCY REFUGEE) and adopts emergency measures if insufficient blood withdrawal or circuit occlusion are detected. After that, control quickly returns to the original control routine.

#### 4.2.3.1. *Pump start regulation*

The fundamental regulatory behavior consists of systemic arterial pressure (BP), diastolic pulmonary arterial pressure (PAD), venous reservoir height (BRH), blood volume within the venous reservoir (BRW), and change in blood volume within the venous reservoir (dBRW). Based on these factors, the change in flow volume through the blood flow pump (dPF), change in venous reservoir height (dBRH), and phase change of the occluder opening (dOcc) can be determined.

The target regulatory value for the pump flow volume is at least 90% of the optimal pump flow volume, which is approximately 3000 mm liter for an adult male. Other necessary conditions are systemic arterial pressure of 50–100 mm Hg and a diastolic pulmonary arterial pressure which is equal to the initial value plus approximately less than 14 mm Hg.

In order for the regulation routine to fulfill the regulatory objectives and to achieve the necessary conditions, the system automatically shifts to normal operation. In Fig. 10, the conditions of systemic arterial pressure and diastolic pulmonary arterial pressure for pump start regulation are classified into 4 phenomena. The ideal condition is condition 1. In this condition, the rotation frequency of the blood flow pump increases and the supply volume increases. In addition, the line occluder slowly opens. In condition 2, the ECC blood flow volume is insufficient because of too much withdrawal blood. In this condition, the blood supply volume slowly increases by trace amounts. In condition 3, the ECC blood flow volume is too high because of an excess blood supply. In this condition, the supply blood volume slowly increases due to trace amounts of withdrawal. Condition 4 is the worst. In this condition, the blood supply volume quickly increases due to trace amounts of withdrawal.

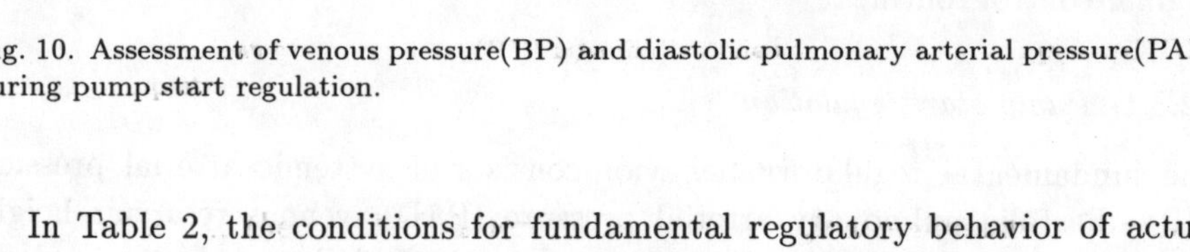


Fig. 10. Assessment of venous pressure(BP) and diastolic pulmonary arterial pressure(PAD) during pump start regulation.

In Table 2, the conditions for fundamental regulatory behavior of actual ECC management by an operator (ECC technician) are stated in the form of numerical expressions. PAD(0) is the initial value of the diastolic pulmonary arterial pressure.

BRH, BRW, dBRW, as well as dPF and dBRH are expressed in percentages which represent the ratio of their respective ranges.

Table 2. Expert's fundamental behavior during pump start regulation.

| | Antecedents | | | | | Consequences | | |
|---|---|---|---|---|---|---|---|---|
| | 1. | 2. | 3. | 4. | 5. | 1. | 2. | 3. |
| Rule | BP (mm Hg) | PAD (mm Hg) | BRH (%) | BRW (%) | dBRW (%) | dPF (%) | dBRH (%) | dOcc |
| 1 | 50⩽BP | PAD<PAD(0)+14 | BRH<50 | 20⩽BRW<50 | 1⩽dBRW | +5 | 0 | 0 |
| 2 | 50⩽BP | PAD<PAD(0)+14 | BRH<50 | 20⩽BRW<50 | −1 ⩽dBRW<1 | +5 | 0 | −1 |
| 3 | 50⩽BP | PAD<PAD(0)+14 | BRH<50 | 20⩽BRW<50 | dBRW<−1 | −5 | −10 | −1 |
| 4 | 50⩽BP | PAD<PAD(0)+14 | BRH<50 | BRW<20 | Unknown | −25 | −20 | +1 |
| 5 | 50⩽BP | PAD<PAD(0)+14 | BRH<50 | 80⩽BRW | Unknown | +10 | +20 | +1 |
| 6 | 50⩽BP | PAD<PAD(0)+14 | BRH<50 | 50⩽BRW<80 | 1⩽dBRW | +5 | +10 | 0 |
| 7 | 50⩽BP | PAD<PAD(0)+14 | BRH<50 | 50⩽BRW<80 | −1 ⩽dBRW<1 | +5 | 0 | −1 |
| 8 | 50⩽BP | PAD<PAD(0)+14 | BRH<50 | 50⩽BRW<80 | dBRW< −1 | 0 | 0 | −1 |
| 9 | 50⩽BP | PAD<PAD(0)+14 | 50⩽BRH | 20⩽BRW<50 | 1⩽dBRW | +5 | 0 | 0 |
| 10 | 50⩽BP | PAD<PAD(0)+14 | 50⩽BRH | 20⩽BRW<50 | −1 ⩽dBRW<1 | +5 | −10 | −1 |
| 11 | 50⩽BP | PAD<PAD(0)+14 | 50⩽BRH | 20⩽BRW<50 | dBRW<−1 | −5 | −10 | −1 |
| 12 | 50⩽BP | PAD<PAD(0)+14 | 50⩽BRH | BRW<20 | Unknown | −25 | −20 | +1 |
| 13 | 50⩽BP | PAD<PAD(0)+14 | 50⩽BRH | 80⩽BRW | Unknown | +10 | +20 | +1 |
| 14 | 50⩽BP | PAD<PAD(0)+14 | 50⩽BRH | 50⩽BRW<80 | 1⩽dBRW | +5 | 0 | 0 |
| 15 | 50⩽BP | PAD<PAD(0)+14 | 50⩽BRH | 50⩽BRW<80 | −1 ⩽dBRW<1 | +5 | 0 | −1 |
| 16 | 50⩽BP | PAD<PAD(0)+14 | 50⩽BRH | 50⩽BRW<80 | dBRW<−1 | 0 | −10 | −1 |
| 17 | 50⩽BP | PAD(0)+14⩽PAD | BRH<50 | 20⩽BRW<50 | 1⩽dBRW | +5 | 0 | 0 |
| 18 | 50⩽BP | PAD(0)+14⩽PAD | BRH<50 | 20⩽BRW<50 | −1 ⩽dBRW<1 | +5 | 0 | −1 |
| 19 | 50⩽BP | PAD(0)+14⩽PAD | BRH<50 | 20⩽BRW<50 | dBRW<−1 | −5 | −10 | −1 |
| 20 | 50⩽BP | PAD(0)+14⩽PAD | BRH<50 | BRW<20 | Unknown | −25 | −20 | +1 |
| 21 | 50⩽BP | PAD(0)+14⩽PAD | BRH<50 | 80⩽BRW | Unknown | 0 | +20 | +1 |
| 22 | 50⩽BP | PAD(0)+14⩽PAD | BRH<50 | 50⩽BRW<80 | 1⩽dBRW | +5 | +10 | 0 |

Table 2. (*Cont'd*)

| Rule | Antecedents 1. BP (mm Hg) | 2. PAD (mm Hg) | 3. BRH (%) | 4. BRW (%) | 5. dBRW (%) | Consequences 1. dPF (%) | 2. dBRH (%) | 3. dOcc |
|---|---|---|---|---|---|---|---|---|
| 23 | 50≤BP | PAD(0)+14≤PAD | BRH<50 | 50≤BRW<80 | −1 ≤dBRW<1 | +5 | 0 | −1 |
| 24 | 50≤BP | PAD(0)+14≤PAD | BRH<50 | 50≤BRW<80 | dBRW<−1 | −5 | 0 | −1 |
| 25 | 50≤BP | PAD(0)+14≤PAD | 50≤BRH | 20≤BRW<50 | 1≤dBRW | +5 | 0 | 0 |
| 26 | 50≤BP | PAD(0)+14≤PAD | 50≤BRH | 20≤BRW<50 | −1 ≤dBRW<1 | +5 | −10 | −1 |
| 27 | 50≤BP | PAD(0)+14≤PAD | 50≤BRH | 20≤BRW<50 | dBRW<−1 | −5 | −10 | −1 |
| 28 | 50≤BP | PAD(0)+14≤PAD | 50≤BRH | BRW<20 | Unknown | −25 | −20 | +1 |
| 29 | 50≤BP | PAD(0)+14≤PAD | 50≤BRH | 80≤BRW | Unknown | 0 | +20 | +1 |
| 30 | 50≤BP | PAD(0)+14≤PAD | 50≤BRH | 50≤BRW<80 | 1≤dBRW | +5 | 0 | 0 |
| 31 | 50≤BP | PAD(0)+14≤PAD | 50≤BRH | 50≤BRW<80 | −1 ≤dBRW<1 | +5 | 0 | −1 |
| 32 | 50≤BP | PAD(0)+14≤PAD | 50≤BRH | 50≤BRW<80 | dBRW<−1 | 0 | −10 | −1 |
| 33 | BP<50 | PAD<PAD(0)+14 | BRH<50 | 20≤BRW<50 | 1≤dBRW | +5 | 0 | 0 |
| 34 | BP<50 | PAD<PAD(0)+14 | BRH<50 | 20≤BRW<50 | −1 ≤dBRW<1 | +5 | 0 | −1 |
| 35 | BP<50 | PAD<PAD(0)+14 | BRH<50 | 20≤BRW<50 | dBRW<−1 | −5 | −10 | −1 |
| 36 | BP<50 | PAD<PAD(0)+14 | BRH<50 | BRW<20 | Unknown | −25 | −20 | +1 |
| 37 | BP<50 | PAD<PAD(0)+14 | BRH<50 | 80≤BRW | Unknown | +10 | +20 | +1 |
| 38 | BP<50 | PAD<PAD(0)+14 | BRH<50 | 50≤BRW<80 | 1≤dBRW | +5 | +10 | 0 |
| 39 | BP<50 | PAD<PAD(0)+14 | BRH<50 | 50≤BRW<80 | −1 ≤dBRW<1 | +5 | +10 | −1 |
| 40 | BP<50 | PAD<PAD(0)+14 | BRH<50 | 50≤BRW<80 | dBRW<−1 | 0 | 0 | −1 |
| 41 | BP<50 | PAD<PAD(0)+14 | 50≤BRH | 20≤BRW<50 | 1≤dBRW | +5 | 0 | 0 |
| 42 | BP<50 | PAD<PAD(0)+14 | 50≤BRH | 20≤BRW<50 | −1 ≤dBRW<1 | +5 | 0 | −1 |
| 43 | BP<50 | PAD<PAD(0)+14 | 50≤BRH | 20≤BRW<50 | dBRW<−1 | −5 | −10 | −1 |

Table 2. (*Cont'd*)

| Rule | Antecedents 1. BP (mm Hg) | 2. PAD (mm Hg) | 3. BRH (%) | 4. BRW (%) | 5. dBRW (%) | Consequences 1. dPF (%) | 2. dBRH (%) | 3. dOcc |
|---|---|---|---|---|---|---|---|---|
| 44 | BP<50 | PAD<PAD(0)+14 | 50⩽BRH | BRW<20 | Unknown | −25 | −20 | +1 |
| 45 | BP<50 | PAD<PAD(0)+14 | 50⩽BRH | 80⩽BRW | Unknown | +10 | +20 | +1 |
| 46 | BP<50 | PAD<PAD(0)+14 | 50⩽BRH | 50⩽BRW<80 | 1⩽dBRW | +5 | 0 | 0 |
| 47 | BP<50 | PAD<PAD(0)+14 | 50⩽BRH | 50⩽BRW<80 | −1 ⩽dBRW<1 | +5 | 0 | −1 |
| 48 | BP<50 | PAD<PAD(0)+14 | 50⩽BRH | 50⩽BRW<80 | dBRW<−1 | 0 | −10 | 0 |
| 49 | BP<50 | PAD(0)+14⩽PAD | BRH<50 | 20⩽BRW<50 | 1⩽dBRW | +5 | 0 | 0 |
| 50 | BP<50 | PAD(0)+14⩽PAD | BRH<50 | 20⩽BRW<50 | −1 ⩽dBRW<1 | +5 | 0 | −1 |
| 51 | BP<50 | PAD(0)+14⩽PAD | BRH<50 | 20⩽BRW<50 | dBRW<−1 | −5 | −10 | −1 |
| 52 | BP<50 | PAD(0)+14⩽PAD | BRH<50 | BRW<20 | Unknown | −25 | −20 | +1 |
| 53 | BP<50 | PAD(0)+14⩽PAD | BRH<50 | 80⩽BRW | Unknown | 0 | +20 | +1 |
| 54 | BP<50 | PAD(0)+14⩽PAD | BRH<50 | 50⩽BRW<80 | 1⩽dBRW | +5 | +10 | 0 |
| 55 | BP<50 | PAD(0)+14⩽PAD | BRH<50 | 50⩽BRW<80 | −1⩽dBRW<1 | +5 | 0 | −1 |
| 56 | BP<50 | PAD(0)+14⩽PAD | BRH<50 | 50⩽BRW<80 | dBRW<−1 | −5 | 0 | −1 |
| 57 | BP<50 | PAD(0)+14⩽PAD | 50⩽BRH | 20⩽BRW<50 | 1⩽dBRW | +5 | 0 | 0 |
| 58 | BP<50 | PAD(0)+14⩽PAD | 50⩽BRH | 20⩽BRW<50 | −1⩽dBRW<1 | +5 | 0 | −1 |
| 59 | BP<50 | PAD(0)+14⩽PAD | 50⩽BRH | 20⩽BRW<50 | dBRW<−1 | −5 | −10 | −1 |
| 60 | BP<50 | PAD(0)+14⩽PAD | 50⩽BRH | BRW<20 | Unknown | −25 | −20 | +1 |
| 61 | BP<50 | PAD(0)+14⩽PAD | 50⩽BRH | 80⩽BRW | Unknown | 0 | +20 | +1 |
| 62 | BP<50 | PAD(0)+14⩽PAD | 50⩽BRH | 50⩽BRW<80 | 1⩽dBRW | +5 | 0 | 0 |
| 63 | BP<50 | PAD(0)+14⩽PAD | 50⩽BRH | 50⩽BRW<80 | −1⩽dBRW<1 | +5 | −10 | −1 |
| 64 | BP<50 | PAD(0)+14⩽PAD | 50⩽BRH | 50⩽BRW<80 | dBRW<−1 | −5 | −10 | −1 |

4.2.3.2. *Regulation during normal operation*

Fundamental regulation integrates pump flow volume, systemic arterial pressure, venous reservoir height, blood volume inside the reservoir as well as its changes, in addition to pump flow fluctuation, changes in venous reservoir height, and changes in line occluder opening. Furthermore, the target regulatory value of venous pressure is 50–100 mm Hg. Concurrently, an important condition is that the pump flow volume be 90–110% of the optimal pump flow volume. However, if the pump flow volume becomes less than 90% of the optimal volume, then the regulation routine automatically switches to pump start regulation. In order to analyze the basic operation, the venous pressure level is classified into 3 phenomena. The ideal condition is that in which the venous pressure is high. In this condition, the blood pump rotation frequency is raised, the supply blood volume is increased, and the line occluder is slowly opened. Another condition is that the venous pressure is low; the circulatory blood volume is insufficient because of excess withdrawal. In this condition, the supply volume is slowly increased by trace amounts. Finally, in the condition of normal venous pressure, a fixed withdrawal volume is maintained by opening and closing of the line occluder and by adjustment of the venous reservoir height.

Table 3 lists an expert's fundamental behavior during actual normal operation expressed as functional conditional statements.

4.2.3.3. *Regulation during pump weaning*

Fundamental regulation integrates the pump flow volume, systemic arterial pressure, venous reservoir height, blood volume inside reservoir as well as its changes, in addition to pump flow fluctuation, changes in venous reservoir height, and changes in line occluder opening. Furthermore, the target regulatory value of the pump flow volume is zero. At the same time, the important conditions are that the systemic arterial pressure be more than approximately 70 mm Hg and that the diastolic pulmonary arterial pressure be no more than 14 plus the initial value. However, in the condition that the effective individual heart rate (no more 30–200 beats/minute) is not obtained, in addition to SW2 being released which signifies the selection of *weaning*, SW1 is selected which signifies the selection of *starting*. In other words, the regulatory routine automatically switches to pump start regulation.

Table 3. Expert's fundamental behavior during normal operation.

| | Antecedents | | | | | Consequences | | |
|---|---|---|---|---|---|---|---|---|
| | 1. | 2. | 3. | 4. | 5. | 1. | 2. | 3. |
| Rule | PF (%) | PAD (mm Hg) | BRH (%) | BRW (%) | dBRW (%) | dPF (%) | dBRH (%) | dOcc |
| 1 | PF<110 | 50⩽BP<100 | BRH<50 | 20⩽BRW<50 | 1⩽dBRW | +5 | +10 | 0 |
| 2 | PF<110 | 50⩽BP<100 | BRH<50 | 20⩽BRW<50 | −1⩽dBRW<1 | 0 | 0 | 0 |
| 3 | PF<110 | 50⩽BP<100 | BRH<50 | 20⩽BRW<50 | dBRW< −1 | −5 | −10 | −1 |
| 4 | PF<110 | 50⩽BP<100 | BRH<50 | BRW<20 | Unknown | −25 | −20 | +1 |
| 5 | PF<110 | 50⩽BP<100 | BRH<50 | 80⩽BRW | Unknown | 0 | +20 | +1 |
| 6 | PF<110 | 50⩽BP<100 | BRH<50 | 50⩽BRW<80 | 1⩽dBRW | 0 | +20 | +1 |
| 7 | PF<110 | 50⩽BP<100 | BRH<50 | 50⩽BRW<80 | −1⩽dBRW<1 | 0 | +10 | 0 |
| 8 | PF<110 | 50⩽BP<100 | BRH<50 | 50⩽BRW<80 | dBRW< −1 | 0 | −10 | −1 |
| 9 | PF<110 | 50⩽BP<100 | 50⩽BRH | 20⩽BRW<50 | 1⩽dBRW | +5 | 0 | +1 |
| 10 | PF<110 | 50⩽BP<100 | 50⩽BRH | 20⩽BRW<50 | −1⩽dBRW<1 | 0 | −10 | 0 |
| 11 | PF<110 | 50⩽BP<100 | 50⩽BRH | 20⩽BRW<50 | dBRW< −1 | −5 | −20 | −1 |
| 12 | PF<110 | 50⩽BP<100 | 50⩽BRH | BRW<20 | Unknown | −25 | −20 | +1 |
| 13 | PF<110 | 50⩽BP<100 | 50⩽BRH | 80⩽BRW | Unknown | 0 | +20 | +1 |
| 14 | PF<110 | 50⩽BP<100 | 50⩽BRH | 50⩽BRW<80 | 1⩽dBRW | 0 | 0 | +1 |
| 15 | PF<110 | 50⩽BP<100 | 50⩽BRH | 50⩽BRW<80 | −1⩽dBRW<1 | 0 | 0 | 0 |
| 16 | PF<110 | 50⩽BP<100 | 50⩽BRH | 50⩽BRW<80 | dBRW< −1 | 0 | −10 | −1 |
| 17 | PF<110 | 100⩽BP | BRH<50 | 20⩽BRW<50 | 1⩽dBRW | +5 | 0 | 0 |
| 18 | PF<110 | 100⩽BP | BRH<50 | 20⩽BRW<50 | −1⩽dBRW<1 | 0 | −10 | −1 |
| 19 | PF<110 | 100⩽BP | BRH<50 | 20⩽BRW<50 | dBRW< −1 | −5 | −10 | −1 |
| 20 | PF<110 | 100⩽BP | BRH<50 | BRW<20 | Unknown | −25 | −20 | +1 |
| 21 | PF<110 | 100⩽BP | BRH<50 | 80⩽BRW | Unknown | 0 | +20 | +1 |
| 22 | PF<110 | 100⩽BP | BRH<50 | 50⩽BRW<80 | 1⩽dBRW | 0 | +10 | +1 |

Table 3. (*Cont'd*)

| Rule | Antecedents 1. PF (%) | 2. PAD (mm Hg) | 3. BRH (%) | 4. BRW (%) | 5. dBRW (%) | Consequences 1. dPF (%) | 2. dBRH (%) | 3. dOcc |
|---|---|---|---|---|---|---|---|---|
| 23 | PF<110 | 100⩽BP | BRH<50 | 50⩽BRW<80 | −1⩽dBRW<1 | 0 | 0 | −1 |
| 24 | PF<110 | 100⩽BP | BRH<50 | 50⩽BRW<80 | dBRW< −1 | 0 | −10 | −1 |
| 25 | PF<110 | 100⩽BP | 50⩽BRH | 20⩽BRW<50 | 1⩽dBRW | +5 | 0 | 0 |
| 26 | PF<110 | 100⩽BP | 50⩽BRH | 20⩽BRW<50 | −1⩽dBRW<1 | 0 | −10 | 0 |
| 27 | PF<110 | 100⩽BP | 50⩽BRH | 20⩽BRW<50 | dBRW< −1 | −5 | −20 | −1 |
| 28 | PF<110 | 100⩽BP | 50⩽BRH | BRW<20 | Unknown | −25 | −20 | +1 |
| 29 | PF<110 | 100⩽BP | 50⩽BRH | 80⩽BRW | Unknown | 0 | +20 | +1 |
| 30 | PF<110 | 100⩽BP | 50⩽BRH | 50⩽BRW<80 | 1⩽dBRW | +5 | 0 | 0 |
| 31 | PF<110 | 100⩽BP | 50⩽BRH | 50⩽BRW<80 | −1⩽dBRW<1 | 0 | −10 | −1 |
| 32 | PF<110 | 100⩽BP | 50⩽BRH | 50⩽BRW<80 | dBRW< −1 | 0 | −20 | −1 |
| 33 | PF<110 | BP<50 | BRH<50 | 20⩽BRW<50 | 1⩽dBRW | +5 | +10 | 0 |
| 34 | PF<110 | BP<50 | BRH<50 | 20⩽BRW<50 | −1⩽dBRW<1 | 0 | 0 | 0 |
| 35 | PF<110 | BP<50 | BRH<50 | 20⩽BRW<50 | dBRW< −1 | −5 | −10 | −1 |
| 36 | PF<110 | BP<50 | BRH<50 | BRW<20 | Unknown | −25 | −20 | −1 |
| 37 | PF<110 | BP<50 | BRH<50 | 80⩽BRW | Unknown | +10 | +20 | +1 |
| 38 | PF<110 | BP<50 | BRH<50 | 50⩽BRW<80 | 1⩽dBRW | +5 | +10 | 0 |
| 39 | PF<110 | BP<50 | BRH<50 | 50⩽BRW<80 | −1⩽dBRW<1 | 0 | +10 | +1 |
| 40 | PF<110 | BP<50 | BRH<50 | 50⩽BRW<80 | dBRW< −1 | 0 | 0 | −1 |
| 41 | PF<110 | BP<50 | 50⩽BRH | 20⩽BRW<50 | 1⩽dBRW | 0 | −10 | +1 |
| 42 | PF<110 | BP<50 | 50⩽BRH | 20⩽BRW<50 | −1⩽dBRW<1 | 0 | 0 | 0 |

Table 3. (*Cont'd*)

| Rule | Antecedents 1. PF (%) | 2. PAD (mm Hg) | 3. BRH (%) | 4. BRW (%) | 5. dBRW (%) | Consequences 1. dPF (%) | 2. dBRH (%) | 3. dOcc |
|---|---|---|---|---|---|---|---|---|
| 43 | PF<110 | BP<50 | 50⩽BRH | 20⩽BRW<50 | dBRW< −1 | −5 | −20 | −1 |
| 44 | PF<110 | BP<50 | 50⩽BRH | BRW<20 | Unknown | −25 | −20 | −1 |
| 45 | PF<110 | BP<50 | 50⩽BRH | 80⩽BRW | Unknown | 0 | +20 | +1 |
| 46 | PF<110 | BP<50 | 50⩽BRH | 50⩽BRW<80 | 1⩽dBRW | +5 | 0 | 0 |
| 47 | PF<110 | BP<50 | 50⩽BRH | 50⩽BRW<80 | −1⩽dBRW<1 | 0 | 0 | +1 |
| 48 | PF<110 | BP<50 | 50⩽BRH | 50⩽BRW<80 | dBRW< −1 | 0 | −10 | −1 |
| 49 | 110⩽PF | Unknown | BRH<50 | 20⩽BRW<50 | 1⩽dBRW | 0 | +10 | 0 |
| 50 | 110⩽PF | Unknown | BRH<50 | 20⩽BRW<50 | −1⩽dBRW<1 | −5 | 0 | 0 |
| 51 | 110⩽PF | Unknown | BRH<50 | 20⩽BRW<50 | dBRW< −1 | −5 | −10 | −1 |
| 52 | 110⩽PF | Unknown | BRH<50 | BRW<20 | Unknown | −25 | −20 | −1 |
| 53 | 110⩽PF | Unknown | BRH<50 | 80⩽BRW | Unknown | 0 | +20 | +1 |
| 54 | 110⩽PF | Unknown | BRH<50 | 50⩽BRW<80 | 1⩽dBRW | 0 | +10 | +1 |
| 55 | 110⩽PF | Unknown | BRH<50 | 50⩽BRW<80 | −1⩽dBRW<1 | 0 | +10 | 0 |
| 56 | 110⩽PF | Unknown | BRH<50 | 50⩽BRW<80 | dBRW< −1 | −5 | 0 | 0 |
| 57 | 110⩽PF | Unknown | 50⩽BRH | 20⩽BRW<50 | 1⩽dBRW | 0 | +10 | 0 |
| 58 | 110⩽PF | Unknown | 50⩽BRH | 20⩽BRW<50 | −1⩽dBRW<1 | −5 | 0 | 0 |
| 59 | 110⩽PF | Unknown | 50⩽BRH | 20⩽BRW<50 | dBRW<−1 | −5 | −20 | −1 |
| 60 | 110⩽PF | Unknown | 50⩽BRH | BRW<20 | Unknown | −25 | −20 | −1 |
| 61 | 110⩽PF | Unknown | 50⩽BRH | 80⩽BRW | Unknown | 0 | +20 | +1 |
| 62 | 110⩽PF | Unknown | 50⩽BRH | 50⩽BRW<80 | 1⩽dBRW | 0 | +10 | +1 |
| 63 | 110⩽PF | Unknown | 50⩽BRH | 50⩽BRW<80 | −1⩽dBRW<1 | −5 | 0 | 0 |
| 64 | 110⩽PF | Unknown | 50⩽BRH | 50⩽BRW<80 | dBRW< −1 | −5 | −10 | 0 |

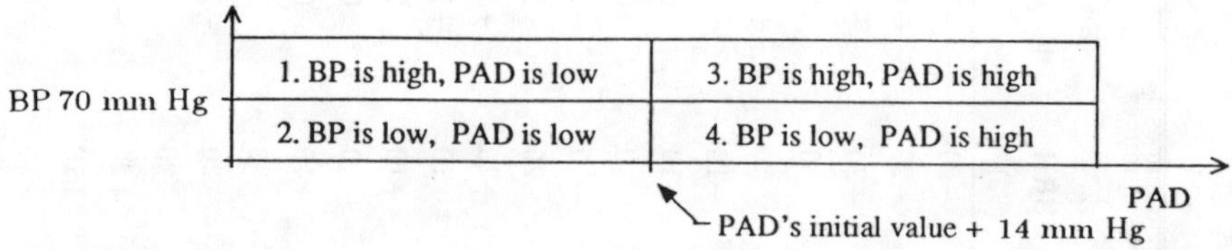


Fig. 11. Assessment of venous pressure (BP) and diastolic pulmonary arterial pressure (PAD) during pump weaning regulation.

Meanwhile, Fig. 11 lists the conditions for systemic arterial pressure and diastolic pulmonary arterial pressure during pump weaning regulation as classified into 4 phenomena.

The ideal condition is condition 1. In this condition, the rotation frequency of the blood flow pump is decreased and the blood supply volume is reduced. In addition, the line occluder is slowly closed. In other words, *weaning* is completed when the blood flow pump stops while the systemic arterial pressure and the diastolic pulmonary arterial pressure are maintained in a favorable condition. Condition 2 is that the ECC blood volume is insufficient by excess withdrawal. In this condition, the supply blood volume slowly increases by trace amounts. Condition 3 is that the ECC blood volume is too high because of an excess supply. In this condition, the supply volume slowly increases by trace amounts of withdrawal. Finally, condition 4 is the worst condition; *weaning* cannot be executed. In this condition, it is important to quickly increase the pump flow volume to the optimal value.

Table 4 lists an expert's fundamental behavior during actual pump weaning expressed as functional conditional statements.

#### 4.2.3.4. *Regulation during safe operation*

Fundamental regulation integrates line pressure (LP), venous reservoir height, blood volume inside reservoir as well as its changes, in addition to pump flow fluctuation, change in venous reservoir height, and changes in line occluder opening. This regulatory routine differs from the three ECC regulatory routines of pump start, normal operation, and pump weaning in that this routine is only passed through no more than once. In other words, after completion the system quickly returns to the original regulatory routine. However, in manual-based operation, returning to the original regulatory routine takes precedence

Table 4. Expert's fundamental behavior during pump weaning.

| | Antecedents | | | | | Consequences | | |
|---|---|---|---|---|---|---|---|---|
| | 1. | 2. | 3. | 4. | 5. | 1. | 2. | 3. |
| Rule | BP (mm Hg) | PAD (mm Hg) | BRH (%) | BRW (%) | dBRW (%) | dPF (%) | dBRH (%) | dOcc |
| 1 | 70⩽BP | PAD < PAD(0)+14 | BRH < 50 | 20⩽BRW < 50 | 1⩽dBRW | 0 | +10 | +1 |
| 2 | 70⩽BP | PAD < PAD(0)+14 | BRH < 50 | 20⩽BRW < 50 | −1⩽dBRW < 1 | −5 | 0 | 0 |
| 3 | 70⩽BP | PAD < PAD(0)+14 | BRH < 50 | 20⩽BRW < 50 | dBRW < −1 | −10 | −10 | −1 |
| 4 | 70⩽BP | PAD < PAD(0)+14 | BRH < 50 | BRW < 20 | Unknown | −25 | −20 | +1 |
| 5 | 70⩽BP | PAD < PAD(0)+14 | BRH < 50 | 80⩽BRW | Unknown | +10 | +20 | +1 |
| 6 | 70⩽BP | PAD < PAD(0)+14 | BRH < 50 | 50⩽BRW < 80 | 1⩽dBRW | 0 | +20 | +1 |
| 7 | 70⩽BP | PAD < PAD(0)+14 | BRH < 50 | 50⩽BRW < 80 | −1⩽dBRW < 1 | −5 | +10 | 0 |
| 8 | 70⩽BP | PAD < PAD(0)+14 | BRH < 50 | 50⩽BRW < 80 | dBRW < −1 | −5 | 0 | 0 |
| 9 | 70⩽BP | PAD < PAD(0)+14 | 50⩽BRH | 20⩽BRW < 50 | 1⩽dBRW | 0 | 0 | +1 |
| 10 | 70⩽BP | PAD < PAD(0)+14 | 50⩽BRH | 20⩽BRW < 50 | −1⩽dBRW < 1 | −5 | 0 | 0 |
| 11 | 70⩽BP | PAD < PAD(0)+14 | 50⩽BRH | 20⩽BRW < 50 | dBRW < −1 | −10 | −10 | −1 |
| 12 | 70⩽BP | PAD < PAD(0)+14 | 50⩽BRH | BRW < 20 | Unknown | −25 | −20 | +1 |
| 13 | 70⩽BP | PAD < PAD(0)+14 | 50⩽BRH | 80⩽BRW | Unknown | +10 | +20 | +1 |
| 14 | 70⩽BP | PAD < PAD(0)+14 | 50⩽BRH | 50⩽BRW < 80 | 1⩽dBRW | 0 | 0 | +1 |
| 15 | 70⩽BP | PAD < PAD(0)+14 | 50⩽BRH | 50⩽BRW < 80 | −1⩽dBRW < 1 | −5 | 0 | 0 |
| 16 | 70⩽BP | PAD < PAD(0)+14 | 50⩽BRH | 50⩽BRW < 80 | dBRW < −1 | −5 | −10 | 0 |
| 17 | 70⩽BP | PAD(0)+14⩽PAD | BRH < 50 | 20⩽BRW < 50 | 1⩽dBRW | −5 | +10 | +1 |
| 18 | 70⩽BP | PAD(0)+14⩽PAD | BRH < 50 | 20⩽BRW < 50 | −1⩽dBRW < 1 | −5 | 0 | 0 |
| 19 | 70⩽BP | PAD(0)+14⩽PAD | BRH < 50 | 20⩽BRW < 50 | dBRW < −1 | −10 | 0 | −1 |
| 20 | 70⩽BP | PAD(0)+14⩽PAD | BRH < 50 | BRW < 20 | Unknown | −25 | −20 | +1 |
| 21 | 70⩽BP | PAD(0)+14⩽PAD | BRH < 50 | 80⩽BRW | Unknown | 0 | +20 | +1 |
| 22 | 70⩽BP | PAD(0)+14⩽PAD | BRH < 50 | 50⩽BRW < 80 | 1⩽dBRW | −5 | +10 | +1 |

Table 4. (*Cont'd*)

| Rule | Antecedents 1. BP (mm Hg) | 2. PAD (mm Hg) | 3. BRH (%) | 4. BRW (%) | 5. dBRW (%) | Consequences 1. dPF (%) | 2. dBRH (%) | 3. dOcc |
|---|---|---|---|---|---|---|---|---|
| 23 | 70⩽BP | PAD(0)+14⩽PAD | BRH<50 | 50⩽BRW<80 | −1⩽dBRW<1 | −5 | 0 | 0 |
| 24 | 70⩽BP | PAD(0)+14⩽PAD | BRH<50 | 50⩽BRW<80 | dBRW< −1 | −5 | 0 | −1 |
| 25 | 70⩽BP | PAD(0)+14⩽PAD | 50⩽BRH | 20⩽BRW<50 | 1⩽dBRW | −5 | 0 | +1 |
| 26 | 70⩽BP | PAD(0)+14⩽PAD | 50⩽BRH | 20⩽BRW<50 | −1⩽dBRW<1 | −5 | 0 | 0 |
| 27 | 70⩽BP | PAD(0)+14⩽PAD | 50⩽BRH | 20⩽BRW<50 | dBRW< −1 | −10 | −10 | −1 |
| 28 | 70⩽BP | PAD(0)+14⩽PAD | 50⩽BRH | BRW<20 | Unknown | −25 | −20 | +1 |
| 29 | 70⩽BP | PAD(0)+14⩽PAD | 50⩽BRH | 80⩽BRW | Unknown | 0 | +20 | +1 |
| 30 | 70⩽BP | PAD(0)+14⩽PAD | 50⩽BRH | 50⩽BRW<80 | 1⩽dBRW | −5 | 0 | +1 |
| 31 | 70⩽BP | PAD(0)+14⩽PAD | 50⩽BRH | 50⩽BRW<80 | −1⩽dBRW<1 | −5 | 0 | 0 |
| 32 | 70⩽BP | PAD(0)+14⩽PAD | 50⩽BRH | 50⩽BRW<80 | dBRW< −1 | −5 | −10 | 0 |
| 33 | BP<70 | PAD<PAD(0)+14 | BRH<50 | 20⩽BRW<50 | 1⩽dBRW | 0 | +20 | +1 |
| 34 | BP<70 | PAD<PAD(0)+14 | BRH<50 | 20⩽BRW<50 | −1⩽dBRW<1 | −5 | 0 | 0 |
| 35 | BP<70 | PAD<PAD(0)+14 | BRH<50 | 20⩽BRW<50 | dBRW< −1 | −10 | 0 | −1 |
| 36 | BP<70 | PAD<PAD(0)+14 | BRH<50 | BRW<20 | Unknown | −25 | −20 | +1 |
| 37 | BP<70 | PAD<PAD(0)+14 | BRH<50 | 80⩽BRW | Unknown | +10 | +30 | +1 |
| 38 | BP<70 | PAD<PAD(0)+14 | BRH<50 | 50⩽BRW<80 | 1⩽dBRW | 0 | +20 | +1 |
| 39 | BP<70 | PAD<PAD(0)+14 | BRH<50 | 50⩽BRW<80 | −1⩽dBRW<1 | −5 | +10 | +1 |
| 40 | BP<70 | PAD<PAD(0)+14 | BRH<50 | 50⩽BRW<80 | dBRW< −1 | −5 | 0 | 0 |
| 41 | BP<70 | PAD<PAD(0)+14 | 50⩽BRH | 20⩽BRW<50 | 1⩽dBRW | 0 | 0 | +1 |
| 42 | BP<70 | PAD<PAD(0)+14 | 50⩽BRH | 20⩽BRW<50 | −1⩽dBRW<1 | −5 | +10 | 0 |
| 43 | BP<70 | PAD<PAD(0)+14 | 50⩽BRH | 20⩽BRW<50 | dBRW< −1 | −5 | −20 | −1 |

Table 4. (*Cont'd*)

| | Antecedents | | | | | Consequences | | |
|---|---|---|---|---|---|---|---|---|
| | 1. | 2. | 3. | 4. | 5. | 1. | 2. | 3. |
| Rule | BP (mm Hg) | PAD (mm Hg) | BRH (%) | BRW (%) | dBRW (%) | dPF (%) | dBRH (%) | dOcc |
| 44 | BP<70 | PAD<PAD(0)+14 | 50⩽BRH | BRW<20 | Unknown | −25 | −20 | +1 |
| 45 | BP<70 | PAD<PAD(0)+14 | 50⩽BRH | 80⩽BRW | Unknown | +10 | +20 | +1 |
| 46 | BP<70 | PAD<PAD(0)+14 | 50⩽BRH | 50⩽BRW<80 | 1⩽dBRW | 0 | 0 | +1 |
| 47 | BP<70 | PAD<PAD(0)+14 | 50⩽BRH | 50⩽BRW<80 | −1⩽dBRW<1 | −5 | 0 | +1 |
| 48 | BP<70 | PAD<PAD(0)+14 | 50⩽BRH | 50⩽BRW<80 | dBRW< −1 | −5 | −10 | 0 |
| 49 | BP<70 | PAD(0)+14⩽PAD | BRH<50 | 20⩽BRW<50 | 1⩽dBRW | 0 | 0 | 0 |
| 50 | BP<70 | PAD(0)+14⩽PAD | BRH<50 | 20⩽BRW<50 | −1⩽dBRW<1 | −5 | 0 | 0 |
| 51 | BP<70 | PAD(0)+14⩽PAD | BRH<50 | 20⩽BRW<50 | dBRW< −1 | −10 | 0 | −1 |
| 52 | BP<70 | PAD(0)+14⩽PAD | BRH<50 | BRW<20 | Unknown | −25 | −20 | +1 |
| 53 | BP<70 | PAD(0)+14⩽PAD | BRH<50 | 80⩽BRW | Unknown | 0 | +20 | +1 |
| 54 | BP<70 | PAD(0)+14⩽PAD | BRH<50 | 50⩽BRW<80 | 1⩽dBRW | 0 | +10 | +1 |
| 55 | BP<70 | PAD(0)+14⩽PAD | BRH<50 | 50⩽BRW<80 | −1⩽dBRW<1 | −5 | 0 | 0 |
| 56 | BP<70 | PAD(0)+14⩽PAD | BRH<50 | 50⩽BRW<80 | dBRW< −1 | −5 | 0 | 0 |
| 57 | BP<70 | PAD(0)+14⩽PAD | 50⩽BRH | 20⩽BRW<50 | 1⩽dBRW | 0 | 0 | +1 |
| 58 | BP<70 | PAD(0)+14⩽PAD | 50⩽BRH | 20⩽BRW<50 | −1⩽dBRW<1 | −5 | −10 | 0 |
| 59 | BP<70 | PAD(0)+14⩽PAD | 50⩽BRH | 20⩽BRW<50 | dBRW< −1 | −5 | −10 | −1 |
| 60 | BP<70 | PAD(0)+14⩽PAD | 50⩽BRH | BRW<20 | Unknown | −25 | −20 | +1 |
| 61 | BP<70 | PAD(0)+14⩽PAD | 50⩽BRH | 80⩽BRW | Unknown | 0 | 0 | +1 |
| 62 | BP<70 | PAD(0)+14⩽PAD | 50⩽BRH | 50⩽BRW<80 | 1⩽dBRW | 0 | 0 | +1 |
| 63 | BP<70 | PAD(0)+14⩽PAD | 50⩽BRH | 50⩽BRW<80 | −1⩽dBRW<1 | −5 | 0 | 0 |
| 64 | BP<70 | PAD(0)+14⩽PAD | 50⩽BRH | 50⩽BRW<80 | dBRW< −1 | −5 | −10 | −1 |

Table 5. Expert's fundamental behavior during safety operation.

| | Antecedents | | | | Consequences | | |
|---|---|---|---|---|---|---|---|
| Rule | 1. LP (mm Hg) | 2. BRH (%) | 3. BRW (%) | 4. dBRW (%) | 1. dPF (%) | 2. dBRH (%) | 3. dOcc |
| 1 | 400⩽LP | BRH<50 | 20⩽BRW<50 | 1⩽dBRW | −PF/2 | +10 | +1 |
| 2 | 400⩽LP | BRH<50 | 20⩽BRW<50 | −1⩽dBRW<1 | −PF/2 | +10 | +1 |
| 3 | 400⩽LP | BRH<50 | 20⩽BRW<50 | dBRW<−1 | −PF/2 | 0 | 0 |
| 4 | 400⩽LP | BRH<50 | BRW<20 | Unknown | −PF/2 | −30 | −1 |
| 5 | 400⩽LP | BRH<50 | 80⩽BRW | Unknown | −20 | +30 | +1 |
| 6 | 400⩽LP | BRH<50 | 50⩽BRW<80 | 1⩽dBRW | −PF/2 | +30 | +1 |
| 7 | 400⩽LP | BRH<50 | 50⩽BRW<80 | −1⩽dBRW<1 | −PF/2 | +30 | +1 |
| 8 | 400⩽LP | BRH<50 | 50⩽BRW<80 | dBRW<−1 | −PF/2 | +30 | +1 |
| 9 | 400⩽LP | 50⩽BRH | 20⩽BRW<50 | 1⩽dBRW | −PF/2 | 0 | +1 |
| 10 | 400⩽LP | 50⩽BRH | 20⩽BRW<50 | −1⩽dBRW<1 | −PF/2 | 0 | +1 |
| 11 | 400⩽LP | 50⩽BRH | 20⩽BRW<50 | dBRW<−1 | −PF/2 | 0 | +1 |
| 12 | 400⩽LP | 50⩽BRH | BRW<20 | Unknown | −PF/2 | −30 | −1 |
| 13 | 400⩽LP | 50⩽BRH | 80⩽BRW | Unknown | −20 | +20 | +2 |
| 14 | 400⩽LP | 50⩽BRH | 50⩽BRW<80 | 1⩽dBRW | −PF/2 | +20 | +2 |
| 15 | 400⩽LP | 50⩽BRH | 50⩽BRW<80 | −1⩽dBRW<1 | −PF/2 | +20 | +2 |
| 16 | 400⩽LP | 50⩽BRH | 50⩽BRW<80 | dBRW<−1 | −PF/2 | +20 | +2 |
| 17 | LP<400 | Unknown | Unknown | Unknown | 0 | 0 | 0 |

even during regulation. In other words, after this operation, it quickly returns to the original regulatory routine. However, the return to the original regulatory routine according to manual operation receives precedence even during passage. In the condition that the line pressure is extremely high (more than apporoximately 400 mm Hg), generally, while controlling the withdrawal volume by the closing of the line occluder and the fluctuation of venous reservoir, the supply volume is reduced by half. In addition, in the condition that the line pressure is somewhat high(no more than 300–400 mm Hg) the circumstances are checked.

Table 5 lists an expert's fundamental behavior during actual safety operation (including conditions) regulation expressed as functional conditional statements.

## 5. Simplification of Control Rules

ECC incorporates four main control routines: pump start control, normal operating control, pump weaning control, and safe operating (and conditions) control. In this paper, we do not provide the set of complex control rules that are necessary for each regulatory behavior. In Table 2, the 64 control rules for pump start regulatory behavior are listed as conditions in the form of numerical expressions. For example, the 64th control rule is expressed as the following conditional statement:

"*IF the systemic arterial pressure is less than 50 mm Hg, AND the diastolic pulmonary arterial pressure is greater than the initial value plus 14 mm Hg, AND the height of the venous reservoir is greater than 50% of the total range, AND the blood volume inside the venous reservoir is 50–80% of the total range, AND the change in blood volume inside the venous reservoir is less than −1%; THEN the pump flow volume decreases by 5% of the total range, AND the height of the venous reservoir drops by 10% of the total range, AND the opening of the occluder increases by one phase.*"

In the following, the conditional statements will be simplified to a set of logic expressions.

### 5.1. *Integration of control rules*

By combining the antecedent phrases of the complex conditional statements which share the same consequence, the antecedents can be simplified to form a single conditional statement. For example, Eq. (11) expresses "*IF the systemic arterial pressure is low, THEN the pump flow volume increases; and IF the systemic arterial pressure is normal, THEN the pump flow volume increases.*"

$$(a \to r) * (b \to r) \tag{11}$$

$$\begin{aligned} \text{where } a &= \langle\langle \text{the systemic arterial pressure is low} \rangle\rangle \\ b &= \langle\langle \text{the systemic arterial pressure is normal} \rangle\rangle \\ r &= \langle\langle \text{increase the pump flow volume} \rangle\rangle. \end{aligned}$$

Furthermore, from the formula related to the conditional statements, Eq. (12) (the right side) is obtained. The logic symbols are '$\sim$', which signifies negation (NOT), '$*$', which signifies product (AND), and '$\vee$', which signifies sum (OR).

$$\begin{aligned} (a \to r) * (b \to r) \equiv & (\sim a \vee r) * (\sim b \vee r) \\ & (\sim a * \sim b) \vee r \\ & \sim (a \vee b) \vee r \\ & (a \vee b) \to r \end{aligned} \tag{12}$$

## 5.2. *Definition of logic variables*

We define two complete sets, $BP$ and $Y_1 = \{0, 1\}$. Then, the mapping of $f_a$ and $f_b$ from $BP$ to $Y_1$ is defined by Eq. (13). In other words, $BP$ is the fixed range of the characteristic functions $f_a$ and $f_b$. In addition, $Y_1 = \{0, 1\} = \{y_1', y_1\}$ is the range of values for $f_a$ and $f_b$.

$$f_a,\ f_b : BP \to \{0, 1\} \tag{13}$$

Where $BP$ is the measured possible range of the systemic arterial pressure, $A$ is the range in which the systemic arterial pressure is low, and $B$ is the range in which the systemic arterial pressure is not low. The fixed range of $BP$ and the partial sets of $A$ and $B$ are expressed in Eqs. (14)–(16).

$$BP = \{X \mid 0 \leqslant X < 100 \text{ mm Hg} \tag{14}$$

$$A = f_a^{-1}(1) = \{x \mid 0 \leqslant x < 50 \text{ mm Hg},\ x \in BP\} \tag{15}$$

$$B = f_b^{-1}(1) = \{x \mid 50 \leqslant x < 100 \text{ mm Hg},\ x \in BP\} \tag{16}$$

The relationships among $BP$, $A$, and $B$ are expressed in Eqs. (17) and (18). In other words, $B$ is the complement of $A$, called $A^c$.

$$A \cap B (= A^c) = \phi \tag{17}$$

$$A \cup B (= A^c) = BP \tag{18}$$

As a rule, an algebraic set containing $n$ elements which form a finite set, called a *power set*, has the same form as a finite Boolean expression, $B_2^n$, which originates from $2^n$ elements. In addition, the logic equation has the same form as the Boolean expression, $B_2 = \{0, 1\}$. Therefore, the antecedents in Table 2 are replaced by logic variables of crisp functions. Examples of consequences for the conditional statements {dPF = +5, dBRH = 0, dOcc = 0} are listed in Table 6.

Table 6. Minimal terms for Boolean expressions.

| Rule | BP | PAD | BRH | BRW | dBRW |
|---|---|---|---|---|---|
| 1 | $y_1'$ | $*y_2$ | $*y_3$ | $*(y_4' * y_5 * y_6' * y_7')$ | $*(y_8' * y_9' * y_{10})$ |
| 9 | $y_1'$ | $*y_2$ | $*y_3'$ | $*(y_4' * y_5 * y_6' * y_7')$ | $*(y_8' * y_9' * y_{10})$ |
| 14 | $y_1'$ | $*y_2$ | $*y_3'$ | $*(y_4' * y_5' * y_6 * y_7')$ | $*(y_8' * y_9' * y_{10})$ |
| 17 | $y_1'$ | $*y_2'$ | $*y_3$ | $*(y_4' * y_5 * y_6' * y_7')$ | $*(y_8' * y_9' * y_{10})$ |
| 25 | $y_1'$ | $*y_2'$ | $*y_3'$ | $*(y_4' * y_5 * y_6' * y_7')$ | $*(y_8' * y_9' * y_{10})$ |
| 30 | $y_1'$ | $*y_2'$ | $*y_3'$ | $*(y_4' * y_5' * y_6 * y_7')$ | $*(y_8' * y_9' * y_{10})$ |
| 33 | $y_1$ | $*y_2$ | $*y_3$ | $*(y_4' * y_5 * y_6' * y_7')$ | $*(y_8' * y_9' * y_{10})$ |
| 41 | $y_1$ | $*y_2$ | $*y_3'$ | $*(y_4' * y_5 * y_6' * y_7')$ | $*(y_8' * y_9' * y_{10})$ |
| 46 | $y_1$ | $*y_2$ | $*y_3'$ | $*(y_4' * y_5' * y_6 * y_7')$ | $*(y_8' * y_9' * y_{10})$ |
| 49 | $y_1$ | $*y_2'$ | $*y_3$ | $*(y_4' * y_5 * y_6' * y_7')$ | $*(y_8' * y_9' * y_{10})$ |
| 57 | $y_1$ | $*y_2'$ | $*y_3'$ | $*(y_4' * y_5 * y_6' * y_7')$ | $*(y_8' * y_9' * y_{10})$ |
| 62 | $y_1$ | $*y_2'$ | $*y_3'$ | $*(y_4' * y_5' * y_6 * y_7')$ | $*(y_8' * y_9' * y_{10})$ |

### 5.3. *Simplification of logic expressions*

The $n$-variable logic expression $f(y_1, \ldots, y_n)$ is represented by the mapping from $B_2^n$ to $B_2$, $f : B_2^n \to B_2$. For example, the 10-variable logic expression $f_1$ is represented by Eq. (19).

$$\begin{aligned} f_1(y_1, y_2, \ldots, y_{10}) = {} & y_1' * y_2 * y_3 * (y_4' * y_5 * y_6' * y_7') * (y_8' * y_9' * y_{10}) \\ & + y_1' * y_2 * y_3' * (y_4' * y_5 * y_6' * y_7') * (y_8' * y_9' * y_{10}) + \cdots \\ & + y_1 * y_2' * y_3' * (y_4' * y_5' * y_6 * y_7') * (y_8' * y_9' * y_{10}) . \end{aligned} \tag{19}$$

When simplifying logic expressions, similar terms are grouped together. If, within the context of logical OR, two terms differ in that one is negative and the other is positive, then they can be simplified. For example, if $P$ is the common logic expression of the two terms $y * P$ and $y' * P$, then the simplified form is the following:

$$y * P + y' * P = (y + y') * P = P \tag{20}$$

Among the conventional methods for obtaining the simplest form of the logic expressions, there is the Karnaugh map method and the Quine–McClusky method. In the Karnaugh map method,the simplification process of rewriting the truth-value expression of the function as a two-dimensional expression is intuitively easy to understand. However, if there are more than 5 variables, then the expression is rather difficult to express with a Karnaugh map.

Therefore, for obtaining the simplest form of the 10-variable logic expression $f_1$, as shown in Eq. (19), after various trials, the Quine–McClusky method is used which gathers only the necessary terms.

### 5.4. *Simplification of logic expressions based on the Quine–McClusky method*

As for the minimal terms of Equation 19 (the expression formed by a number of logic variables or their negative product), the number of non-negative variables is called the Hamming distance of that term. For example, the Hamming distance of $y_1' * y_2 * y_3 * y_5 * y_{10}$ is 4. Simplification is done within the interval of two terms whose Hamming distances differ only by 1. Possible combinations are completely simplified. The result is that all candidate terms (principal terms), which can be expressed in the simplest form, are obtained. The targeted simplest form can be obtained by selecting the most important terms from among the principal terms.

The simplest form of Eq. (19) is expressed in Eq. (21).

$$f_1(y_1, y_2, \ldots, y_{10}) = y_3' * y_4' * y_5' * y_6 * y_7' * y_8' * y_9' * y_{10} + y_4' * y_5 * y_6' * y_7' * y_8' * y_9' * y_{10} \tag{21}$$

### 5.5. *Simplified fundamental regulatory behavior*

Based on the equality of the set algebra and Boolean algebra in Table 2, the simplified logic expressions are rewritten from logic variables to a crisp set. Table 7 lists the 20 elements which were condensed from 64 elements based on simplified fundamental regulatory behavior of ECC. In a similar manner,

Table 7. Simplified fundamental regulatory behavior during pump start regulation.

| | Antecedents | | | | | Consequences | | |
|---|---|---|---|---|---|---|---|---|
| Rule | 1. BP (mm Hg) | 2. PAD (mm Hg) | 3. BRH (%) | 4. BRW (%) | 5. dBRW (%) | 1. dPF (%) | 2. dBRH (%) | 3. dOcc |
| 1 | Unknown | Unknown | 50⩽BRH | 50⩽BRW<80 | 1⩽dBRW | +5 | 0 | 0 |
| 2 | Unknown | Unknown | Unknown | 20⩽BRW<50 | 1⩽dBRW | +5 | 0 | 0 |
| 3 | Unknown | PAD<PAD(0)+14 | 50⩽BRH | 50⩽BRW<80 | −1⩽dBRW<1 | +5 | 0 | −1 |
| 4 | Unknown | PAD(0)+14⩽PAD | BRH<50 | 50⩽BRW<80 | −1⩽dBRW<1 | +5 | 0 | −1 |
| 5 | 50⩽BP | Unknown | Unknown | 50⩽BRW<80 | −1⩽dBRW<1 | +5 | 0 | −1 |
| 6 | Unknown | Unknown | BRH<50 | 20⩽BRW<50 | −1⩽dBRW<1 | +5 | 0 | −1 |
| 7 | BP<50 | Unknown | Unknown | 20⩽BRW<50 | −1⩽dBRW<1 | +5 | 0 | −1 |
| 8 | BP<50 | PAD(0)+14⩽PAD | 50⩽BRH | 50⩽BRW<80 | dBRW< −1 | −5 | −10 | −1 |
| 9 | Unknown | Unknown | Unknown | 20⩽BRW<50 | dBRW< −1 | −5 | −10 | −1 |
| 10 | Unknown | Unknown | Unknown | BRW<20 | Unknown | −25 | −20 | +1 |
| 11 | Unknown | PAD<PAD(0)+14 | Unknown | 80⩽BRW | Unknown | +10 | +20 | +1 |
| 12 | Unknown | Unknown | BRH<50 | 50⩽BRW<80 | 1⩽dBRW | +5 | +10 | 0 |
| 13 | Unknown | PAD<PAD(0)+14 | BRH<50 | 50⩽BRW<80 | dBRW< −1 | 0 | 0 | −1 |
| 14 | BP<50 | PAD(0)+14⩽PAD | 50⩽BRH | 50⩽BRW<80 | −1⩽dBRW<1 | +5 | −10 | −1 |
| 15 | 50⩽BP | Unknown | 50⩽BRH | 20⩽BRW<50 | −1⩽dBRW<1 | +5 | −10 | −1 |
| 16 | 50⩽BP | Unknown | 50⩽BRH | 50⩽BRW<80 | dBRW< −1 | 0 | −10 | −1 |
| 17 | Unknown | PAD(0)+14⩽PAD | Unknown | 80⩽BRW | Unknown | 0 | +20 | +1 |
| 18 | Unknown | PAD(0)+14⩽PAD | BRH<50 | 50⩽BRW<80 | dBRW< −1 | −5 | 0 | −1 |
| 19 | BP<50 | PAD<PAD(0)+14 | BRH<50 | 50⩽BRW<80 | −1⩽dBRW<1 | +5 | +10 | −1 |
| 20 | BP<50 | PAD<PAD(0)+14 | 50⩽BRH | 50⩽BRW<80 | dBRW< −1 | 0 | −10 | 0 |

as in Tables 8–10, the regulation rules for normal operation, pump weaning, and safety operation (as well as conditions) are summarized in 41, 36, and 9 rules, respectively. In other words, by simplification, the set of regulation rules related to ECC are reduced from 209 to 106.

## 6. Verification Based on Previous Regulatory Method

The number of control rules related to ECC was reduced from 209 to 106; however, the complexity of control involving 5 (or 4) inputs and 3 outputs is still evident. Some labor is necessary in the case of applying the previous PID regulation method to a multiple input/multiple output system, such as dividing it into several single input/single output systems.

However, a logical analysis of ECC operation is rather difficult. Certainly, many researchers have attempted to automate the ECC operation processes. The regulatory objective during ECC is to maintain a fixed blood volume inside the body as well as to maintain a fixed systemic arterial pressure and diastolic pulmonary arterial pressure. The method of maintaining a fixed blood volume inside the body involves maintaining a fixed weight for the individual, maintaining a fixed blood volume inside the venous reservoir, and regulating central venous pressure(CVP) at a fixed value. However, it is difficult to continue to measure the body weight during the operation of the artificial heart–lung apparatus. The method for stabilizing the venous reservoir blood volume has a problem when the target regulatory value changes with factors such as blood loss due to surgical operation, the exchange of water between blood and body fluid, blood transfusion, and injection of cardioplegic solution. Finally, in the method of controlling fixed value of points such as relevance of regulation index test accuracy of target regulatory value are problems. Changes in central venous pressure are limited in its conformity. Therefore, the indices of this type are available at any time that the regulatory system is insufficient.

Also, depending upon the measurement location, sucking pressure and other factors from the withdrawal line easily affect the measurement during ECC and render the credibility of the measured values as questionable. Therefore, even though ordinarily the regulation index and regulation principle are experimentally investigated, they are not theoretically established.

As for an expert's regulatory operation, static regulation method based on conditions under the specified domain, i.e., sequence regulation, can be executed. Before the animal experiments, the regulatory rules and regulatory method were verified. The verification method is a short circuit experiment

Table 8. Simplified fundamental regulatory behavior during normal operation.

| | Antecedents | | | | | Consequences | | |
|---|---|---|---|---|---|---|---|---|
| Rule | 1. BF (%) | 2. BP (mm Hg) | 3. BRH (%) | 4. BRW (%) | 5. dBRW (%) | 1. dPF (%) | 2. dBRH (%) | 3. dOcc |
| 1 | PF<110 | BP<100 | BRH<50 | 20⩽BRW<50 | 1⩽dBRW | +5 | +10 | 0 |
| 2 | PF<110 | BP<50 | BRH<50 | 50⩽BRW<80 | 1⩽dBRW | +5 | +10 | 0 |
| 3 | PF<110 | 50⩽BP<100 | 50⩽BRH | 50⩽BRW<80 | −1⩽dBRW<1 | 0 | 0 | 0 |
| 4 | PF<110 | 50⩽BP<100 | BRH<50 | 20⩽BRW<50 | −1⩽dBRW<1 | 0 | 0 | 0 |
| 5 | PF<110 | BP<50 | Unknown | 20⩽BRW<50 | −1⩽dBRW<1 | 0 | 0 | 0 |
| 6 | Unknown | Unknown | BRH<50 | 20⩽BRW<50 | dBRW< −1 | −5 | −10 | −1 |
| 7 | PF<110 | 50⩽BP | Unknown | BRW<20 | Unknown | −25 | −20 | +1 |
| 8 | PF<110 | BP<50 | 50⩽BRH | 80⩽BRW | Unknown | 0 | +20 | +1 |
| 9 | PF<110 | 50⩽BP<100 | BRH<50 | 50⩽BRW<80 | 1⩽dBRW | 0 | +20 | +1 |
| 10 | 110⩽PF | Unknown | 50⩽BRH | 80⩽BRW | Unknown | 0 | +20 | +1 |
| 11 | PF<110 | 50⩽BP | Unknown | 80⩽BRW | Unknown | 0 | +20 | +1 |
| 12 | PF<110 | 50⩽BP<100 | BRH<50 | 50⩽BRW<80 | −1⩽dBRW<1 | 0 | +10 | 0 |
| 13 | 110⩽PF | Unknown | Unknown | 20⩽BRW<50 | 1⩽dBRW | 0 | +10 | 0 |
| 14 | 110⩽PF | Unknown | BRH<50 | 50⩽BRW<80 | −1⩽dBRW<1 | 0 | +10 | 0 |
| 15 | PF<110 | 50⩽BP<100 | Unknown | 50⩽BRW<80 | dBRW< −1 | 0 | −10 | −1 |
| 16 | PF<110 | 100⩽BP | BRH<50 | 20⩽BRW<50 | −1⩽dBRW<1 | 0 | −10 | −1 |
| 17 | PF<110 | 100⩽BP | BRH<50 | 50⩽BRW<80 | dBRW< −1 | 0 | −10 | −1 |
| 18 | PF<110 | 100⩽BP | 50⩽BRH | 50⩽BRW<80 | −1⩽dBRW<1 | 0 | −10 | −1 |
| 19 | PF<110 | BP<50 | 50⩽BRH | 50⩽BRW<80 | dBRW< −1 | 0 | −10 | −1 |
| 20 | PF<110 | 50⩽BP<100 | 50⩽BRH | 20⩽BRW<50 | 1⩽dBRW | +5 | 0 | +1 |
| 21 | PF<110 | 50⩽BP | 50⩽BRH | 20⩽BRW<50 | −1⩽dBRW<1 | 0 | −10 | 0 |
| 22 | Unknown | Unknown | 50⩽BRH | 20⩽BRW<50 | dBRW< −1 | −5 | −20 | −1 |

Table 8. (*Cont'd*)

| Rule | Antecedents 1. BF (%) | 2. BP (mm Hg) | 3. BRH (%) | 4. BRW (%) | 5. dBRW (%) | Consequences 1. dPF (%) | 2. dBRH (%) | 3. dOcc |
|---|---|---|---|---|---|---|---|---|
| 23 | PF<110 | 50⩽BP<100 | 50⩽BRH | 50⩽BRW<80 | 1⩽dBRW | 0 | 0 | +1 |
| 24 | PF<110 | BP<50 | 50⩽BRH | 50⩽BRW<80 | −1⩽dBRW<1 | 0 | 0 | +1 |
| 25 | PF<110 | 100⩽BP | Unknown | 20⩽BRW<50 | 1⩽dBRW | +5 | 0 | 0 |
| 26 | PF<110 | 100⩽BP | 50⩽BRH | 50⩽BRW<80 | 1⩽dBRW | +5 | 0 | 0 |
| 27 | PF<110 | BP<50 | 50⩽BRH | 50⩽BRW<80 | 1⩽dBRW | +5 | 0 | 0 |
| 28 | PF<110 | 100⩽BP | BRH<50 | 50⩽BRW<80 | 1⩽dBRW | 0 | +10 | +1 |
| 29 | PF<110 | BP<50 | BRH<50 | 50⩽BRW<80 | −1⩽dBRW<1 | 0 | +10 | +1 |
| 30 | 110⩽PF | Unknown | Unknown | 50⩽BRW<80 | 1⩽dBRW | 0 | +10 | +1 |
| 31 | PF<110 | 100⩽BP | BRH<50 | 50⩽BRW<80 | −1⩽dBRW<1 | 0 | 0 | −1 |
| 32 | PF<110 | BP<50 | BRH<50 | 50⩽BRW<80 | dBRW< −1 | 0 | 0 | −1 |
| 33 | PF<110 | 100⩽BP | 50⩽BRH | 50⩽BRW<80 | dBRW< −1 | 0 | −20 | −1 |
| 34 | PF<110 | BP<50 | Unknown | BRW<20 | Unknown | −25 | −20 | −1 |
| 35 | 110⩽PF | Unknown | Unknown | BRW<20 | Unknown | −25 | −20 | −1 |
| 36 | PF<110 | BP<50 | BRH<50 | 80⩽BRW | Unknown | +10 | +20 | +1 |
| 37 | PF<110 | BP<50 | 50⩽BRH | 20⩽BRW<50 | 1⩽dBRW | 0 | −10 | +1 |
| 38 | 110⩽PF | Unknown | Unknown | 20⩽BRW<50 | −1⩽dBRW<1 | −5 | 0 | 0 |
| 39 | 110⩽PF | Unknown | BRH<50 | 50⩽BRW<80 | dBRW< −1 | −5 | 0 | 0 |
| 40 | 110⩽PF | Unknown | 50⩽BRH | 50⩽BRW<80 | −1⩽dBRW<1 | −5 | 0 | 0 |
| 41 | 110⩽PF | Unknown | 50⩽BRH | 50⩽BRW<80 | dBRW< −1 | −5 | −10 | 0 |

Table 9. Simplified fundamental regulatory behavior during pump weaning.

| | Antecedents | | | | | Consequences | | |
|---|---|---|---|---|---|---|---|---|
| | 1. | 2. | 3. | 4. | 5. | 1. | 2. | 3. |
| Rule | BP (mm Hg) | PAD (mm Hg) | BRH (%) | BRW (%) | dBRW (%) | dPF (%) | dBRH (%) | dOcc |
| 1 | 70≤BP | PAD<PAD(0)+14 | BRH<50 | 20≤BRW<50 | 1≤dBRW | 0 | +10 | +1 |
| 2 | BP<70 | PAD(0)+14≤PAD | BRH<50 | 50≤BRW<80 | 1≤dBRW | 0 | +10 | +1 |
| 3 | 70≤BP | Unknown | 50≤BRH | 50≤BRW<80 | −1≤dBRW<1 | −5 | 0 | 0 |
| 4 | Unknown | PAD<PAD(0)+14 | BRH<50 | 50≤BRW<80 | dBRW< −1 | −5 | 0 | 0 |
| 5 | BP<70 | Unknown | BRH<50 | 50≤BRW<80 | dBRW< −1 | −5 | 0 | 0 |
| 6 | 70≤BP | Unknown | Unknown | 20≤BRW<50 | −1≤dBRW<1 | −5 | 0 | 0 |
| 7 | Unknown | PAD(0)+14≤PAD | Unknown | 50≤BRW<80 | −1≤dBRW<1 | −5 | 0 | 0 |
| 8 | Unknown | Unknown | BRH<50 | 20≤BRW<50 | −1≤dBRW<1 | −5 | 0 | 0 |
| 9 | 70≤BP | Unknown | 50≤BRH | 20≤BRW<50 | dBRW< −1 | −10 | −10 | −1 |
| 10 | 70≤BP | PAD<PAD(0)+14 | Unknown | 20≤BRW<50 | dBRW< −1 | −10 | −10 | −1 |
| 11 | Unknown | Unknown | Unknown | BRW<20 | Unknown | −25 | −20 | +1 |
| 12 | 70≤BP | PAD<PAD(0)+14 | Unknown | 80≤BRW | Unknown | +10 | +20 | +1 |
| 13 | Unknown | PAD<PAD(0)+14 | 50≤BRH | 80≤BRW | Unknown | +10 | +20 | +1 |
| 14 | BP<70 | PAD<PAD(0)+14 | BRH<50 | 20≤BRW<50 | 1≤dBRW | 0 | +20 | +1 |
| 15 | 70≤BP | PAD(0)+14≤PAD | Unknown | 80≤BRW | Unknown | 0 | +20 | +1 |
| 16 | Unknown | PAD(0)+14≤PAD | BRH<50 | 80≤BRW | Unknown | 0 | +20 | +1 |
| 17 | Unknown | PAD<PAD(0)+14 | BRH<50 | 50≤BRW<80 | 1≤dBRW | 0 | +20 | +1 |
| 18 | 70≤BP | PAD<PAD(0)+14 | BRH<50 | 50≤BRW<80 | −1≤dBRW<1 | −5 | +10 | 0 |
| 19 | BP<70 | PAD<PAD(0)+14 | 50≤BRH | 20≤BRW<50 | −1≤dBRW<1 | −5 | +10 | 0 |
| 20 | BP<70 | PAD(0)+14≤PAD | 50≤BRH | 80≤BRW | Unknown | 0 | 0 | +1 |
| 21 | Unknown | PAD<PAD(0)+14 | 50≤BRH | 20≤BRW<80 | 1≤dBRW | 0 | 0 | +1 |
| 22 | BP<70 | Unknown | 50≤BRH | 20≤BRW<80 | 1≤dBRW | 0 | 0 | +1 |

Table 9. (*Cont'd*)

| | Antecedents | | | | | Consequences | | |
|---|---|---|---|---|---|---|---|---|
| Rule | 1. BP (mm Hg) | 2. PAD (mm Hg) | 3. BRH (%) | 4. BRW (%) | 5. dBRW (%) | 1. dPF (%) | 2. dBRH (%) | 3. dOcc |
| 23 | BP<70 | PAD(0)+14⩽PAD | 50⩽BRH | 20⩽BRW<50 | −1⩽dBRW<1 | −5 | −10 | 0 |
| 24 | 70⩽BP | Unknown | 50⩽BRH | 50⩽BRW<80 | dBRW< −1 | −5 | −10 | 0 |
| 25 | Unknown | PAD<PAD(0)+14 | 50⩽BRH | 50⩽BRW<80 | dBRW< −1 | −5 | −10 | 0 |
| 26 | 70⩽BP | PAD(0)+14⩽PAD | BRH<50 | 20⩽BRW<80 | 1⩽dBRW | −5 | +10 | +1 |
| 27 | BP<70 | PAD<PAD(0)+14 | BRH<50 | 50⩽BRW<80 | −1⩽dBRW<1 | −5 | +10 | +1 |
| 28 | Unknown | PAD(0)+14⩽PAD | BRH<50 | 20⩽BRW<50 | dBRW< −1 | −10 | 0 | −1 |
| 29 | BP<70 | Unknown | BRH<50 | 20⩽BRW<50 | dBRW< −1 | −10 | 0 | −1 |
| 30 | 70⩽BP | PAD(0)+14⩽PAD | BRH<50 | 50⩽BRW<80 | dBRW< −1 | −5 | 0 | −1 |
| 31 | 70⩽BP | PAD(0)+14⩽PAD | 50⩽BRH | 20⩽BRW<80 | 1⩽dBRW | −5 | 0 | +1 |
| 32 | BP<70 | PAD<PAD(0)+14 | 50⩽BRH | 50⩽BRW<80 | −1⩽dBRW<1 | −5 | 0 | +1 |
| 33 | BP<70 | PAD<PAD(0)+14 | BRH<50 | 80⩽BRW | Unknown | +10 | +30 | +1 |
| 34 | BP<70 | PAD<PAD(0)+14 | 50⩽BRH | 20⩽BRW<50 | dBRW< −1 | −5 | −20 | −1 |
| 35 | BP<70 | PAD(0)+14⩽PAD | BRH<50 | 20⩽BRW<50 | 1⩽dBRW | 0 | 0 | 0 |
| 36 | BP<70 | PAD(0)+14⩽PAD | 50⩽BRH | 20⩽BRW<80 | dBRW< −1 | −5 | −10 | −1 |

Table 10. Simplified fundamental behavior during safety operation.

| Rule | Antecedents | | | | Consequences | | |
|---|---|---|---|---|---|---|---|
| | 1. LP (mm Hg) | 2. BRH (%) | 3. BRW (%) | 4. dBRW (%) | 1. dPF (%) | 2. dBRH (%) | 3. dOcc |
| 1 | 400⩽LP | BRH<50 | 20⩽BRW<50 | −1⩽dBRW | −PF/2 | +10 | +1 |
| 2 | 400⩽LP | BRH<50 | 20⩽BRW<50 | dBRW<−1 | −PF/2 | 0 | 0 |
| 3 | 400⩽LP | Unknown | BRW<20 | Unknown | −PF/2 | −30 | −1 |
| 4 | 400⩽LP | BRH<50 | 80⩽BRW | Unknown | −20 | +30 | +1 |
| 5 | 400⩽LP | BRH<50 | 50⩽BRW<80 | Unknown | −PF/2 | +30 | +1 |
| 6 | 400⩽LP | 50⩽BRH | 20⩽BRW<50 | Unknown | −PF/2 | 0 | +1 |
| 7 | 400⩽LP | 50⩽BRH | 80⩽BRW | Unknown | −20 | +20 | +2 |
| 8 | 400⩽LP | 50⩽BRH | 50⩽BRW<80 | Unknown | −PF/2 | +20 | +2 |
| 9 | LP<400 | Unknown | Unknown | Unknown | 0 | 0 | 0 |

including a water tank which models the living body and an auto-regulation system. Actual clinical data for ECC was used. A water circulation experiment was conducted to evaluate the control sequence; however, due to problems concerning slow controller operation, favorable results could not be obtained.

## 7. Application of Fuzzy Control

In an attempt to improve the inadequate control characteristics of the static regulation method, fuzzy logic was applied. Membership functions characterize the fuzzy sets using the standard functions of Fig. 5. The functional expressions are functions $Z$, $P$, $S$, and impulse function $SF$ which were defined in Eqs. (7)–(10).

### 7.1. *Fuzzification of control of pump-start*

The characteristics of the fuzzy set corresponding to each input/output variable is expressed by the conditional expressions for Table 7. In Table 11, the fuzzy sets of Table 7 are defined in terms of the standard functions of Eqs. (7)–(10). Using fuzzy inference, dPF and dBRH are detemined by the MAX–MIN composition/center-of-gravity method, and dOcc is determined by the MAX–MIN composition/maximum-height method.

In addition, the labels of the indices which modify the membership functions are expressed in the hexadecimal system (in Table 12). The processor manages the fuzzy inference board in addition to the labels of the indices.

### 7.2. *Fuzzification of control of stable control*

In Table 13, the characterized fuzzy sets of Table 8 are defined by the standard functions of Eqs. (7)–(10). Using fuzzy inference, dPF and dBRH are determined by the MAX–MIN composition/center-of-gravity method, and dOcc is determined by the MAX–MIN composition/maximum-height method.

In addition, the labels of the indices which modify the membership functions are expressed in hexadecimal in Table 14.

### 7.3. *Fuzzification of control of pump weaning*

Similarly, the fuzzy sets of Table 9 are defined by the standard functions of Eqs. (7)–(10), Using fuzzy inference, dPF and dBRH are determined by the MAX–MIN composition/center-of-gravity method, and dOcc is determined by the MAX–MIN composition/maximum-height method.

Table 11. Fuzzy set descriptions as related to input/output variable for pump start regulation.

| Antecedents | | | Consequences | | |
|---|---|---|---|---|---|
| Var. | Fuzzy set | Translation to base Membership function | Var. | Fuzzy set | Translation to base Membership function |
| BP | BP<50 | Z(u;40,50,60) | dPF | −25 | Z(u;−30, −26.25, −22.5) |
| | 50⩽BP | S(u;40,50,60) | | −5 | P(u;−15, −11.25, −7.5, −7.5, 3.75, 0) |
| PAD | PAD<PAD(0)+14 | Z(u-PAD(0);12,14,16) | | 0 | P(u;−7.5, −3.75, 0, 0, 3.75, 7.5) |
| | PAD(0)+14⩽PAD | S(u-PAD(0);12,14,16) | | +5 | P(u;0,3.75,7.5,7.5,11.25,15) |
| BRH | BRH<50 | Z(u;40,50,60) | | +10 | P(u;7.5,11.25,15,15,18.75,22.5) |
| | 50⩽BRH | S(u;40,50,60) | dBRH | −20 | Z(u;−20, −15, −10) |
| BRW | BRW<20 | Z(u;14,20,26) | | −10 | P(u;−20, −15, −10, −10, −5, 0) |
| | 20⩽BRW<50 | P(u;14,20,26,44,50,56) | | 0 | P(u;−10, −5, 0, 0, 5, 10) |
| | 50⩽BRW<80 | P(u;44,50,56,74,80,86) | | +10 | P(u;0,5,10,10,15,20) |
| | 80⩽BRW | S(u;74,80,86) | | +20 | S(u;10,15,20) |
| dBRW | dBRW< −1 | Z(u;−2, −1,0) | | | |
| | −1⩽dBRW<1 | P(u;−2, −1, 0, 0, 1, 2) | | | |
| | 1⩽dBRW | S(u;0,1,2) | | | |

Table 12. Hexadecimal representation of the membership functions.

| | Antecedents | | | | | Consequences | | | | Antecedents | | | | | Consequences | | |
|---|---|---|---|---|---|---|---|---|---|---|---|---|---|---|---|---|---|
| Rule | 1 | 2 | 3 | 4 | 5 | 1 | 2 | 3 | Rule | 1 | 2 | 3 | 4 | 5 | 1 | 2 | 3 |
| sr01 | $01 | $01 | $03 | $5e | $12 | $3b | $17 | $80 | sr02 | $01 | $01 | $01 | $5d | $12 | $3b | $17 | $82 |
| sr03 | $01 | $02 | $03 | $5e | $11 | $3b | $17 | $84 | sr04 | $01 | $03 | $02 | $5e | $11 | $3b | $17 | $86 |
| sr05 | $03 | $01 | $01 | $5e | $11 | $3b | $17 | $88 | sr06 | $01 | $01 | $02 | $5d | $11 | $3b | $17 | $8a |
| sr07 | $02 | $01 | $01 | $5d | $11 | $3b | $17 | $8c | sr08 | $02 | $03 | $03 | $5e | $10 | $39 | $16 | $8e |
| sr09 | $01 | $01 | $01 | $5d | $10 | $39 | $16 | $90 | sr10 | $01 | $01 | $01 | $21 | $01 | $36 | $15 | $92 |
| sr11 | $01 | $02 | $01 | $27 | $01 | $3c | $19 | $94 | sr12 | $01 | $01 | $02 | $5e | $12 | $3b | $18 | $96 |
| sr13 | $01 | $02 | $02 | $5e | $10 | $3a | $17 | $98 | sr14 | $02 | $03 | $03 | $5e | $11 | $3b | $16 | $9a |
| sr15 | $03 | $01 | $03 | $5d | $11 | $3b | $16 | $9c | sr16 | $03 | $01 | $03 | $5e | $10 | $3a | $16 | $9e |
| sr17 | $01 | $03 | $01 | $27 | $01 | $3a | $19 | $a0 | sr18 | $01 | $03 | $02 | $5e | $10 | $39 | $17 | $a2 |
| sr19 | $02 | $02 | $02 | $5e | $11 | $3b | $18 | $a4 | sr20 | $02 | $02 | $03 | $5e | $10 | $3a | $16 | $a6 |

Table 13. Fuzzy set descriptions as related to input/output variable for normal operation.

| Antecedents | | | Consequences | | |
|---|---|---|---|---|---|
| Var. | Fuzzy set | Translation to base Membership function | Var. | Fuzzy set | Translation to base Membership function |
| PF | PF<110 | Z(u;100,110,120) | dPF | −25 | Z(u;−30, −26.25, −22.5) |
| | 110⩽PF | S(u;100,110,120) | | −5 | P(u;−15, −11.25, −7.5, −7.5, −3.75, 0) |
| BP | BP<50 | Z(u;43,51,59) | | 0 | P(u;−7.5, −3.75, 0, 0, 3.75, 7.5) |
| | BP<100 | Z(u;91,99,107) | | +5 | P(u;0,3.75,7.5,7.5,11.25,15) |
| | 50⩽BP<100 | P(u;43,51,59,91,99,107) | | +10 | P(u;7.5,11.25,15,15,18.75,22.5) |
| | 50⩽BP | S(u;43,51,59) | dBRH | −20 | Z(u;−20, −15, −10) |
| | 100⩽BP | S(u;91,99,107) | | −10 | P(u;−20, −15, −10, −10, −5, 0) |
| BRH | BRH<50 | Z(u;40,50,60) | | 0 | P(u;−10, −5, 0, 0, 5, 10) |
| | 50⩽BRH | S(u;40,50,60) | | +10 | P(u;0,5,10,10,15,20) |
| BRW | BRW<20 | Z(u;14,20,26) | | +20 | S(u;10,15,20) |
| | 20⩽BRW<50 | P(u;14,20,26,44,50,56) | | | |
| | 50⩽BRW<80 | P(u;44,50,56,74,80,86) | | | |
| | 80⩽BRW | S(u;74,80,86) | | | |
| dBRW | dBRW< −1 | Z(u;−2, −1, 0) | | | |
| | −1⩽dBRW<1 | P(u;−2, −1, 0, 0, 1, 2) | | | |
| | 1⩽dBRW | S(u;0,1,2) | | | |

Table 14. Hexadecimal representation of the membership functions.

| | Antecedents | | | | | Consequences | | | | Antecedents | | | | | Consequences | | |
|---|---|---|---|---|---|---|---|---|---|---|---|---|---|---|---|---|---|
| Rule | 1 | 2 | 3 | 4 | 5 | 1 | 2 | 3 | Rule | 1 | 2 | 3 | 4 | 5 | 1 | 2 | 3 |
| cr01 | $02 | $1c | $02 | $5d | $12 | $3b | $18 | $80 | cr02 | $02 | $15 | $02 | $5e | $12 | $3b | $18 | $82 |
| cr03 | $02 | $20 | $03 | $5e | $11 | $3a | $17 | $84 | cr04 | $02 | $20 | $02 | $5d | $11 | $3a | $17 | $86 |
| cr05 | $02 | $15 | $01 | $5d | $11 | $3a | $17 | $88 | cr06 | $01 | $01 | $02 | $5d | $10 | $39 | $16 | $8a |
| cr07 | $02 | $1d | $01 | $21 | $01 | $36 | $15 | $8c | cr08 | $02 | $15 | $03 | $27 | $01 | $3a | $19 | $8e |
| cr09 | $02 | $20 | $02 | $5e | $12 | $3a | $19 | $90 | cr10 | $03 | $01 | $03 | $27 | $01 | $3a | $19 | $92 |
| cr11 | $02 | $1d | $01 | $27 | $01 | $3a | $19 | $94 | cr12 | $02 | $20 | $02 | $5e | $11 | $3a | $18 | $96 |
| cr13 | $03 | $01 | $01 | $5d | $12 | $3a | $18 | $98 | cr14 | $03 | $01 | $02 | $5e | $11 | $3a | $18 | $9a |
| cr15 | $02 | $20 | $01 | $5e | $10 | $3a | $16 | $9c | cr16 | $02 | $19 | $02 | $5d | $11 | $3a | $16 | $9e |
| cr17 | $02 | $19 | $02 | $5e | $10 | $3a | $16 | $a0 | cr18 | $02 | $19 | $03 | $5e | $11 | $3a | $16 | $a2 |
| cr19 | $02 | $15 | $03 | $5e | $10 | $3a | $16 | $a4 | cr20 | $02 | $20 | $03 | $5d | $12 | $3b | $17 | $a6 |
| cr21 | $02 | $1d | $03 | $5d | $11 | $3a | $16 | $a8 | cr22 | $01 | $01 | $03 | $5d | $10 | $39 | $15 | $aa |
| cr23 | $02 | $20 | $03 | $5e | $12 | $3a | $17 | $ac | cr24 | $02 | $15 | $03 | $5e | $11 | $3a | $17 | $ae |
| cr25 | $02 | $19 | $01 | $5d | $12 | $3b | $17 | $b0 | cr26 | $02 | $19 | $03 | $5e | $12 | $3b | $17 | $b2 |
| cr27 | $02 | $15 | $03 | $5e | $12 | $3b | $17 | $b4 | cr28 | $02 | $19 | $02 | $5e | $12 | $3a | $18 | $b6 |
| cr29 | $02 | $15 | $02 | $5e | $11 | $3a | $18 | $b8 | cr30 | $03 | $01 | $01 | $5e | $12 | $3a | $18 | $ba |
| cr31 | $02 | $19 | $02 | $5e | $11 | $3a | $17 | $bc | cr32 | $02 | $15 | $02 | $5e | $10 | $3a | $17 | $be |
| cr33 | $02 | $19 | $03 | $5e | $10 | $3a | $15 | $c0 | cr34 | $02 | $15 | $01 | $21 | $01 | $36 | $15 | $c2 |
| cr35 | $03 | $01 | $01 | $21 | $01 | $36 | $15 | $c4 | cr36 | $02 | $15 | $02 | $27 | $01 | $3c | $19 | $c6 |
| cr37 | $02 | $15 | $03 | $5d | $12 | $3a | $16 | $c8 | cr38 | $03 | $01 | $01 | $5d | $11 | $39 | $17 | $ca |
| cr39 | $03 | $01 | $02 | $5e | $10 | $39 | $17 | $cc | cr40 | $03 | $01 | $03 | $5e | $11 | $39 | $17 | $ce |
| cr41 | $03 | $01 | $03 | $5e | $10 | $39 | $16 | $d0 | | | | | | | | | |

Table 15. Fuzzy set descriptions as related to input/output variable.

| Antecedents | | | Consequences | | |
|---|---|---|---|---|---|
| Var. | Fuzzy set | Translation to base Membership function | Var. | Fuzzy set | Translation to base Membership function |
| BP | BP<70 | Z(u;60,70,80) | dPF | −25 | Z(u;−30, −26.25, −22.5) |
| | 70⩽BP | S(u;60,70,80) | | −10 | P(u;−22.5, −18.75, −15, −15, −11.25, −7.5) |
| PAD | PAD<PAD(0)+14 | Z(u-PAD(0);12,14,16) | | −5 | P(u;−15, −11.25, −7.5, −7.5, −3.75, 0) |
| | PAD(0)+14⩽PAD | S(u-PAD(0);12,14,16) | | 0 | P(u;−7.5, −3.75, 0, 0, 3.75, 7.5) |
| BRH | BRH<50 | Z(u;40,50,60) | | +10 | P(u;7.5,11.25,15,15,18.75,22.5) |
| | 50⩽BRH | S(u;40,50,60) | dBRH | −20 | P(u;−30, −25, −20, −20, −15, −10) |
| BRW | BRW<20 | Z(u;14,20,26) | | −10 | P(u;−20, −15, −10, −10, −5, 0) |
| | 20⩽BRW<50 | P(u;14,20,26,44,50,56) | | 0 | P(u;−10, −5, 0, 0, 5, 10) |
| | 20⩽BRW<80 | P(u;14,20,26,74,80,86) | | +10 | P(u;0,5,10,10,15,20) |
| | 50⩽BRW<80 | P(u;44,50,56,74,80,86) | | +20 | P(u;10,15,20,20,25,30) |
| | 80⩽BRW | S(u;74,80,86) | | +30 | S(u;20,25,30) |
| dBRW | dBRW<−1 | Z(u;−2, −1, 0) | | | |
| | −1⩽dBRW<1 | P(u;−2, −1, 0, 0, 1, 2) | | | |
| | 1⩽dBRW | S(u;0,1,2) | | | |

Table 16. Hexadecimal representation of the membership functions.

| | Antecedents | | | | | Consequences | | | | Antecedents | | | | | Consequences | | |
|---|---|---|---|---|---|---|---|---|---|---|---|---|---|---|---|---|---|
| Rule | 1 | 2 | 3 | 4 | 5 | 1 | 2 | 3 | Rule | 1 | 2 | 3 | 4 | 5 | 1 | 2 | 3 |
| wr01 | $03 | $02 | $02 | $5d | $12 | $3a | $25 | $80 | wr02 | $02 | $03 | $02 | $5e | $12 | $3a | $25 | $82 |
| wr03 | $03 | $01 | $03 | $5e | $11 | $39 | $24 | $84 | wr04 | $01 | $02 | $02 | $5e | $10 | $39 | $24 | $86 |
| wr05 | $02 | $01 | $02 | $5e | $10 | $39 | $24 | $88 | wr06 | $03 | $01 | $01 | $5d | $11 | $39 | $24 | $8a |
| wr07 | $01 | $03 | $01 | $5e | $11 | $39 | $24 | $8c | wr08 | $01 | $01 | $02 | $5d | $11 | $39 | $24 | $8e |
| wr09 | $03 | $01 | $03 | $5d | $10 | $38 | $23 | $90 | wr10 | $03 | $02 | $01 | $5d | $10 | $38 | $23 | $92 |
| wr11 | $01 | $01 | $01 | $21 | $01 | $36 | $22 | $94 | wr12 | $03 | $02 | $01 | $27 | $01 | $3c | $26 | $96 |
| wr13 | $01 | $02 | $03 | $27 | $01 | $3c | $26 | $98 | wr14 | $02 | $02 | $02 | $5d | $12 | $3a | $26 | $9a |
| wr15 | $03 | $03 | $01 | $27 | $01 | $3a | $26 | $9c | wr16 | $01 | $03 | $02 | $27 | $01 | $3a | $26 | $9e |
| wr17 | $01 | $02 | $02 | $5e | $12 | $3a | $26 | $a0 | wr18 | $03 | $02 | $02 | $5e | $11 | $39 | $25 | $a2 |
| wr19 | $02 | $02 | $03 | $5d | $11 | $39 | $25 | $a4 | wr20 | $02 | $03 | $03 | $27 | $01 | $3a | $24 | $a6 |
| wr21 | $01 | $02 | $03 | $34 | $12 | $3a | $24 | $a8 | wr22 | $02 | $01 | $03 | $34 | $12 | $3a | $24 | $aa |
| wr23 | $02 | $03 | $03 | $5d | $11 | $39 | $23 | $ac | wr24 | $03 | $01 | $03 | $5e | $10 | $39 | $23 | $ae |
| wr25 | $01 | $02 | $03 | $5e | $10 | $39 | $23 | $b0 | wr26 | $03 | $03 | $02 | $34 | $12 | $39 | $25 | $b2 |
| wr27 | $02 | $02 | $02 | $5e | $11 | $39 | $25 | $b4 | wr28 | $01 | $03 | $02 | $5d | $10 | $38 | $24 | $b6 |
| wr29 | $02 | $01 | $02 | $5d | $10 | $38 | $24 | $b8 | wr30 | $03 | $03 | $02 | $5e | $10 | $39 | $24 | $ba |
| wr31 | $03 | $03 | $03 | $34 | $12 | $39 | $24 | $bc | wr32 | $02 | $02 | $03 | $5e | $11 | $39 | $24 | $be |
| wr33 | $02 | $02 | $02 | $27 | $01 | $3c | $27 | $c0 | wr34 | $02 | $02 | $03 | $5d | $10 | $39 | $22 | $c2 |
| wr35 | $02 | $03 | $02 | $5d | $12 | $3a | $24 | $c4 | wr36 | $02 | $03 | $03 | $34 | $10 | $39 | $23 | $c6 |

In addition, the labels of the indices which modify the membership functions are expressed in hexadecimal in Table 16.

### 7.4. *Fuzzification of conditions and control for safe operation*

In Table 17, the fuzzy sets of Table 10 are defined by the standard functions of Eqs. (7)–(10). Using fuzzy inference, dPF, dBRH and dOcc are determined by the MAX–MIN composition/maximum-height method.

In addition, the labels of the indices which modify the membership functions are expressed in hexadecimal in Table 18.

Table 17. Fuzzy set descriptions as related to input/output variable.

| Var. | Fuzzy set | Antecedents<br>Translation to base membership function |
|---|---|---|
| LP | LP<400 | Z(u;350,400,450) |
| | 400⩽LP | S(u;350,400,450) |
| BRH | BRH<50 | Z(u;40,50,60) |
| | 50⩽BRH | S(u;40,50,60) |
| BRW | BRW<20 | Z(u;14,20,26) |
| | 20⩽BRW<50 | P(u;14,20,26,44,50,56) |
| | 50⩽BRW<80 | P(u;44,50,56,74,80,86) |
| | 80⩽BRW | S(u;74,80,86) |
| dBRW | dBRW<−1 | Z(u;−2, −1, 0) |
| | −1⩽dBRW | S(u;−2, −1, 0) |

## 8. Evaluation of Animal Experiment

For infant's ECC, the optimal pump flow volume is low (approximately 1200 mm-lit./min.). Furthermore, even if the corporeal blood flow volume changes only slightly, the effects on the infant's body are great. In animal experiments, the clinical conditions for infants are assumed; mongrel dogs are used. The experimental preparations begin in the order of post-anesthesia, artificial respiration, and right open-heart surgery. In addition, in the regulatory system, after attaining normal operation regulation for the commencement of pump start regulation, cooling of the blood flow to the heart–lung apparatus begins based on the operation of the blood flow pump and the connection of the heart–lung circuit. The heart soon loses its effective heart beat. After

Table 18. Hexadecimal representation of the membership functions.

| Rule | Antecedents | | | | Consequence | Rule | Antecedents | | | | Consequence |
|---|---|---|---|---|---|---|---|---|---|---|---|
| | 1 | 2 | 3 | 4 | 1 | | 1 | 2 | 3 | 4 | 1 |
| er01 | \$03 | \$02 | \$5d | \$03 | \$80 | er02 | \$03 | \$02 | \$5d | \$02 | \$82 |
| er03 | \$03 | \$01 | \$21 | \$01 | \$84 | er04 | \$03 | \$02 | \$27 | \$01 | \$86 |
| er05 | \$03 | \$02 | \$5e | \$01 | \$88 | er06 | \$03 | \$03 | \$5d | \$01 | \$8a |
| er07 | \$03 | \$03 | \$27 | \$01 | \$8c | er08 | \$03 | \$03 | \$5e | \$01 | \$8e |
| er09 | \$02 | \$01 | \$01 | \$01 | \$90 | | | | | | |

that, in the body, the heart is stopped by prompt isolation of the greater vein, followed by injection of the cardioplegic solution and localized cooling. The heart is stopped for approximately 30 minutes. Finally, the shift from mechanical ECC to physiological body circulation proceeds in the order of restoring temperature and deisolation of the greater vein, provision of effective heart rate, and shut-off of the blood flow pump from the commencement of pump weaning.

Table 19 lists the infant clinical data based on actual manual operation, and Table 20 lists the experimental data for mongrel dogs based on auto-regulation.

Table 19. The data of clinical manual circulation.

| Case | BW (Kg) | preBP (mm Hg) | minBP (mm Hg) | init.drp.BP (%) | prePAD (mm Hg) | maxPAD (mm Hg) | minPAD (mm Hg) | var.PAD (%) |
|---|---|---|---|---|---|---|---|---|
| 1 | 14.20 | 76 | 43 | 43.42 | 6.0 | 15.0 | 5.0 | 166.67 |
| 2 | 15.20 | 92 | 50 | 45.65 | 11.0 | 16.0 | 2.0 | 127.27 |
| 3 | 6.58 | 82 | 42 | 48.78 | 3.0 | 18.0 | 4.0 | 466.67 |
| 4 | 21.00 | 104 | 72 | 30.77 | 7.0 | 10.0 | 3.0 | 100.00 |
| 5 | 11.60 | 75 | 42 | 44.00 | 4.0 | 6.0 | 3.0 | 75.00 |
| 6 | 13.00 | 120 | 42 | 65.00 | 5.0 | 9.0 | 5.0 | 80.00 |
| 7 | 12.90 | 118 | 56 | 52.54 | 4.0 | 9.0 | 4.0 | 125.00 |
| 8 | 11.80 | 100 | 40 | 60.00 | 5.2 | 8.0 | 3.0 | 96.15 |
| 9 | 28.60 | 118 | 44 | 62.71 | 5.0 | 14.0 | 4.0 | 200.00 |
| 10 | 8.49 | 122 | 42 | 65.57 | 6.0 | 15.0 | 0.0 | 250.00 |
| 11 | 7.74 | 96 | 46 | 52.08 | 4.0 | 22.0 | 0.0 | 550.00 |
| 12 | 14.00 | 100 | 46 | 54.00 | 4.0 | 9.0 | 2.0 | 175.00 |
| 13 | 8.06 | 84 | 44 | 47.62 | 3.0 | 15.0 | 2.0 | 433.33 |

The headings in the table are: Case, the index for the clinical example (or animal experiment); BW, the weight; preBP, the value of the systemic arterial pressure immediately before the start of ECC; minBP, the minimum value of systemic arterial pressure immediately after the start of ECC; init.drp.BP, the percentage change in systemic arterial pressure immediately after the start of ECC; prePAD, the value of diastolic pulmonary arterial pressure immediately before the start of ECC; maxPAD, the maximum value of diastolic pulmonary arterial pressure during ECC; minPAD, the minimum value of pulmonary diastolic arterial pressure during ECC; and var.PAD, the percentage change in pulmonary diastolic arterial pressure after the start of ECC. Evaluation was made

Table 20. The data of experimental automatic circulation.

| Case | BW (Kg) | preBP (mm Hg) | minBP (mm Hg) | init.drp.BP (%) | prePAD (mm Hg) | maxPAD (mm Hg) | minPAD (mm Hg) | var.PAD (%) |
|---|---|---|---|---|---|---|---|---|
| 1 | 10.70 | 240 | 118 | 50.83 | 17.0 | 25.0 | 12.0 | 76.47 |
| 2 | 12.30 | 117 | 73 | 37.61 | 16.0 | 21.0 | 12.0 | 56.25 |
| 3 | 11.60 | 112 | 74 | 33.93 | 26.0 | 46.0 | 23.0 | 88.46 |
| 4 | 14.50 | 150 | 82 | 45.33 | 16.0 | 24.0 | 3.0 | 131.25 |
| 5 | 9.80 | 90 | 130 | 44.44 | 22.0 | 26.0 | 18.0 | 36.36 |
| 6 | 15.80 | 220 | 116 | 47.27 | 12.0 | 18.0 | 10.0 | 66.67 |
| 7 | 8.70 | 280 | 142 | 49.29 | 25.0 | 32.0 | 15.0 | 68.00 |
| 8 | 10.50 | 72 | 58 | 19.44 | 9.0 | 16.0 | 7.0 | 100.00 |
| 9 | 9.60 | 178 | 110 | 38.20 | 25.0 | 42.0 | 20.0 | 88.00 |
| 10 | 12.80 | 156 | 89 | 42.95 | 23.0 | 31.0 | 16.0 | 65.22 |

based on two characteristic data which expresses the stability of hemodynamic behavior. In order words, a comparison was made between the percentage change in systemic arterial pressure immediately after the start of ECC (init.drp.BP) and the percentage change in diastolic pulmonary arterial pressure after the start of ECC (var.PAD). In Eq. (22), the systemic arterial pressure is used to express the smooth transition from physiological body circulation to mechanical ECC. In addition, in Eq. (23), the percentage change in diastolic pulmonary arterial pressure expresses the change in blood flow volume during ECC.

$$\text{init.drp.BP} = (|\text{preBP-minBP}|)/\text{preBP} \times 100 \tag{22}$$

$$\text{var.PAD} = (|\text{maxPAD-minPAD}|)/\text{prePAD} \times 100 \tag{23}$$

In Figs. 12 and 13, the trends for the percentage change for systemic arterial pressure (BP) and diastolic pulmonary arterial pressure (PAD), respectively, are illustrated. The percentage change for the auto-regulation group (10 cases, average weight: 11.63 ± 2.25 kg) show a slighter trend than that for manual-regulation group (13 cases, average weight: 13.32 ± 5.99 kg).

## 9. Discussion

When attempting the auto-regulation of ECC, two serious problems arise. One is whether or not it is necessary to automate the safe ECC circulatory operation

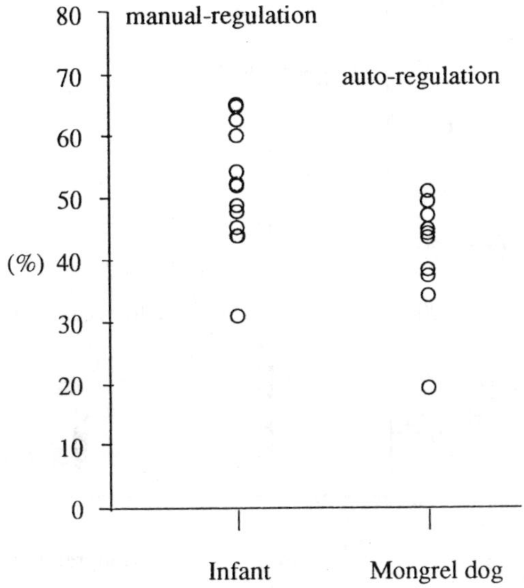


Fig. 12. Comparison of percentage change in BP before and immediately after ECC.

based on an expert's operation; the other is the establishment of essential regulatory principles for auto-regulation. An expert creates a pattern for all of the instantaneous data during circulation. Each parameter in the series of regulatory processes is classified according to condition. Expert operation is based upon a rational pattern recognition process with accurate decision making. On the other hand, the theory of control is based on the adequacy of the regulatory operation selected specifically for each pattern.

The significance of achieving the auto-regulation of ECC is not merely in labor saving but in the embodiment of an expert's knowledge and in the establishment of grounds for such an argument. In this experiment, each pattern is classified, and auto-regulation was possible due to the establishment of all regulatory rules.[4]

This time, we applied fuzzy logic to compensate for the gaps among the regulation rules. The result was that a highly reliable regulation system was smoothly established. This implies that the uncertain elements of analysis of expertise and theory-making concerning expertise are sufficiently conceivable.

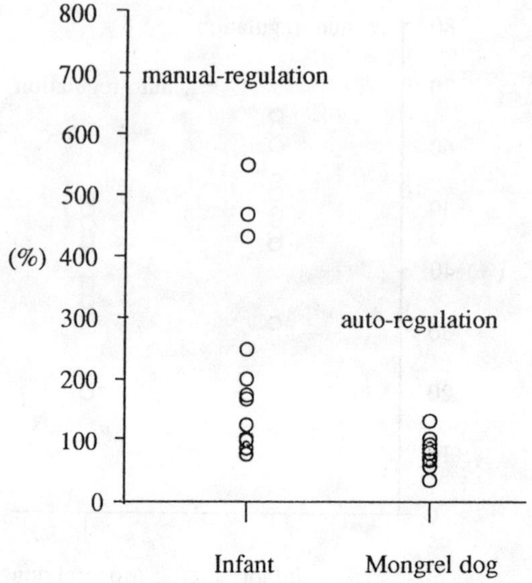


Fig. 13. Comparison of percentage change in PAD before and after ECC.

## 10. Conclusion

While providing for fixed levels of systemic arterial pressure and diastolic pulmonary arterial pressure, we have designed and tested an auto-regulation system for ECC which automatically controls blood flow volume. We analyzed manual operation methods for the principal regulatory functions of ECC and expressed them as a set of simple, basic rules. Experimental results were obtained for 10 mongrel dogs. Based on these results, we were able to confirm that our dynamic regulatory method based on fuzzy theory stabilizes circulatory blood flow volume and provides a practical means for oxygenation.

Furthermore, we were able to condense the set of control rules, which were obtained from an expert, from 209 to 106 rules. However, if the antecedent of the control rule is comprised of input variables which contain complex partial sets, then simplification of the logic functions based on the Quine–McClusky method becomes intuitively hard to understand. In calculation by fuzzy inference, it is characteristic that the complementary rule is not realized. In other words, the law of the excluded middle and the law of contradiction are not recognizable. Therefore, direct simplification by fuzzy inference is desirable; however, it is not simple.

In theoretical research related to fuzzy inference functions, the methods of standardized expression,[11] minimization,[12] and simplification[13–15] of $n$-fuzzy-variable inference functions are studied. In this field, the adequacy among the regulatory rule, as well as the evaluation of the redundancy of the rules, is an important research topic.

## References

1. B. J. Miller, J. H. Gibbon, and M. H. Gibbon, "Recent advance in the development of a mechanical and lung apparatus," *Ann. Surg.* **134** (1951) pp. 694–706.
2. J. H. Gibbon, Jr.,"Application of a heart lung apparatus to cardiac surgery," *Med.* **37** (1954) p. 171.
3. A. Kantrowitz, S. Reiner, and D. Abelsen,"An automatically controlled, inexpensive pump oxygenator," *J. Thorac Cardiovasc. Surg.* **38** (1959) p. 586.
4. J. Anbe *et al.*,"The development of a microcomputer-based auto-regulation system for ECC — Investigation of the control algorithm and control device," *Artificial Organs* **20**, 3 (1990), pp. 1201–1205.
5. L. A. Zadeh,"Fuzzy sets," *Information and Control* **8** (1965), pp. 338–353.
6. L. A. Zadeh,"Fuzzy sets as a basis for a theory of possibility," *Fuzzy Sets and Systems* **1** (1978), pp. 3–28.
7. L. A. Zadeh,"Fuzzy algorithms," *Information and Control* **12** (1968), pp. 94–104.
8. L. A. Zadeh,"Outline of a new approach to the analysis of complex systems and decision processes," *IEEE Trans. Systems, Man, and Cybernetics* **SMC-3** (1973), pp. 28–44.
9. S. Yasunobu and S. Miyamoto,"Automatic train operation system by predictive fuzzy control," *Industrial Applications of Fuzzy Control*, ed. M. Sugeno (North-Holland, Amsterdam, 1985), pp. 1–18.
10. M. Mizumoto and H.-J. Zimmermann,"Comparison of fuzzy reasoning methods," *Fuzzy Sets and Systems* **8**, 3 (1982), pp. 253–283.
11. M. Mukaidono,"About fuzzy logic functions," *The Road to Fuzzy Inference, Science* (1988), pp. 85–92 (in Japanese).
12. A. Kandel and S. C. Lee, *Fuzzy Switching and Automata, Crane* (Russak & Company, New York, 1979).
13. K. Asai and C. V. Negoita, *Introduction to Ambiguous Fuzzy Inference* (OHM, 1978) (in Japanese).
14. A. Kandel,"On minimization of fuzzy functions," *IEEE Trans. Computers* **C–22**, 9 (1973), pp. 826–832.
15. M. Mukaidono,"The algebraic structure of fuzzy logic functions and their simplest and rule form," *Shingakuron(D)* **58-D**, 12 (1975), pp. 748–755 (in Japanese).

# AUTOMATIC CRANE OPERATION USING FUZZY COOPERATIVE CONTROL METHOD

OSAMU ITOH, HIROHISA MIGITA
*Fuji Electric Corporate Research and Development Ltd., 1, Fuji-machi, Hino-city, 191 Tokyo, Japan*

JUN ITOH
*Fuji Electric Co., Ltd. 1, Fuji-machi, Hino-city, 191 Tokyo, Japan*

YASUFUMI IRIE
*KAWADEN Co., Ltd. 1, Kawasaki Street, Mizushima, Kurashiki-city, 712 Okayama, Japan*

## 1. Introduction

A lot of researches have been done on the development of automated overhead-traveling crane. One of conventional control methods proposes to apply the optimal control theory and the phase plane under the conditions of definite rope length, in order to create the target speed patterns to be followed by the crane. However, this method belongs to the open system in which the angle of swing both at the initial position and destination is assumed to be zero, and cannot stand against the problems of friction and mechanical external disturbance.

In another research aimed at solving the above-mentioned problems, researchers are trying to make the crane tracking speed patterns with the feedback control which utilizes the optimum track obtained by using accurate mathematical models and the symmetry of time acceleration and deceleration. On the other hand, the following proposals were also presented: the predictive fuzzy control method in which control results obtained using dynamic system models are evaluated with the fuzzy set, the target speed pattern control method based on fuzzy inference, the model reference adaptive control method which has the fuzzy adaptation rules, and other control method based on the evaluation function minimization algorithm which uses the fuzzy inference model.

However, either report of these researches is still at the level of simulation or experiment. Moreover, these paper are dependent either on the swing prevention or the positioning, and no paper is reported about the cooperation control.

In order to solve the foregoing problems, we have used the fuzzy control and realized the cooperative operation which satisfies at the same time the positioning and the swing prevention that contradict each other, for possible automation of crane operation. The control method and the evaluation result by a practical machine are reported in Ref. 1.

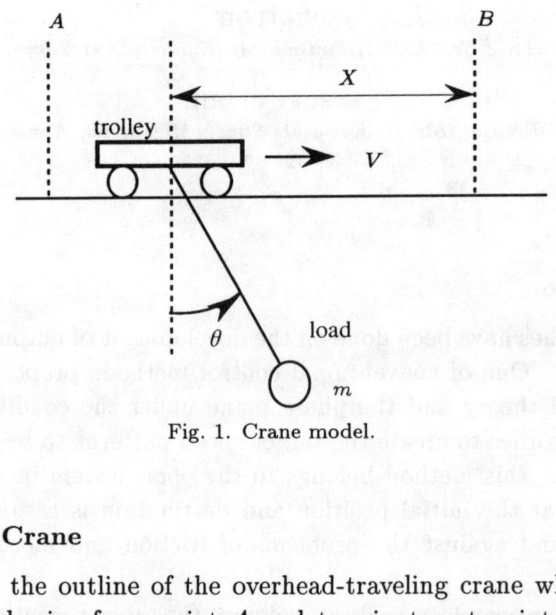


Fig. 1. Crane model.

## 2. Outline of Crane

Figure 1 shows the outline of the overhead-traveling crane which was used. The movement begins from point $A$ and terminates at point $B$. With this movement, the traveling (traverse motion) pattern of crane is determined based on the distance $X$ between the crane truck and the destination and the velocity $V$ of trolley. Further, the information of swing angle $\theta$ of hanging load and the wire rope length is added to obtain the setting velocity (or both acceleration and deceleration) for each traveling (or traverse motion) pattern in order to control the positioning and the swing prevention.

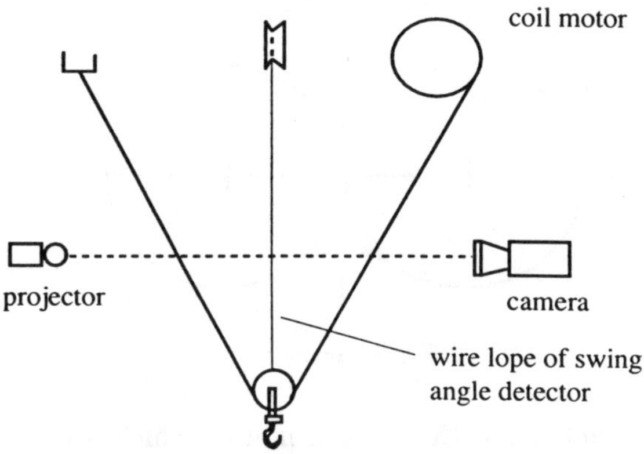


Fig. 2. Detection of swing angle.

Composition of swing angle detector is as shown in Fig. 2. A wire rope is hanged perpendicularly from the trolley at the hook of hanging load and the extent of swing is monitored with a camera so as to calculate the swing angle. The swing angle detector is installed for the traverse motion as well as for the traveling.

## 3. Basic Principle of Crane Model

The crane model equation of motion can be obtained using Lagrange equation as

$$L = \frac{1}{2}m\left\{\left(-l\dot{\theta}\cos\theta - \dot{l}\sin\theta - \dot{X}\right)^2 + \left(l\dot{\theta}\sin\theta - \dot{l}\cos\theta\right)^2\right\} + lgm\cos\theta \quad (1)$$

$$\frac{d}{dt}\left(\frac{\partial L}{\partial\dot{\theta}}\right) - \frac{\partial L}{\partial\theta} = m\left(-2l\dot{l}\dot{\theta} + l^2\ddot{\theta} - l\ddot{X}\cos\theta - gl\sin\theta\right) = 0\,. \quad (2)$$

Using $\sin\theta \approx \theta$ & $\cos\theta \approx 1$ (when $0 \approx \theta$), $\dot{l} = 0$; the following equation can be obtained.

$$\ddot{\theta} + \frac{g}{l}\theta + \ddot{X} = 0\,. \quad (3)$$

Where we integrated Eq. (3), using $\omega = \sqrt{g/l}$, $\alpha = \ddot{X}$ the following equation can be obtained.

$$\left(\theta + \frac{\alpha}{g}\right)^2 + \left(\frac{\dot{\theta}}{\omega}\right)^2 = \text{constant}\,. \quad (4)$$

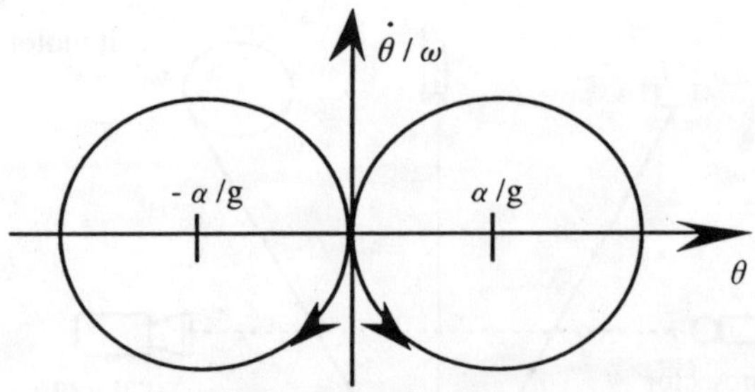


Fig. 3. Swing state.

The swing can be defined as the circular motion which is made with its center at $(-\alpha/g, 0)$ on the coordinate space shown in Fig. 3. Based on this definition, it is considered that the swing can be returned to the initial swing state during the acceleration (or deceleration) if a constant acceleration (or deceleration) is applied for the duration equivalent to the natural period of hanging load which is multiplied by an integer.

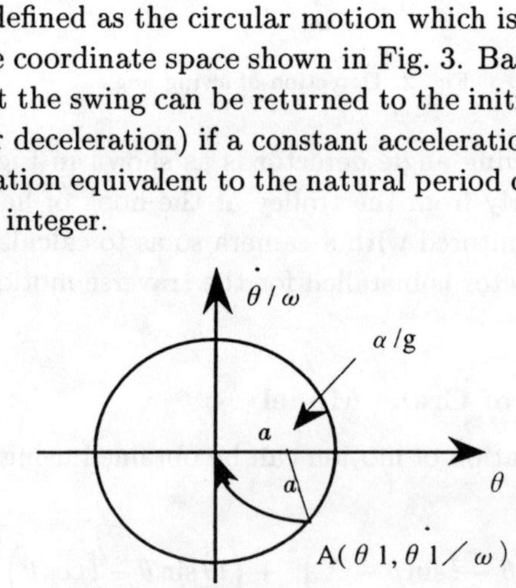


Fig. 4. Swing control method 1.

Figure 4 shows an example of calculation of swing prevention. Acceleration applied to the trolley exists at a point on the $\theta$-axis. Therefore, it can be calculated with the following equation on the condition that the point $(\alpha/g)$ on $\theta$-axis is at the position equal to the distance $(a)$ between the zero point and the swing position $(\theta_1, \dot{\theta}_1/\omega)$.

$$\alpha = -g\frac{(\theta^2 + (\dot{\theta}/\omega)^2)}{2\theta}. \tag{5}$$

The calculation shown in Fig. 4 is in the range of ($\theta > 0$, $\dot{\theta}/\omega < 0$) and the same calculation is made also in the range of ($\theta < 0$, $\dot{\theta}/\omega > 0$). By the way, it is possible only to reduce the swing in the ranges of ($\theta > 0$, $\dot{\theta}/\omega > 0$) and ($\theta < 0$, $\dot{\theta}/\omega < 0$) as shown in Fig. 5. In other words, when the hanging load is swinging, the swing can be stopped in the ranges of ($\theta > 0$, $\dot{\theta}/\omega < 0$) and ($\theta < 0$, $\dot{\theta}/\omega > 0$) but the swing can only be reduced in the ranges of ($\theta > 0$, $\dot{\theta}/\omega > 0$) and ($\theta < 0$, $\dot{\theta}/\omega < 0$). Furthermore, in order to stop the swing, it is necessary to accelerate in the range ($\dot{\theta}/\omega > 0$) while it must be decelerated in the range ($\dot{\theta}/\omega > 0$) and besides the deceleration is necessary to simultaneously conduct the positioning as the final purpose.

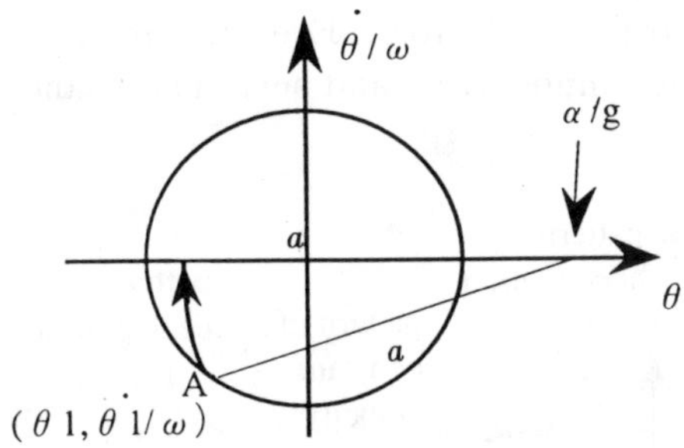


Fig. 5. Swing control method 2.

The positioning of Eq. (8) can be calculated with the equation of motion of Eq. (6) and Eq. (7).

$$V = \beta t \tag{6}$$

$$X = \frac{1}{2}\beta t^2 \tag{7}$$

$$V = \sqrt{2\beta X}\,. \tag{8}$$

The $\beta$ can be calculated with the following equation when it entered in the positioning and swing prevention pattern.

$$\beta = \frac{V}{2X}\,. \tag{9}$$

Due to the foregoing reasons, it is known that there is no general solution that can satisfy both the swing control and the positioning simultaneously.

## 4. Control Method

### 4.1. *Basic control pattern*

Figure 6 shows the basic traveling pattern of a conventional crane, which consists of factors largely divided as an acceleration, a fixed velocity and a deceleration. During the acceleration and the deceleration patterns, the control is made in accordance with the natural period of wire rope. There will be no swing of the hanging load when reaching a target position theoretically, if it is assumed that there is neither the initial swing nor any disturbance. However, because there are actually such factors as the initial swing, looseness of components, time lag, friction, etc., it is very difficult to develop an ideally precise model to control such factors. Even if a model can be made, a fault still remains in that it cannot withstand such disturbance as gusts.

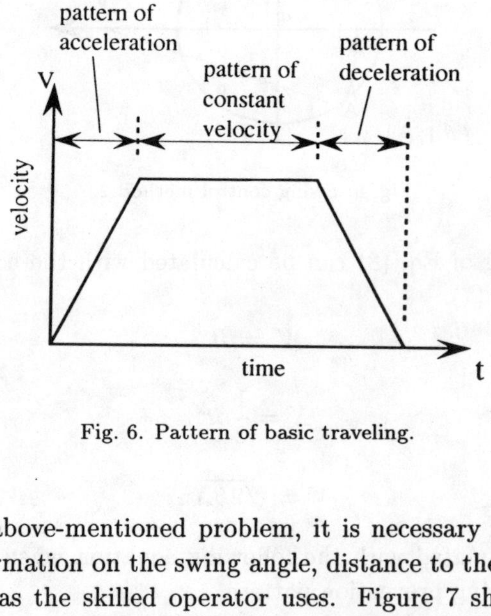


Fig. 6. Pattern of basic traveling.

To solve the above-mentioned problem, it is necessary to adopt the control in which information on the swing angle, distance to the destination, etc., is provided, just as the skilled operator uses. Figure 7 shows the traveling pattern of the crane which was used for our study, and it consists of the following factors: acceleration, fixed velocity swing prevention, deceleration and positioning/swing prevention. When there is an initial swing, it is suppressed during the fixed velocity traveling, and the later swing raised by any disturbance is controlled by the swing prevention and the positioning pattern.

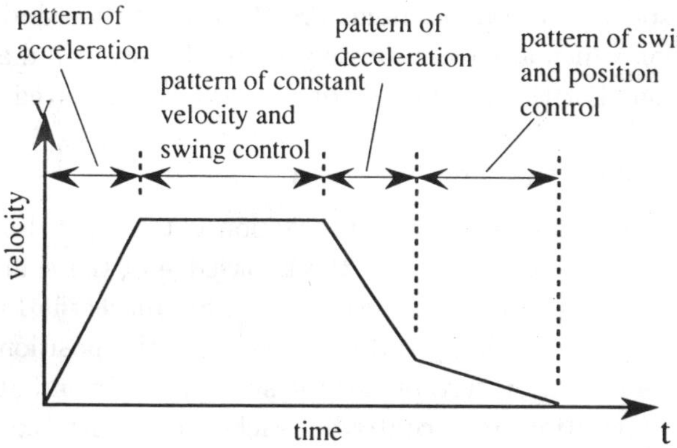


Fig. 7. Pattern of new control method.

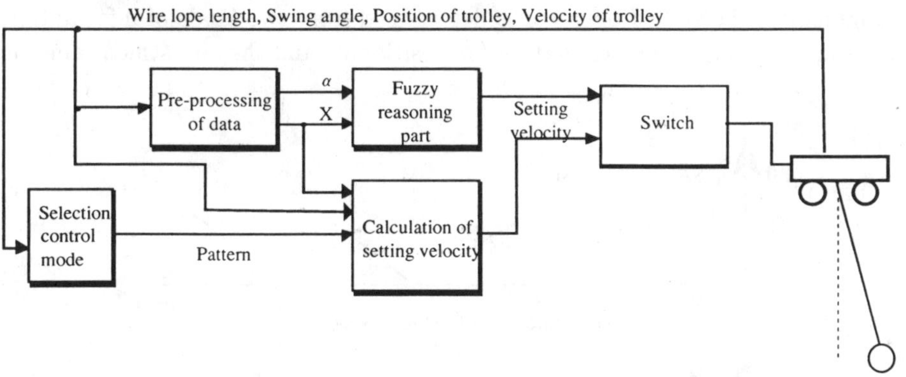


Fig. 8. Block diagram.

## 4.2. *Control method*

Figure 8 shows the control block diagram. Patterns of acceleration, high velocity swing prevention, deceleration and positioning/swing prevention are determined by the target position, the trolley velocity, the trolley position and the wire rope length. The trolley velocity during the acceleration, high velocity swing prevention and deceleration patterns is calculated with the setting velocity calculator and the data preprocessor. The input value of fuzzy reasoning, necessary for the positioning and swing prevention control, is made from the data preprocessor, and the input items are the acceleration and the distance to the destination which are necessary to prevent the swing and are calculated with the equation. For the switch a setting velocity is produced.

### 4.3. *Design of control rule*

Control of the positioning and swing prevention is the most difficult element and requires more than anything else the knowledge of the skilled operators. The operation demands a difficult technique to approach the target position suppressing the swing. Such operation suitable for the positioning does not necessarily match operation suitable for the swing prevention but, on the contrary, these two operations may contradict each other. These complicated operations are skillfully coordinated by the experienced operator depending on the situation. The positioning and the swing prevention are converged rather forcibly with the fuzzy control rule with which the setting velocity equal to the judgment of the skilled operator is determined from two kinds of information, i.e., the distance to the destination for positioning and the adjustment amount

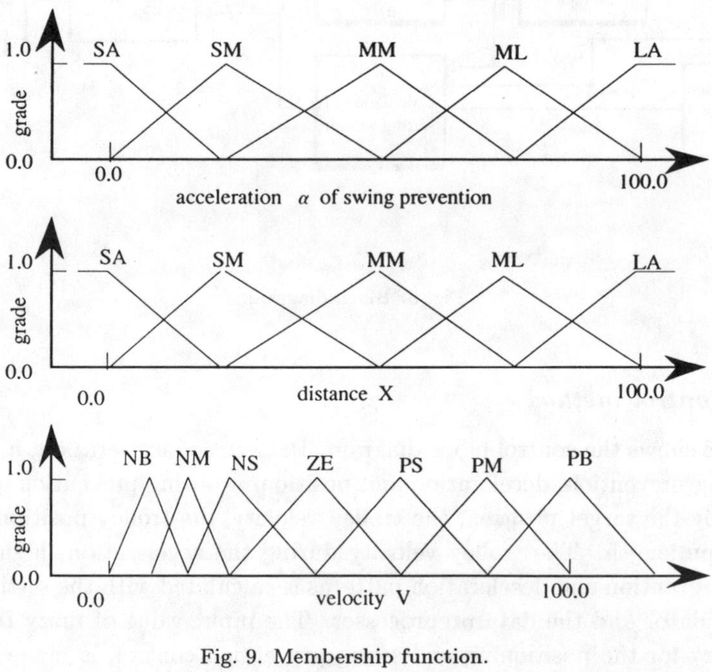


Fig. 9. Membership function.

Table 1. Fuzzy rules.

| | | |
|---|---|---|
| 1 | IF X1=SA,X2=SA | THEN V=NB |
| 2 | IF X1=SM,X2=MM | THEN V=NS |
| ⋮ | ⋮ | ⋮ |
| 16 | IF X1=LA,X2=ML | THEN V=PB |

of acceleration and deceleration to stop the swing. Figure 9 and Table 1 show an example of the membership function and the control rule which were used in this case.

## 5. Analysis by Simulation

Figure 10 shows the result of simulation where, under the positioning and swing prevention pattern, the control cycle is 0.1 sec, the maximum acceleration is $0.25\,\mathrm{m/sec^2}$ and the motor velocity control is of the time lag of first order.

Conditions of simulation are that the initial swing angle is 2°, the distance to the destination is 0.2 m and the length of wire rope is 6 m. It is shown from this figure that the swing prevention is inhibited under the positioning control while the positioning is disabled under the swing prevention control. On the other hand, with the positioning and swing prevention control based on the fuzzy reasoning which is shown by the solid line, it converges within the time of approximately 2.5 times of natural period which is determined by the wire rope length, although a slight residual swing remains around the point close to the destination.

## 6. Result of Actual Crane Operation

Table 2 shows the particulars of the crane actually operated during the experiment. Figure 11 shows an example of the experiment result. In this case, a 6 m long wire rope is used and it has entered in the positioning and swing prevention pattern after the velocity was reduced to 10 % of the fixed velocity, which is the velocity when entering in the fixed velocity swing prevention pattern, under the deceleration pattern.

In Fig. 11, it is shown orderly from the top: the crane velocity, the swing angle of wire rope and the distance to the target position. Since there is no initial swing, the acceleration and deceleration have been conducted with the

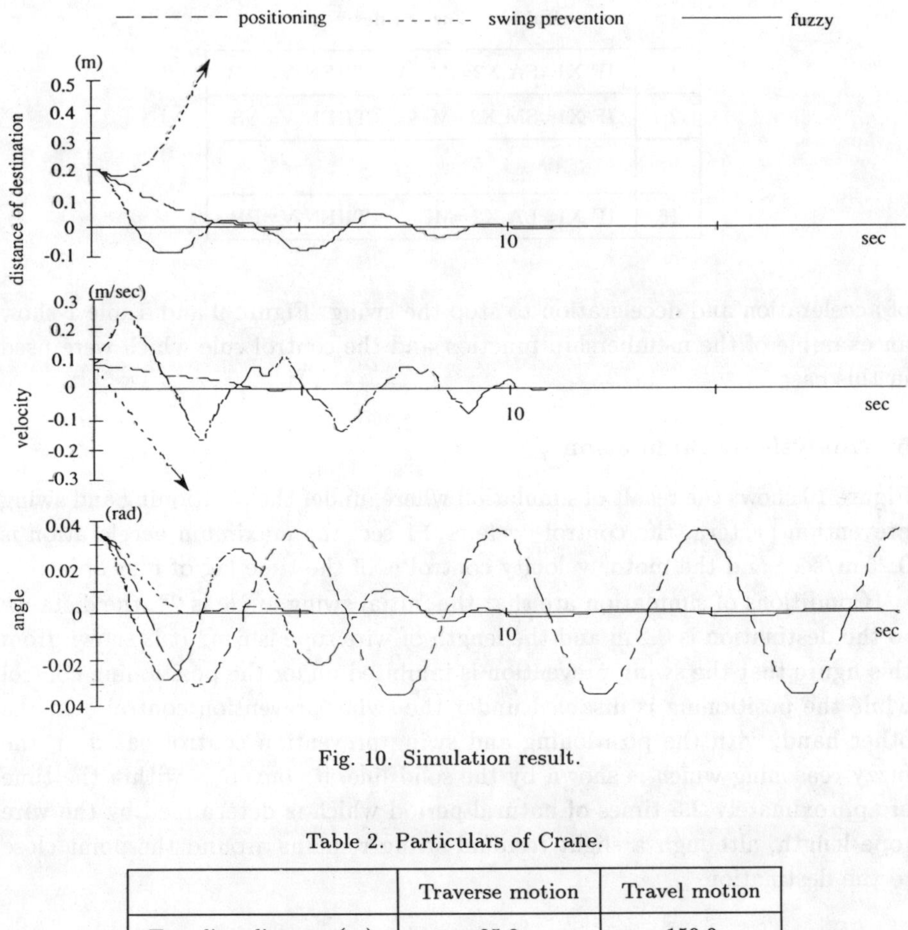


Fig. 10. Simulation result.

Table 2. Particulars of Crane.

| | Traverse motion | Travel motion |
|---|---|---|
| Traveling distance (m) | 25.0 | 150.0 |
| Rated velocity (m/sec) | 50.0 | 120.0 |
| Weight (t) | 2.8 | 25.1 |

natural period of 3 m long wire rope in order to raise the swing forcibly under the acceleration pattern and the deceleration pattern. The data of manual operation were obtained by the skilled operator. Fixed velocity swing prevention, positioning and swing prevention control was not made under the velocity pattern control. The control was conducted solely in accordance with the pattern calculated in advance. The fuzzy control pattern was applied to the controls of

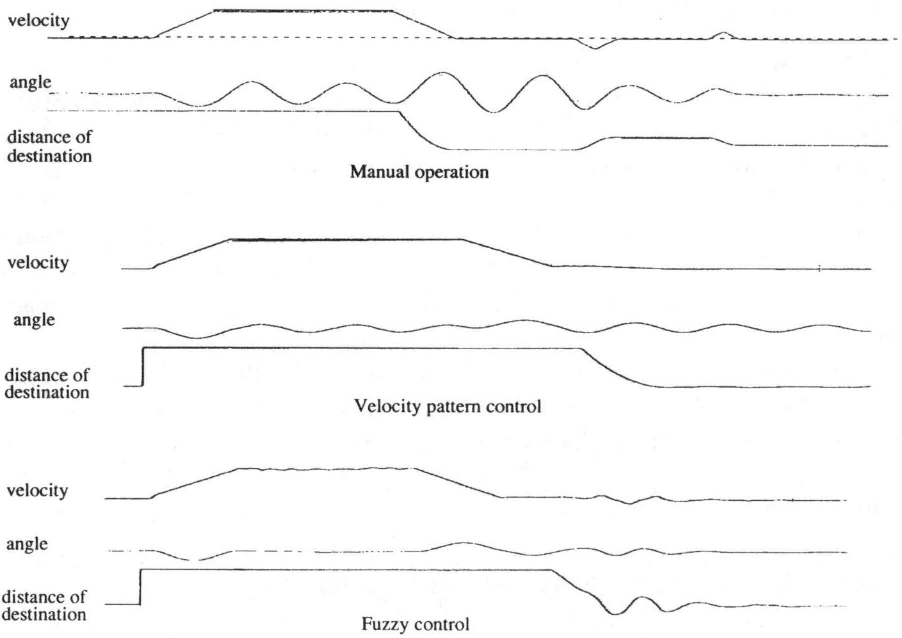


Fig. 11. Experiment result.

fixed velocity swing prevention, positioning and swing prevention in addition to the velocity pattern control.

The swing occurred during the acceleration or the deceleration is suppressed under the fuzzy control, but it persists during the manual operation and under the velocity pattern control. During the manual operation, the positioning at the target is dependent on the visual observation; a better control can be made only from where a good timing can be taken, and the positioning accuracy is not necessarily high. Under the velocity control pattern, the positioning accuracy improves but the swing cannot be eliminated. With the fuzzy control, on the other hand, the swing raised during the acceleration is suppressed and the swing can be converged when it stops at the destination. Under the velocity pattern control, the swing of the fixed velocity traveling crane is about $\pm 1.0°$ and the same extent is reserved at the final phase while, under the fuzzy control, it is about $\pm 0.1°$ and less for both during the fixed velocity traveling and at the final phase.

## 7. Conclusion

Residual swing will often be generated in the crane drive by factors such as the delay and the friction of the machine even if it is controlled along the pattern. Therefore we have confirmed, by computer simulation, the effect of the fuzzy control method which combines the means of the control of positioning and swing prevention. As a result of the experiment using a practical crane, it was shown that, under the positioning and swing prevention control, the position of the trolley converges at about 2.3 times of natural period (4.9 sec.), while the trolley moves back and forth and approaches the object point with the load's swinging around the angle 0° (maximum swinging angle ±1.0°). Moreover, we compared the result obtained by the fuzzy control with that of the performance obtained by the skilled operator and also that of the velocity pattern control and confirmed that the fuzzy control is effective as well.

## References

1. O. Itoh, H. Migita, J. Itoh and Y. Irie, "Application of fuzzy control to automatic crane operation", **IECON'93**, Vol. 1 of 3, pp. 161–164.

# INTEGRATION OF KNOWLEDGE-BASED CONFIGURATION WITH FUZZY LOGIC AND OPTIMIZATION

A. GÜNTER

*FB Informatik, Universität Hamburg, Vogt-Koelln-Str. 30 Hamburg, D-22527, Germany*

M. KOPISCH

*FB Informatik, Universität Hamburg, Vogt-Koelln-Str. 30 Hamburg, D-22527, Germany*

H.-J. SEBASTIAN

*Lehrstuhl für Unternehmensforschung (Operations Research), RWTH Aachen, Templergraben 64 Aachen, D-52056, Germany*

The application of fuzzy technology to control problems, data analysis, decision problems and many other application-oriented problems was done successfully during the last years. There are, however, relatively few papers dealing with the application of fuzzy set theory in the area of engineering design and configuration problems. In particular, we acknowledge the Refs. 23–26. Consider real-world design or configuration problems, e.g. design or configuration of computer systems, communication systems, special machines or vehicles, buildings, logistic- or manufacturing systems layout, we come to the conclusion that dealing with imprecision will become more and more important, if we want to approach intelligent Decision Support Systems. More precisely, integration of fuzzy set theory and methods with the successfully applied techniques of the Knowledge-Based Systems will generate a new application oriented topic of the fuzzy logic — fuzzy design.

This review paper describes some of the approaches to integrate Knowledge-Based Configuration with Fuzzy Logic and Optimization which have been elaborated in the frame of the research project PROKON[a] in Germany from 1991 to 1995. The paper focuses on those methods, software tools and applications considered mainly by the PROKON groups of the Hamburg University and the Aachen University of Technology.

## 1. Introduction

Several special programming languages and toolboxes support the development of expert systems. Recently toolboxes are developed that are restricted to a

[a]This work was supported by the German Federal Ministry of Research and Technology (BMFT) under grant number 01 IW 202 F.

certain application class or problem class (see Ref. 13). These toolboxes intend to cut the time for the development of an expert system by providing the knowledge engineer with methods that are especially suited for his application area.

This review article introduces a toolbox that aims on configuration and design problems. It integrates fuzzy techniques as well as optimization methods with knowledge-based type modeling. In order to show the capabilities of our toolbox we will use two applications and their needs for fuzzy techniques and optimization methods as examples.

First we want to give a definition of design tasks. Tong[36] defines design as follows:

> "What is design? Design is the process of constructing a description of an artifact that satisfies a (possibly informal) functional specification, meets certain performance criteria and resource limitations, is realizable in a given target technology, and satisfies criteria such as simplicity, testability, manufacturability, reusability etc.; the design process itself may also be subject to certain restrictions such as time, human power, cost, etc."

In addition to this general definition we understand engineering design as design of "*physical artifacts or physical processes of various kinds.*" Generation of structure from a functional description of requirements is central to this kind of design.

Design tasks are a subclass of synthesis tasks. In contrast to classification tasks they are usually solved by stepwise refinement. One subclass of synthesis tasks — configuration — is well investigated. In Ref. 10 it is described as follows:

"Configuration — as well as planning — is a synthesis task. A configuration task is understood as the assembling of a configuration from objects. Given are:

- a task description that defines requirements that have to be obeyed by the solution.
- a set of objects and their properties.
- a set of relations and restrictions between objects. In particular the 'part-of' relation is important in configuration tasks.
- "control knowledge"

Navinchandra[22] sees the significance of configuration methods for design tasks as follows:

> "The first is that some useful design tasks can be modeled as configuration problems, and that search techniques can be used to find solutions to these problems."

A configuration task may be defined as a simple design task. Many methods that have been developed for configuration problems can successfully be used for design problems as well. These methods are the basis for our toolbox KONWERK (Sec. 4). On top of that we integrated additional methods like fuzzy and optimization techniques for solving design tasks, knowing that there is a small number of publications dealing with these topics only.[1,8,20,25,26,28,30]

The main reasons to do this are briefly described as follows:

The task description which defines the requirements of the user to the goal system becomes more and more complicated. The user requirements might be expressed using imprecise linguistic terms, might be maximization or minimization aims for several, possibly conflicting criteria or might be imprecise restrictions. As a consequence we have to formulate the design task as Fuzzy Optimization or as a Fuzzy MCDM (Multi Criteria Decision Making) problem.

The work we introduce in this review paper has been done in PROKON — a project that was supported by the German Ministry for Research and Technology (BMFT).[9] The project's intention was to enlarge configuration methods for the purpose of using them for design tasks. Under consideration of this paper's thematic it is important, that in KONWERK

- it is possible to represent and process fuzzy knowledge
- optimization methods are considered in order to be able to find optimal solutions for a given task.

The paper is organized as follows: First we briefly describe the basic concepts of the integration of knowledge-based methods with fuzzy technology and optimization. For this introduction of concepts we use very simple examples from different areas for illustrational purposes only (Sec. 2). After this, we introduce our applications — the cabin layout of a passenger aircraft of the type AIRBUS A340 and the preliminary design and evaluation of future Space Launch Systems (Sec. 3). Our workbench KONWERK is described in Sec. 4. The integration of fuzzy techniques and optimization methods into the design task formulation of the two real world examples is shown in Sec. 5. The use of KONWERK for the realization of the first of our applications is described in

Sec. 6 where we again pay special attention to fuzzy techniques and optimization methods. We close this article with a summary and an assessment of the results.

## 2. Integration of Knowledge-Based Design Methods with Fuzzy Technologies and Optimization

To realize optimization or multi-objective decision making or even fuzzy optimization or fuzzy multi-criteria decision making by an Expert System we propose that there is an underlying model of the design or the configuration problem. This model (CONM) mainly consists of a "knowledge model" and a "requirement model." It is published in Refs. 2, 27, 31, 32. Unless we want to consider the aspect of "imprecision" mainly, in the following we have to give a short summary of the model CONM first.

### 2.1. *The knowledge-model*

For the representation of the domain knowledge we use hierarchies to model

- the world of systems under consideration (is-a-hierarchy)
- the world of elements (components) of these systems (is-a-hierarchy) and
- the relations between systems and elements (part-of-hierarchy).

The part-of-hierarchies may be considered as multilevel hierarchies, because a component may also be considered as a system consisting of components of the 2nd level.

It is possible to use only one "conceptual hierarchy" to represent all the hierarchies mentioned above. In Fig. 1 such a conceptual hierarchy is introduced where

$S$ – denotes the most general system (related to the world of systems),
$Si$ – are the system variants (conceptual level),
$OS_{i,k}$ – are system instances (individuals),
$E_j$ – denotes the components of $S_j$ (conceptual level),
$E_{j,r}$ – are component variants (conceptual level) and
$OE_{j,r,e}$ – are the component instances (individuals).

In Fig. 1, for the representation of the basic domain knowledge, we use abstract entities (concepts, generic objects, classes) $S$, $S_i$, $E_j$, $E_{j,r}$ and instances (individuals, specified objects) as well. These objects are connected by relations of the types:

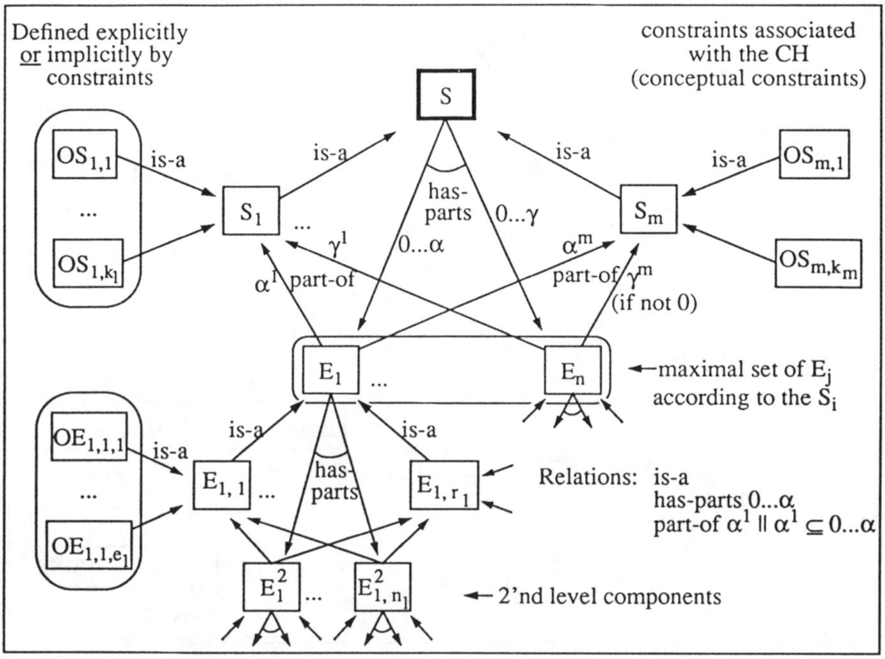


Fig. 1. Conceptual hierarchy.

is-a (specialization) (1–1 relation)
has-parts (composition) (1–$n$ relation) with arc-quantification $0 \ldots \alpha$
part-of (1–1 relation) with arc-quantification $\alpha^1$

The KONWERK system which will be introduced in Chap. 4 allows even N–M-relations for has-parts and part-of.

Figure 2 shows an illustrative example for such kind of a conceptual hierarchy introduced in Fig. 1.

Design or, as a special case configuration, means specification of abstract entities by more precise entities or composition of the precise entities, called specified objects or object classes to build a system.

The specification process starts with the abstract entities, so-called components and runs over several stages of object classes, terminating with precise entities, the so-called object-instances or individual-instances.

For a formal description we denote by $Y$ either a system $S$, a component $E$, a specialized concept or an individual $O$.

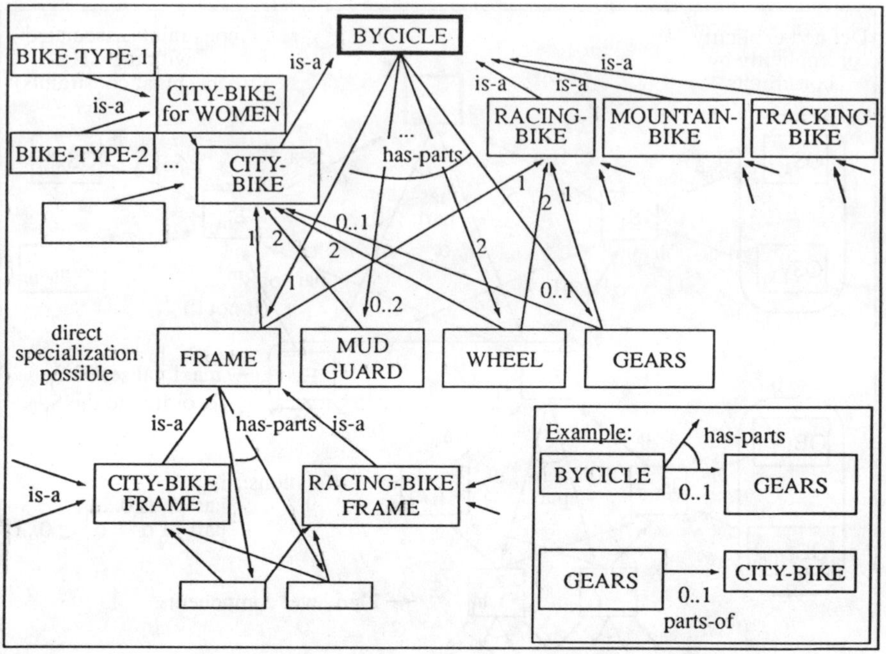


Fig. 2. Conceptual hierarchy —- example.

Then, $Y$ is described by

- a name $N(Y)$
- a vector $y = (p, e)$ with

  $p$ – vector of numerical parameters

  $e$ – vector of linguistic parameters

  $p \in \mathbb{P}, e \in \mathbb{E}$ ($\mathbb{P}$, $\mathbb{E}$ given sets)

  $Y = [N(Y), y]$

The components of $p$ and $e$ are either crisp or fuzzy values.

If $E_i$ denotes a component belonging to the set $E = \{E_1, E_2, \ldots, E_n\}$ of elements of $S$, then $\tilde{O}_i(S)$ denotes the set of all object-instances of $E_i$.

- *Routine design* means, that there is a finite number of predefined object-instances $O_i = O_{i,\alpha}$ for $E_i$, $\alpha = 1, 2, \ldots, \alpha_i$ which are *explicitly* known to the configuration system ($O_{i,\alpha} = N(Y)$, $Y$ is an object from $\tilde{O}_i(S)$)

$$\tilde{O}_i = (O_{i,1}, O_{i,2}, \ldots, O_{i,\alpha_i})$$

- *Innovative design* (special case) means that not all object-instances are explicitly predefined and known. But there are finite sets of well-known *constraints*, which describe feasible object-instances *implicitly*.

$$g_j : \mathbb{P} \times \mathbb{E} \to \mathbb{R}^1 , \qquad j = 1, 2, \ldots , j_0$$
$$g_j(y) \le 0 \qquad \wedge \quad y \in \mathbb{P} \times \mathbb{E}$$

The implementation of the specification process is realized by a FRAME-representation language combined with a Constraint-language (for example PLAKON).[4]

Furthermore in our model, we have to describe the relations between the components of $S$. Therefore a structure graph $G(S) = [E(S), R(S)]$ models these relations on an abstract level and the object-structure graph $GO(S)$ models these relations on the object-instance level. For more precise definition of these graphs see Ref. 2.

The (local) consistency of a configuration of $S$ is characterized by an equivalent to $G(S)$ subgraph of $GO(S)$.

## 2.2. *The requirement — model*

One topic of central importance for a design or configuration model is an adequate representation of the user requirements. We consider three types of requirements:

- functional properties
- global constraints
- objectives

### 2.2.1. *Necessary functional requirements*

With functional properties we model some requirements of functional abilities of the system, for example:

"The car should be fast and safe."
"The computer system should have graphic ability."

Now, we consider this linguistic expressions as crisp knowledge and we use such kind of requirements to specify the necessary components, object-instances the system must have. In 2.3.3, alternatively we will model such kind of requirements taking into account the imprecision of the attributes used.

In the following, we give a syntactic definition of a necessary functional property $F_j$

$$F_j = (f_j, [\{E_{j1}, \dots, E_{jk}\}, \{O^*_{j1}, \dots, O^*_{jk}\}, \text{key}(f_j)]) \qquad j = 1, 2, \dots, \mu$$

with

| | |
|---|---|
| $f_j$: | name of $F_j$ |
| $E_{j1}, \dots, E_{jk}$ | components, which must belong to $S$ to realize $F_j$ |
| $O^*_{j1}, \dots, O^*_{jk}$ | corresponding to $E_{j\nu}$ object-classes, e.g. object instances (defined explicitly or by constraints) |

Then $F = (F_1, F_2, \dots, F_\mu)$ is called necessary functional requirement.

2.2.2. *Global constraints*

In the following, we use a vector x to describe the feasible objects for $S$ or $E_j$ as well and $X$ to denote the set of feasible objects defined above by $\mathbb{P}$, $\mathbb{E}$ and $GO$ $(S)$.
Furthermore, there are resources of limited availability to the whole system $S$, e.g.:

$$C_1: \quad \text{Weight } (S) \leq G_{\text{Max}}$$

$$C_2: \quad \text{Energy-consumption } (S) \leq En_{\text{Max}}$$

$$C_3: \quad \text{Cost } (S) \leq C_{\text{Max}}$$

These resources are distributed to the system-elements by "global-constraints" for instance:

$$C_1: \quad \text{Weight } [\{E_1, E_2, \dots, E_n\}, R(S)] \leq G_{\text{Max}}$$

A configuration $x$ of $S$ is global consistent, if

- the functional requirement $F$ is fulfilled
- $x$ is local consistent
- $x$ fulfils the global constraints $C_j$, $j = 1, 2, \dots, M$

$X^* \subseteq X$ denotes the set of global consistent configurations. A global constraint $C_j$ is syntactically defined by

$$C_j = \{c_j, (E_{j1}, \dots, E_{jk_j}), g_j\}$$

with $E_{j1}, \ldots, E_{jk_j}$ — the components of $E(S)$ which need the resource modelled by $C_j$

$$g_j : X \to \mathbb{R}^1 \qquad \text{with } g_j(x) \leq 0$$

A constraint $C_j$ may be "hard" or "soft,"

$$C_j \in \{C_\nu^h, C_\delta^w\}.$$

Then, $C = (C_1^h, \ldots, C_{\nu_0}^h, C_1^w, \ldots, C_{\delta_0}^w)$ with $\nu_0 + \delta_0 = M$ models the *global constraint requirement.*

### 2.2.3. *Objectives and fuzzy goals*

Optimization means that there exists at least a preference relation on $X$ or $X^*$, respectively. We consider the following case:

(1) Objectives

— There exist A crisp goals (aims)

$$Z_1, Z_2, \ldots, Z_A,$$

called objectives, with:
If $x \in X$ is local consistent, then $Z_i(x) \in \mathbb{R}^1$ is uniquely defined (for each $i = 1, 2, \ldots, A$).

(2) Fuzzy Goals

— There are $B$ "imprecise aims"

$$W_1, W_2, \ldots, W_B,$$

so called "fuzzy goals" with:
$W_j$ is described by a linguistic variable $LV_j$

$$LV_j \in \{w_{j1}, w_{j2}, \ldots, w_j\rho_0\} \quad w_{jp} \quad \text{linguistic values (terms)},$$

$$\rho = 1, 2, \ldots, \rho_0$$

Then, $M_{j\rho}$ = "$LV_j = w_{j\rho}$" is a fuzzy set on $X$, characterized by the membership-function

$$\mu_{M_{j\rho}}(x) : x \in X.$$

— An aggregated fuzzy goal $W_j^*$ is defined by:

$$W_j^* : X \to \mathbb{R}^1 \text{ with } W_j^*(x) = \phi_j[\mu_{M_{j1}}(x), \ldots, \mu_{M_{j\rho_0}}(x)]$$

$$\phi_j : [0, 1]^{\rho_0} \to \mathbb{R}^1$$

An objective/fuzzy goal requirement for $S$ is a tupel

$$ZW = (Z_1, Z_2, \ldots, Z_A; W_1^*, W_2^*, \ldots, W_B^*) \quad \text{(c.w.a.)}$$

and an objective/fuzzy goal variant for $S$ is a tupel

$$ZW = (Z_{i1}, Z_{i2}, \ldots, Z_{in_1}; \quad W_{j1}^*, W_{j2}^*, \ldots, W_{jn_2}^*)$$

with $n_1 \leq A$, $n_2 \leq B$, $i_\alpha$, $j_\beta$ pairwise different.

From the consideration above we can derive a general class of configuration models described by an 8-tupel

$$\text{CONM} = [\text{X, E(X), G(S), } \tilde{\text{O}} \text{ (S), GO(S), F, C, ZW}]$$

## 2.3. *Imprecision in design and configuration models*

In 2.2 we introduced with "fuzzy-goals" only one specific topic of integration of imprecise knowledge into a knowledge-based system. Also, the terminus "fuzzy-goal" we introduced was specialized.

Now, we want to introduce concepts of imprecision according to the knowledge-model and the requirement-model introduced above more systematically.

In considering applications it becomes obvious that in both models, the knowledge model and the requirement model as well, one part of the knowledge is imprecise by nature or can be considered to be imprecise.

### 2.3.1. *Fuzzy frame hierarchies*

The conceptual hierarchy contains objects with imprecise attributes (called imprecise objects). For instance, an attribute of an object might be defined as a linguistic variable in which the terms are characterized by the corresponding fuzzy sets. Also, an attribute might be a fuzzy number. Objects, consequently, are described by a few crisp or fuzzy attributes as well. Therefore, in such hierarchies with imprecise objects, one has to define carefully what specialization and generalization means. In particular, multiple inheritance and inference in fuzzy conceptual hierarchies have to be defined and implemented. The basic problems as well as implementation oriented aspects connected with these hierarchies are solved in KONWERK (Sec. 4).

In the following, we consider an illustrative example:

| Name of Object | Attributes | | |
|---|---|---|---|
| | Name | Type (linguistic/numeric) | Precision (crisp/fuzzy) |
| TUBE | 1) FORM | $L_1$ = {straight, curved} | crisp |
| | 2) MATERIAL | $L_2$ = {steel, lead, plastics, ceramics, ..} | crisp |
| | 3) THICKNESS | $N_3$ = [$di_{Min}$, $di_{Max}$] real | crisp |
| | | $L_3, L_4$ = {thin, medium, thick} | fuzzy |
| | 4) DIAMETER | $N_4$ = [$AD_{Min}$, $AD_{Max}$] real | crisp |
| | 5) LENGTH | $N_5$ = [$l_{Min}$, $l_{Max}$] | crisp |
| | | $L_5$ = {very-short, short, medium, long, very-long} | fuzzy |

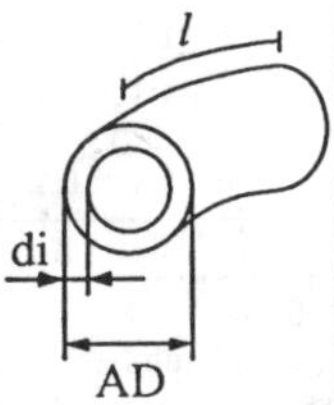


if curved, then additional attribute curvature k (radius)
- crisp
- fuzzy: weak-curvature / medium-curvature / strong-curvature

value sets
in fuzzy cases: membership-functions added

Fig. 3. Conceptual hierarchies (hierarchies of objects described by attributes).

In Fig. 3 FRAME TUBE is generated for the generic object tube (also called class tube) by introducing the attributes (slots) by a name, a type and a characterization of precision or imprecision of the attribute. The subconcept (subclass) TUBE-curved requires the definition of an additional attribute CURVATURE. Figure 4 shows the specialization of a class to a subclass (e.g. an instance or individual). This shows that during the design-process (e.g. specialization process) imprecision will be reduced or even deleted.

Crisp objects in the sense of individuals belong not only to one but in general to several imprecise classes (concepts). The membership-degree is dependent mainly on the interpretation of the fuzzy-“and,” Figs. 5 and 6 show a very simple example. In Figs. 5 and 6 the membership-degrees 0.4 or 0.3 as well are computed by Min-Operator application using the membership-functions of the objects which are on the respective path from the considered object node (complete-specified imprecise class) to the root-node of the tree.

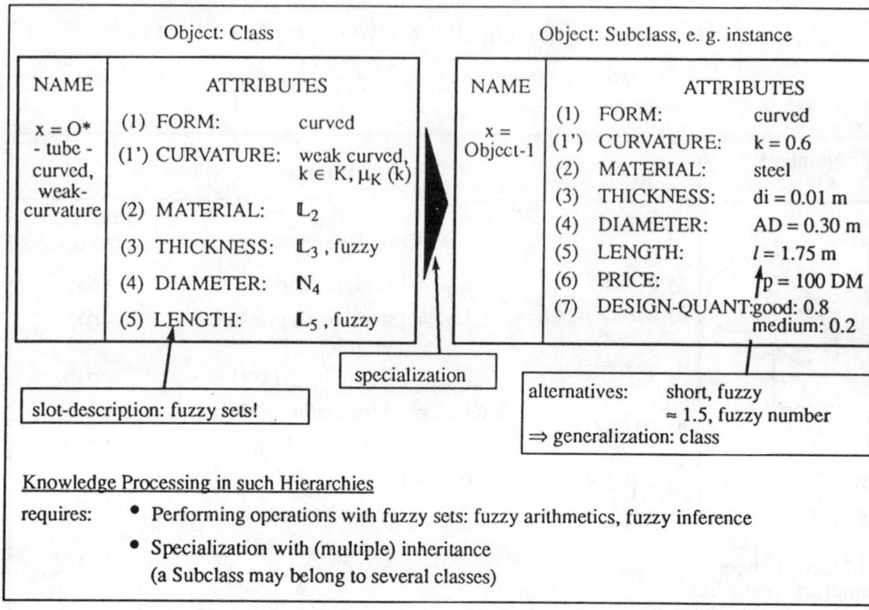


Fig. 4. Classes – subclasses.

$$\mu_{O^*_{4,1,4}}(\text{Object}-1) = \text{Min}\,(\mu_4(K),\, \mu_1^*(\text{di}),\, \bar{\mu}_4(l)) = \text{Min}\,(0.4,\, 0.5,\, 0.3) = 0.3$$

$$\mu_{O^*_{4,1,3}}(\text{Object}-1) = \text{Min}\,(\mu_4(K),\, \mu_1^*(\text{di}),\, \bar{\mu}_3(l)) = \text{Min}\,(0.4,\, 0.5,\, 0.6) = 0.4$$

The membership-values used in these formulas one gets from the respective membership-functions which are available as slot-descriptions of the respective frames.

For a more detailed description of the Fuzzy Frame Hierarchies see Refs. 32 and 35.

### 2.3.2. *Compositional hierarchies with imprecision*

The compositional hierarchies (part-of, has-parts) (Figs. 1, 2) can also be modeled using fuzzy methods, taking into account the problems that the component structure might be partly unknown in earlier design stages or that there are "optional" system components. Therefore, the components of an artifact can

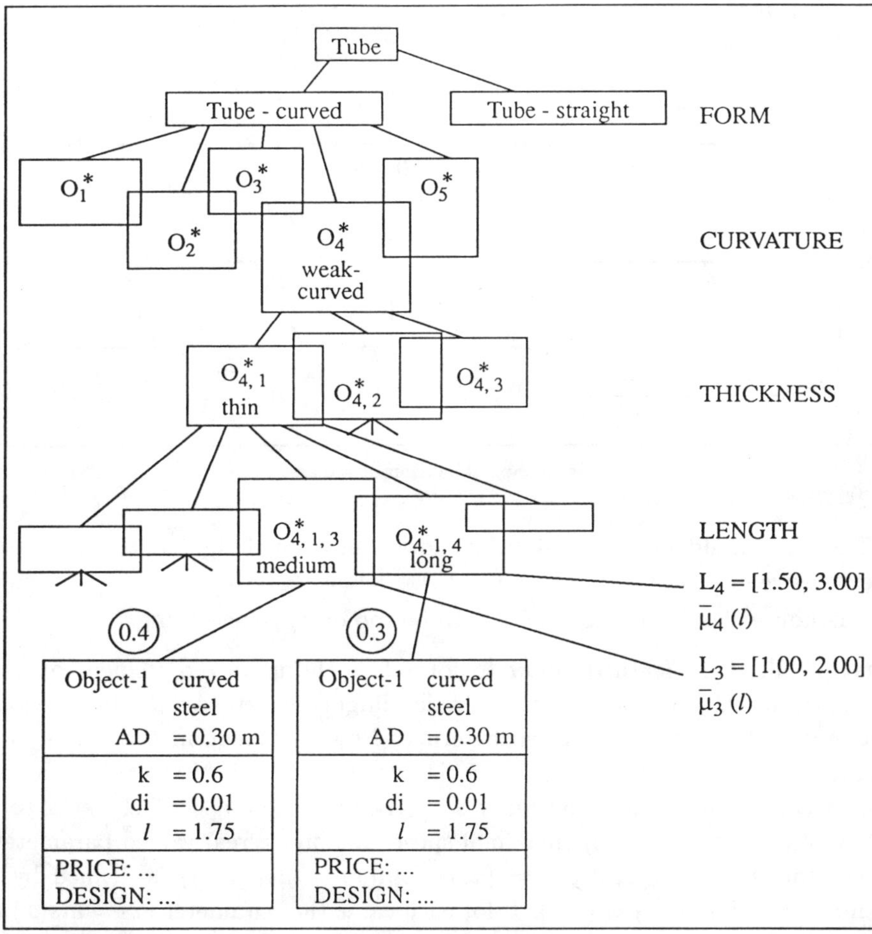


Fig. 5. Example — FRAME TUBE.

be defined as fuzzy sets where the membership function represents the "desirability" of a component to belong to the artifact or not. For more detailed considerations of this topic see Refs. 37 and 40.

### 2.3.3. *Fuzzy-design-requirement*

"Fuzzy-Design-Requirements" are modeling necessary or optional requirements (see 2.2.1, for the crisp case) or functional abilities for the goal-system, for example:

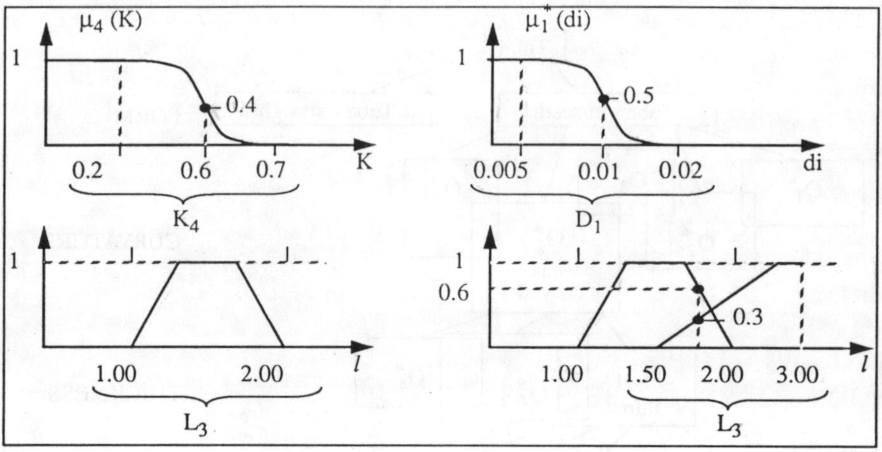


Fig. 6. Specialization of a class.

- The car should be fast and safe, show a modern design, but not be too expensive.
- The house should be designed not to be too energy-consuming.

First of all, these features allow to formulate the users requirement on the linguistic level. In 2.2.1 we used similar linguistic expressions, but we did not model the imprecision of such terms. Now, we mention that there is a linguistic approximation of the imprecise statements, e.g. "fast car," "not too energy-consuming house," by fuzzy sets. Let $x$ denote a car belonging to a set $X$ of cars, $ms = ms(x)$ (maximal speed) denotes a real valued parameter. Then, the fuzzy set, Fast-car $= \{x \in X | \mu(x) = \mu_{\mathrm{FC}}(ms)\}$ is defined by a membership-function $\mu_{\mathrm{FC}}(ms)$. If for each car $x$ the parameter $ms = ms(x)$ is known, we call Fast-Car to be given explicitly. Otherwise $\mu_{\mathrm{FC}}(ms)$ gives the possibility of each value ms, assumed the car is known to be fast.

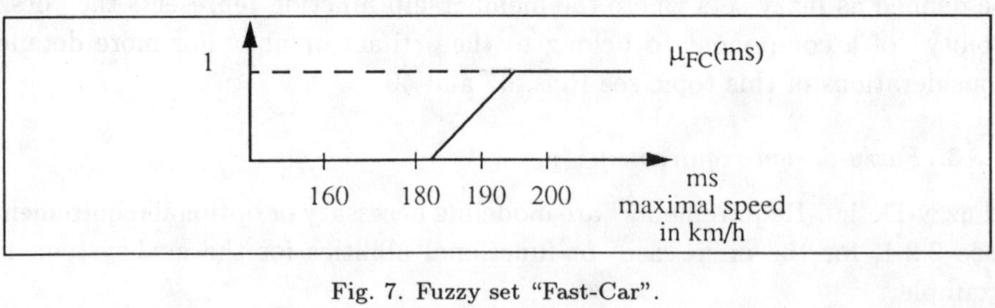


Fig. 7. Fuzzy set "Fast-Car".

This fuzzy set "Fast-Car" gives us a linguistic approximation using a quantification of "Fast" in the context of "Car." Of course, the example is very simple. But it is easily possible to make the example more realistic, e.g. to define Fast-Car by the fuzzy set Fast-Car $= \{x \in X | \mu(x) = \mu^*_{\mathrm{FC}}[ms,\, t(0,\, 100)]\}$ where $t(0,\, 100)$ denotes the time the car needs to reach speed 100 km/h from 0.

It is more difficult but possible by introducing appropriate scales (possibly again multidimensional) to define "Safe-Car," "Modern-Design-Car," "Not-Expensive-Car" by fuzzy sets over X.

But, in addition, e.g. "Safe-Car" is connected with components this car should have and also specializations of these components which are more or less required (see 2.2.1). Therefore, Fuzzy Design Requirements are characterized by compositional hierarchies with imprecise attributes.

For instance, "Safe-Car," in addition to the fuzzy-set description (fuzzy set over $X$), means that the components "two independent brakes," "safety belts," "good tyres," for example, must belong but one or two airbags only should belong to the car with a "desirability measure of 0.8." Also, ABS could be a specialization-requirement for the brakes.

### 2.3.4. *Fuzzy constraints and fuzzy goals*

The most important issue of fuzzy design requirements is to model the dependencies between an imprecise linguistic expression like e.g. "Safe-Car" and the component or specification structure which is required by this expression. However, such a fuzzy set might also be used to model the underlying requirement by a fuzzy-constraint or a fuzzy-goal. Let's consider the fuzzy set Fast-Car $= \{x \in X | \mu(x) = \mu_{\mathrm{FC}}(ms)\}$ once more and assume that $ms = ms(x)$ is known for each $x \in X$.
A crisp constraint could be formulated by

$$ms(x) \geq 190\,\mathrm{km/h}$$

A fuzzy constraint is e.g. (according to the fuzzy set Fast-Car): $x$ is feasible with the degree $\mu_{\mathrm{FC}}(ms)$, that means

$$\mu_{\mathrm{FC}}(ms) = \begin{cases} 0, & \text{if } ms(x) \leq 170 \\ \in [0,\, 1]\,, & \text{if } 170 \leq ms(x) \leq 190 \\ 1\,, & \text{if } ms(x) \geq 190\,. \end{cases}$$

Using this kind of relaxation of the sharp boundary 190 km/h between feasibility and infeasibility we get a smooth transition.

There is another possibility to model the requirements as fuzzy goals, what means to have the idea of maximization of a real-valued function over $X$ in mind, which measures the utility.

Again, let's consider the imprecise requirements Fast-Car, Safe-Car, Modern-Design-Car, Not-Expensive-Car represented by fuzzy sets over the basic set $X$ of cars $x$. There are some different approaches to use such fuzzy sets to model fuzzy goals.

- The fuzzy goal is described by a fuzzy set. Then, we introduce a specific defuzzification by:
  Find $x \in X$ such that $\mu_{\mathrm{FC}_1}(x) \to \mathrm{Max}$ with $\mu_{\mathrm{FC}_1}(x)$ as the membership-function of the fuzzy set

$$\mathrm{FG}_1 = \text{Fast-Car} \wedge \text{Safe-Car} \wedge \text{Modern-Design-Car} \wedge \text{Not-Expensive-Car}\,.$$

e.g. $\mu_{\mathrm{FC}_1} = \mathrm{Min}[\mu_{\text{F-C}}(x),\, \mu_{\text{S-C}}(x),\, \mu_{\text{MD-C}}(x),\, \mu_{\text{NE-C}}(x)]$

Still, there is another possibility to consider an aggregation tree like the following:

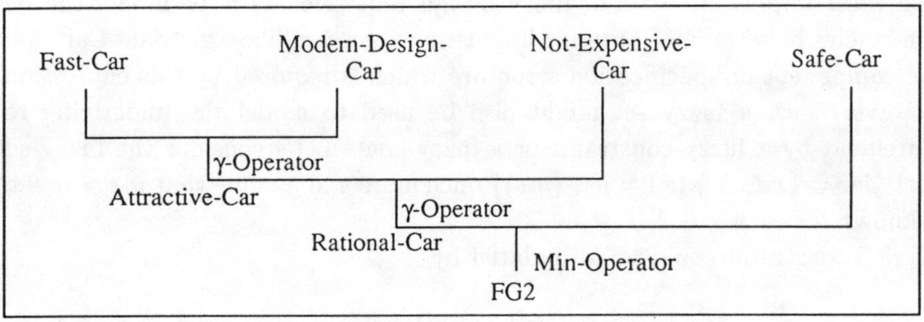


Fig. 8. Example for an aggregation.

We aggregate "fast" and "modern design" to "attractiveness." "Attractiveness" and "inexpensiveness" are aggregated to "rationality" and finally, "rationality" and "safety" will be aggregated to define the goal FG2. This aggregation uses a compensatory operator on the first two levels and the minimum operator on the final level.[38–40]

- Another possibility is the linguistic-variable approach to define fuzzy goals. We introduced this approach in 2.2.3 more mathematically. Here, we give an example. Let Maximum-Speed be a linguistic variable with the terms {slow, medium, high, very high}, then a fuzzy goal can be defined by "medium ∨ high." Then, again as a defuzzification, we are maximizing the membership degree of "medium ∨ high" over X using fuzzy sets "Max.-Speed = medium" and "Max.-Speed = high."

For more detailed considerations of the topics stressed in (3) and (4) see Refs. 40, 41 and in particular of the linguistic variable approach to fuzzy goals see Refs. 27, 29.

## 3. Applications

In the following we describe two of the six real-world applications of the PROKON project.

### 3.1. *Passenger aircraft cabin layout*

A cabin layout (see Fig. 9) defines the passenger cabin of an aircraft by number, arrangement, type and position of the cabin interior components. Cabin interior components are objects like passenger seats, lavatories, galleys or cabin attendant seats which may be installed in the passenger cabin. The entire set of cabin interior components in the passenger cabin defines its capacity, its comfort level, its possible service, etc.

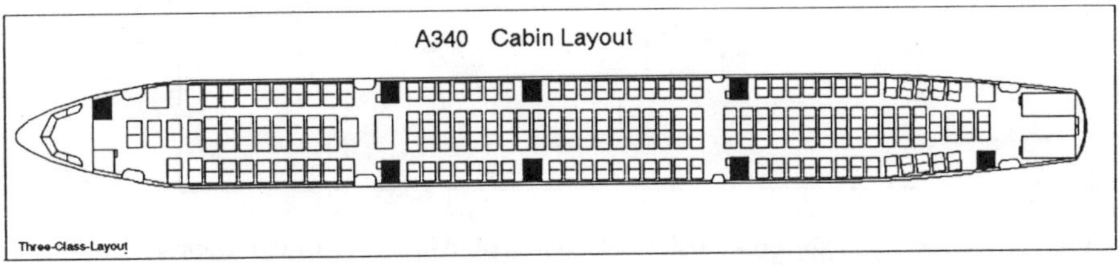


Fig. 9. Cabin layout that was configured with XKL.

Configuring a cabin layout means[17] to place a certain number of cabin interior components optimally in the restricted space of the passenger cabin. Optimality may be defined with respect to factors like the number of seats, the cost, the lead-time or a mixture of these factors. In the process of configuring a cabin layout one has to consider the given restrictions and the requests of the ordering airline. The restrictions may be either certification rules of the approving airworthiness authority or technical requirements of either the aircraft builder or the component supplier. The airline requests concern the division of the passenger cabin into passenger classes, the number of seats, the seat pitch etc.

The difficulty in configuring a cabin layout is to consider all restrictions and requirements at the same time. The following examples give an impression of the restrictions that appear in the cabin layout domain.

Figure 10 illustrates several certification rules: There has to be an assist-space to each side of a type-A-door (width at least 12 inch); the width of a cross aisle between two type-A-doors has to be at least 20 inch; a cross aisle between type-A-doors with two approach paths has to overlap with the 36 inch of the passageway by 50 % (see para 25.807 (a) (7) (IV) and (V) and para. 25.813 (a) of Ref. 5). Another example for a certification rule is the minimum ratio between cabin attendants and passengers.

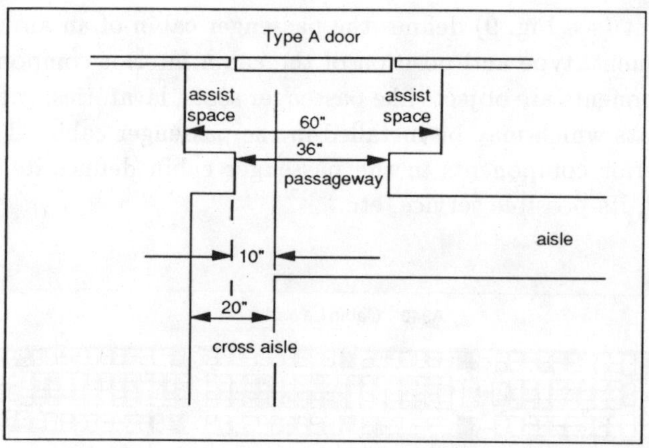


Fig. 10. Restrictions for cross aisles between type-A-doors with two approach paths.

Figure 11 illustrates the technical restrictions that are given for the mounting of a special galley. Another simple technical restriction is the maximum number of lavatories in the passenger cabin.

In addition to the given constraints (legal and technical) one has to consider the customers requirements. Figure 12 shows the seat pitch as an example for such a requirement.

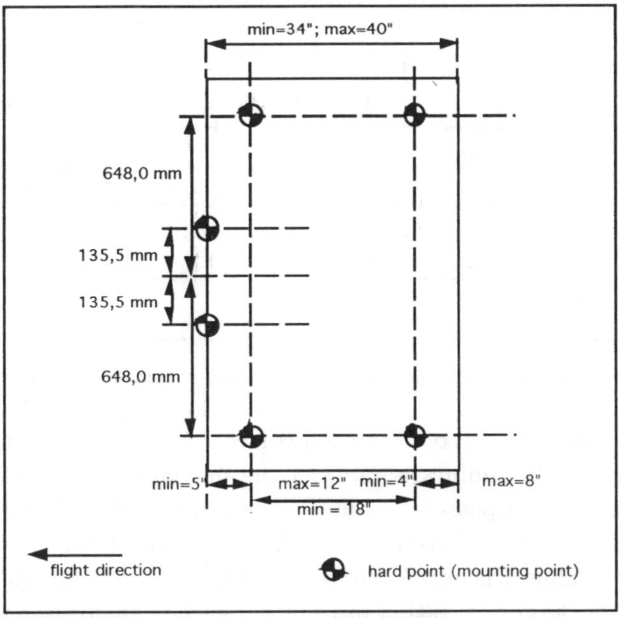


Fig. 11. Mounting restrictions for the transversal galley 25-33.100.

Besides, one has to strive for optimality with regard to criteria like number of seats, comfort level, cost, delivery time etc., which are different for each order. Configuring a cabin layout is getting harder. This is due to the constantly increasing size of aircrafts permitting more and more cabin interior components to be placed and a rising number of certification rules.

In particular, the aircraft builder's representative is no longer able to consider every restriction and thus cannot assure correctness and optimality of his layouts. An expert system might provide assistance by relieving the representative from routine tasks. It could make the necessary tests for correctness and thereby guarantee the consistency of the layout developer's decisions. Such a system should be able to carry out parts of the configuring and optimizing processes on its own. The system XKL,[16] an expert system for cabin layout, supports configuring a passenger cabin for an AIRBUS-A340. This system was build with PLAKON.[4]

Configuration with PLAKON is based on a stepwise refinement of partial solutions. Starting with an initial partial construction which has been derived from a task description a partial construction is refined by a sequence of

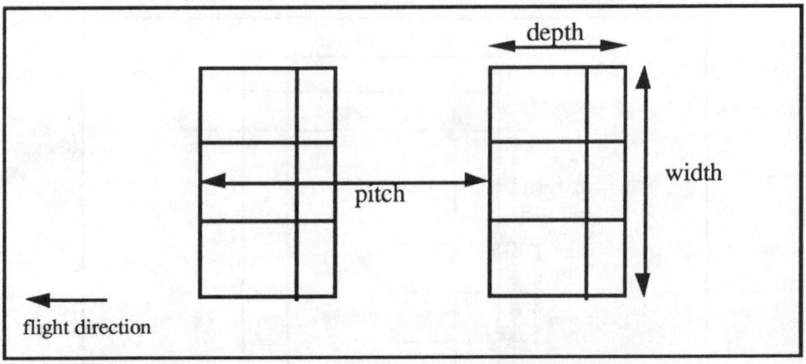


Fig. 12. Customer's requirement for a certain seat pitch.

construction steps until a complete and correct construction is achieved. Each construction step determines a value in the partial construction such as the number of galleys in a passenger class or the position of a lavatory in the passenger cabin. The PLAKON-version of XKL supports its user by considering all constraints. Thus it is possible to show the user the possible values (possible at that point of the configuration process) for each configuration step. Considering all constraints during each configuration step leads finally to a consistent solution of the configuration task. Although the solution is consistent, we do not know whether it is good or even optimal. Thus we need possibilities for the description of criteria for a good solution and mechanisms for their processing. During the project PROKON we built tools for such tasks and implemented a new version of XKL with KONWERK.

### 3.2. *Preliminary design and configuration of future space launch systems*

In space flight a change of paradigm has taken place during the last years. While in former times national prestige and the 'cold war' were the main reasons for space activities, today economical aspects gain more importance. Furthermore, environmental compatibility and other requirements not only act like aims but like restricting arguments. The task of space launch systems (SLS) is to provide the transportation demand for orbital applications like positioning satellites in the orbit and planetary missions. SLS are the key technology for all space activities. Today a large number of different systems and a large amount of experience are available. Future space launch system concepts depend strongly on the mission and become more user-oriented, i.e., aims and requirements are also defined by the user. On principle, different technical solutions are feasible.

The research activities for new space launch systems are a time-consuming process taking years up to decades. In particular the beginning of a project is very important, because the main technical features will be determined in the first phase and cannot be changed in a latter phase. Thus also technical and economical risks and chances are being determined in the early phase. The knowledge of experts helps to select technically feasible systems. The first decisions give the main frame for the further research of the relevant systems.

On principle, the space transportation community is interested in a support system that assists the decision-making process in the first approach towards a new space launch system. In the first approach a preselection of conceivable space transportation systems for a predefined payload with several requirements and one or more goals has to be made. The resulting configurations will be investigated closer in the next step.

The idea of making the attempt or an application of knowledge-based configuration in this technical domain was introduced at the end of 1993 and has been performed with the co-operation of the Institute of Operations Research of RWTH Aachen and the German Aerospace Research Establishment (Space System Analysis Division). The present results are published in Refs. 18, 19. The implementation is presently being done with the KONWERK workbench.

An expert system that assists in the decision process of configuration of future space launch systems is a suitable approach. The highly complicated technical reality must be simplified. For a first approach it is sufficient to describe a space launch system with its components and several technical data. After the first selection, further optimizations have to be done, which is not in the focus of this work but could be an interesting question for an expert system, too.

Space launch systems consist of a limited number of components, that will be aggregated in modules. This yields a good structure of the domain. The components are considered as concepts in a conceptual hierarchy. In this hierarchy both part-of relations and specializations can be described (see Figs. 13 and 14).

A space launch system is the most general concept that may be specialized according the feature propulsion system of the first stage. Here further specialization concerning the number of stages is possible. In this tree-like

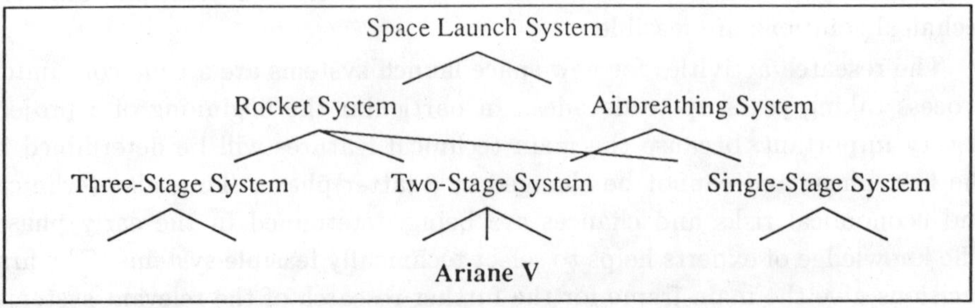


Fig. 13. Specialization hierarchy of space launch systems.

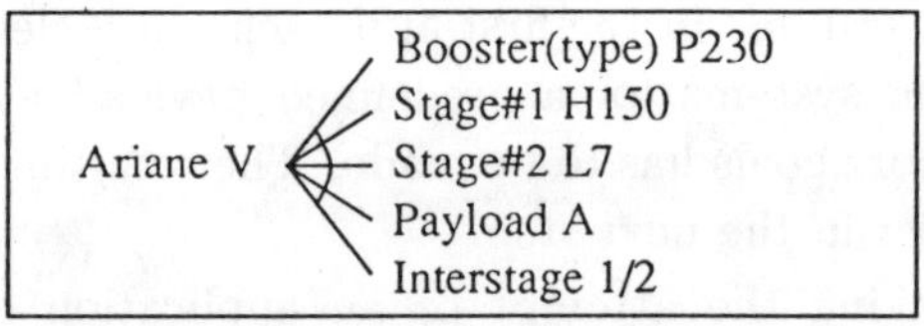


Fig. 14. Part-of hierarchy of instance ariane V.

structure the instances, which are the most specialized objects, are situated at the bottom, like Ariane V, a two-stage, rocket propulsion system.

The concept Ariane V consists of parts that are instances of parts of general concepts, as well. They are specialized for the Ariane V system, and, therefore, determine this concept. The knowledge-base "SLS" is implemented in KONWERK and will be described later more precisely.

The design process starts with the payload mass to be transported, which corresponds to the payload capacity of the launch system. The aim is to configure a space launch system that is best adapted to the problem to bring a specified payload into a prescribed orbit. For simplicity reasons the payload capacity will be normed to the low Earth orbit (LEO).

In this example merely rocket propulsion systems are considered. Airbreathing propulsion systems will be configurated similarly and, thus, will be neglected for this example.

The purpose of the first part of the design algorithm is to select those space systems that are at least capable of transporting the predefined payload into the orbit. This is the main part, since the number of components and the most important modules determining the space launch system are being fixed here. A set of numerical computations have to be performed to obtain the basic parameters, e.g. the take-off mass of the system. The geometric dimensions of the system are not known in the first step, these values will be estimated in a second step.

The algorithm generates a set of alternative SLS which are at least capable of solving the transportation task. Each alternative might be evaluated by several, in general imprecise, criteria. Therefore, a Fuzzy-MADM can be applied to select or rank the alternatives.

For data input of the design algorithm crisp data is used presently. But the imprecise form, e.g. linguistic expressions is a possible alternative. That makes the data handling easier for the user, helps him to formulate his requirements, which are imprecise, in a more natural way.

In the conceptual hierarchy specializations that inherit characteristics of two superconcepts are likely, e.g. there do exist plans for space launch systems with propulsion systems, that are both, airbreathing and a rocket. In addition to this, crisp values are not always wanted, and for the preliminary design phase fuzzy results are sufficient. An example is a fuzzy input of the payload capacity that could be grouped into several weight classes, each class representing a fuzzy set.

## 4. KONWERK — A Configuration Tool

In this section we present KONWERK — a problem-specific expert system shell. The conception and realization of this modular tool aims to support a developer of expert systems for configuration and design tasks. KONWERK consists of several integrated methods and representations mechanisms which are well suited for configuration and design tasks. Some concepts are based on the system PLAKON.[4] During the development of KONWERK six applications from mainly technical domains are considered (see Ref. 12):

- selection and parametrization of slide bearings
- configuration of cardan shafts
- configuration of an optimal layout for a logistics system
- design of liquid crystals
- configuration of cabin layouts for the AIRBUS A340
- preliminary design of space launch systems.

KONWERK is a tool-box for configuration tasks. Its development is part of the project PROKON (see Ref. 9). A central requirement for tools like KONWERK is its applicability in different domains. Thus its concepts have to be universal and must not be too special. On the other hand every domain has its special needs. Therefore we propose a solution of modules that can be chosen by the developer in order to solve his problem.

Figure 15 shows the architecture of KONWERK. SUN stations and Macintosh PCs are used as hardware. The use of standards like CommonLisp,

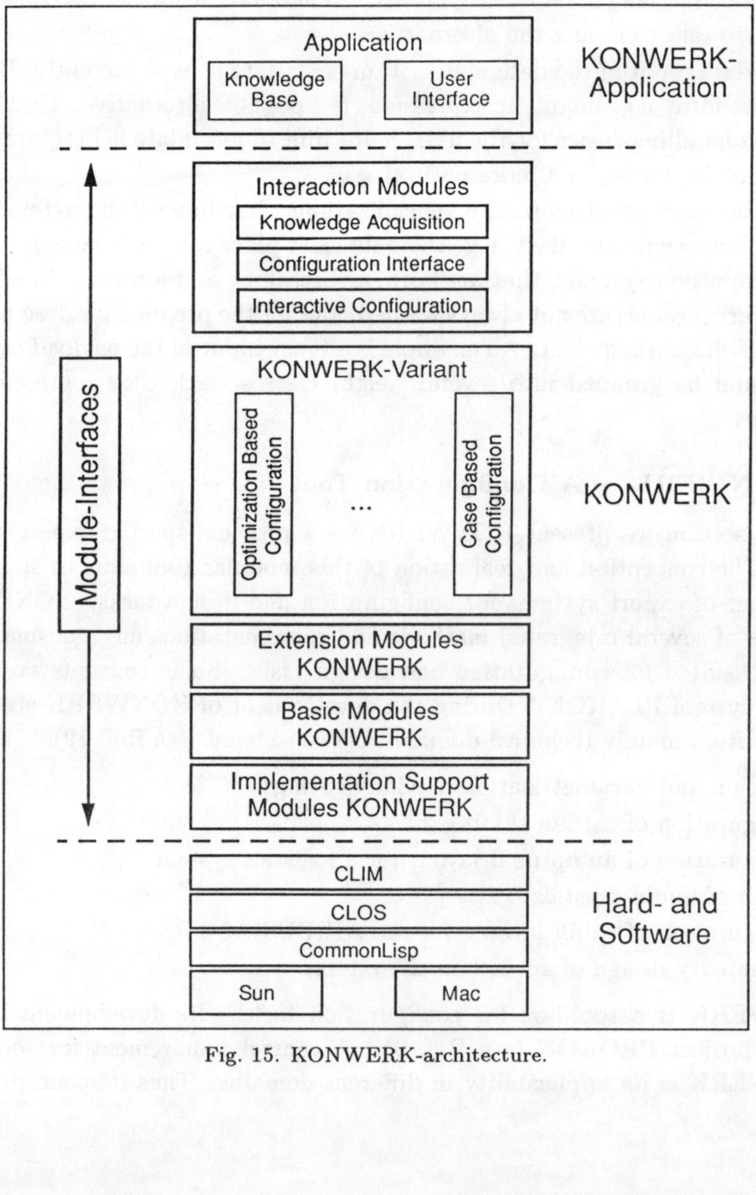


Fig. 15. KONWERK-architecture.

CommonLispObjectSystem (CLOS) and CommonLispInterfaceManager (CLIM) ensures portability. Especially with CLIM the standard was chosen instead of more efficient languages. Basic and extension modules will be explained later on in more detail. For the domain independent user interface we differentiate between knowledge acquisition and configuration. KONWERK supports interactive configuration which, like the explanation module, is an extension module.

In most cases a KONWERK-application consists of a knowledge base and a domain specific user interface. In order to design a KONWERK-application one can

- select a pre-defined problem specific KONWERK-variant
- choose all necessary modules and combine them.

## 4.1. *Modules*

The problem-solving modules are divided in basic modules and extension modules which either enhance the abilities of basic modules or provide new abilities. The basic modules cover the following general tasks in configuration and design systems:

- representation of domain objects
- representation and processing of constraints and heuristics
- formulation of configuration tasks
- control of the configuration process.

We will now introduce the modules (basic and extension modules) which together are the kernel of KONWERK. For the support of their implementation we built, e.g. the following support modules:

- object-oriented enlargement of CommonLispObjectSystem (CLOS)
- definition of interface classes and interface methods between the modules.

The basic modules of KONWERK are necessary in most applications. They offer standard methods for the main tasks in configuration systems. If required by the domain they may be complemented by extension modules. The basic modules of KONWERK are:

- *Conceptual hierarchy for the representation of domain objects*
  Domain objects and their properties are represented in a conceptual hierarchy which comprises a taxonomic (is-a) and a compositional (part-of) hierarchy. The possible values for the properties are pre-defined with help of so-called object descriptors.

- *Constraints for the representation of relations*
  For the representation of configuration restrictions and their processing we use a constraint system. By the use of constraint propagation it is possible to restrict values and to discover inconsistencies.
- *Control*
  The control is based on a depth-first search and supports chronological backtracking. The control knowledge is represented in several so called strategies which control the sequence of configuration steps chosen from an agenda.
- *Specification of the configuration task*
  A configuration task in KONWERK is specified by the selection of a target object which has to be configured. It is possible to select further objects (components of the target object) and to specify their properties (slots).

Extension modules enlarge the functionality of KONWERK in several ways. Due to the concept of a tool-box with a variety of modules the user of KONWERK can build his tool in a very flexible way by selecting the necessary modules. Extension modules extend the basic modules only in a conceptual view. On the implementation level they are often complex modules which contain all the functionality of a basic module and a lot of functional extensions to this basic module.

The conceptual hierarchy may serve as an example for the principle of the modular toolbox. The basic module for representation of a domain object is a "simple" version of a taxonomic hierarchy as a basic module. This enables the user to describe his objects in a strict is-a hierarchy and to define their properties. Possible values for these properties are pre-defined by the user. A special property is the compositional relation between objects. This *n*-ery relation is very important in the configuration process and is especially supported in this module.

There are several extensions to this basic module. We have one extension module for each of them. With this knowledge an engineer can select the functionality he needs and does not have to pay for power he does not want. Possible extensions for the conceptual hierarchy are:

- views and mixins
- fuzzy properties
- linguistic values
- spatial properties (2D and 3D)
- fuzzy is-a hierarchy
- measures.

Table 1. Survey of the basic and extension modules.

| Task | Basic modules | Extension modules |
|---|---|---|
| representation of domain objects | • conceptual hierarchy | • views and mixins in the conceptual hierarchy<br>• fuzzy properties in the conceptual hierarchy<br>• linguistic values in the conceptual hierarchy<br>• spatial properties (2D and 3D)<br>• fuzzy is-a hierarchy<br>• measures |
| representation and processing of relations | • constraints | • uni-directional functions<br>• rules for the representation of heuristics<br>• constraints with fuzzy variables<br>• spatial relations |
| control | • depth-first search with backtracking | • case based configuration<br>• exploration based configuration<br>• interactive configuration<br>• basic optimization module<br>• optimization module MADM<br>• linear optimization<br>• non-linear optimization with constraints<br>• branch-and-bound optimization<br>• knowledge based backtracking<br>• knowledge based backtracking with data acceptance<br>• repair instructions<br>• complete search in an explicit solution space<br>• ATMS-based configuration |
| task specification | • target object, components and their properties | • requirement modelling (functional task specification, fuzzy task specification, goals and wishes for the optimization) |
| special extensions | | • integration of simulation method<br>• compilation of constraints<br>• truth maintenance system<br>• interface to data base systems<br>• interface to CAD systems |

In our terms a KONWERK-variant is a collection of modules (basic and extension modules) which is tailor-made for special requirements. This means there is not a single complex KONWERK-system which solves all configuration problems but there are a lot of KONWERK-variants that are as lean as possible and highly specialized for their type of problem.

A knowledge engineer who wants to implement an application with KONWERK is supported by a description for each module and each variant. These descriptions offer all necessary information about the functionality and map typical application characteristics on modules or variants. Thus the developer can use KONWERK in a way which leads to a solution with all necessary functionality and without any unnecessary complexity.

Due to the focal point of this paper we will now emphasize on fuzzy techniques and optimization methods in KONWERK. Both are integrated in the form of extension modules (Table 1).

### 4.2. *Modeling of the requirements, optimization in KONWERK*

There is an extension module of KONWERK which supports the task specification (Table 1). That means, this module helps to formulate the user requirements in both cases, in the crisp case and in the case of imprecise (e.g. linguistic) requirements. There are several possibilities to model such requirements, e.g.:

- *restrictions*
  required components (generic objects) or individuals, required parameter-values, local constraints (binary relations between components), global constraints (restricted resources of the goal system).
- *goals*
  goals are objectives or criteria for the goal system, which should be maximized or minimized by selecting or constructing an appropriate solution.
- *functional requirements*
  an user often has requirements formulated as functional abilities of the goal system modeled by crisp or linguistic expressions (e.g. the "car" should be safe and fast, the "computer" should have graphic ability).
- *requests*
  requests are optional requirements of the types considered above. That means, such requests have a lower priority and might be relaxed or removed during the configuration process.

All the requirements formulated above might be crisp or imprecise (see Sec. 2).

To be able to deal with the imprecise-case, we need extension modules such fuzzy-is-a-hierarchy, compositional hierarchy with fuzzy attributes, fuzzy constraints and goals as introduced in 2.3.

Using a closed-world assumption, all possible local or global constraints, goals, functional requirements, requests have to be formulated by the knowledge engineer and to be represented in the static knowledge base (requirement model). The modeling of the requirements for a determined case (task) by the user is mainly an interactive process. For this end, there are some requirement-strategies, predefined by the knowledge engineer, which determine e.g. a sequence, in which the requirement types are considered and the requirements will be selected or specified by the user. For example, the user will be asked first to define the functional requirements, then to select goals from a predefined list of goals, after that to select constraints and so on.

For a more detailed description of the KONWERK extension-module which supports the requirement-modeling see Refs. 21, 34.

It is important to remark, that some of the requirements might be modeled in different ways, as goals or constraints, precise (crisp) or imprecise (fuzzy). To illustrate this decision problem see Ref. 33. For example, the user has to choose his requirement model among a set of different alternatives. Therefore, he starts with an initial model. Then, this model is used for the configuration of the target system. If the result is not acceptable, then the requirement model will be modified. Using an explanation module, the process of the modification of the initial requirement model will be supported.

If the requirement model has been established, we have to specify some of the important components of an optimization model, like the constraints and the goals. The decision variables and several additional type constraints might be derived from the conceptual hierarchy.

In general, there are several possibilities to formulate an optimization model using the domain knowledge and the requirement model. The task to choose and to represent the optimization model is done in KONWERK by the extension module "Basic Optimization Module — BOM" of KONWERK.[6,7] The BOM acts as an interface between the knowledge-based configuration subsystem and a set of optimization methods and algorithms which are available using optimization–extension modules like "linear optimization," "non-linear optimization with constraints," "optimization module MADM" and some other more. Figure 16 shows a simple illustration of that idea.

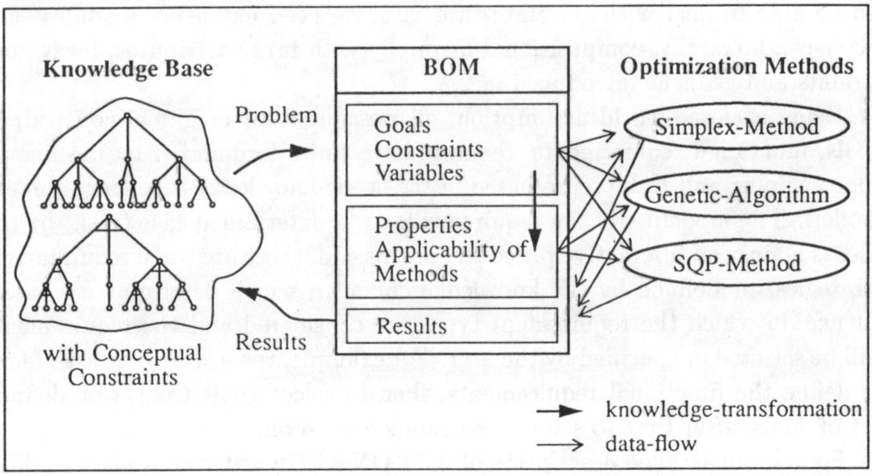


Fig. 16. The basic optimization module — BOM.

In Fig. 16 it is shown that the BOM transforms knowledge from the knowledge-base into such a representation of knowledge which is required for an optimization problem. In particular, the conceptual constraints which are connected with the conceptual hierarchy and "activated" by the requirement model will be transformed into such a representation of constraints which is usually used in the Operations Research.

After this transformation the BOM analyzes the properties of the optimization model and matches this properties with the requirements and the assumptions of the optimization methods which are available. Finally, an agenda of the most appropriate — for the problem — optimization methods is generated and the respective optimization extension module is called to run the chosen optimization algorithm.

The results are given back to the BOM and the BOM has some strategies predefined to feedback these results to the dynamic or static knowledge base. In particular, if an optimization method is used within the configuration process to perform a configuration step, the result will be used for the "partial construction" in the dynamic knowledge base.

Depending on the main-properties of the optimization-models (linearity, non-linearity, finite solution space, unknown properties), currently KONWERK consists of the optimization–extension modules in Fig. 17.

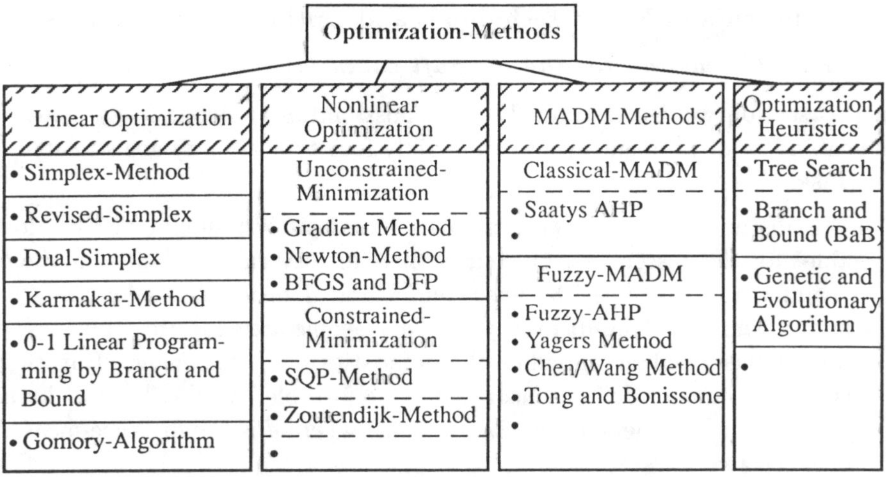


Fig. 17. Optimization–methods.

Each of the blocks in Fig. 17 represents one extension module. The implementation of these modules is still under development. For instance the MADM-Module is almost completely implemented. The conception of each of the modules is such that an additional method can easily be added to the module (open architecture).

Finally, we describe briefly the three main possibilities we have in KONWERK to connect knowledge-based configuration with optimization (multicriteria decision making is included). These different ways to combine configuration with optimization are briefly introduced in Table 2 and discussed in Refs. 6, 14.

Table 2. Configuration by optimization.

| Name | Description of the approach | Available optimization methods |
|---|---|---|
| COOPT | The configuration problem is modelled as an optimization problem | all methods (Fig. 17) |
| COBaB | The control strategy of the configuration method is an optimization method | BaB |
| SPECDEC | The control strategy of the configuration uses optimization as one particular SPEC or DECOMP method | all methods (Fig. 17) |

## 5. Application of Fuzzy Techniques and Optimization Methods

### 5.1. *Fuzzy modeling for the aircraft cabin layout problem*

In this section we describe briefly some possibilities to define fuzzy goals for the cabin layout problem. There are mainly four typical fuzzy goals related to this task:

(1) The distance between two seat-rows ("seat-pitch") should be "comfortable" for the passengers. Of course, there are several possibilities to model "comfortable distance," because comfortable depends on the length of the flights (short- or long-distance), on what people are expecting (business or tourist class) and on individual characteristics of the passengers. But, there is a possibility to define a fuzzy set "Comfortable-Distance" = CD by introducing a membership-function $\mu_{\mathrm{CD}}(d)$ where $d$ is the distance-measure between two rows

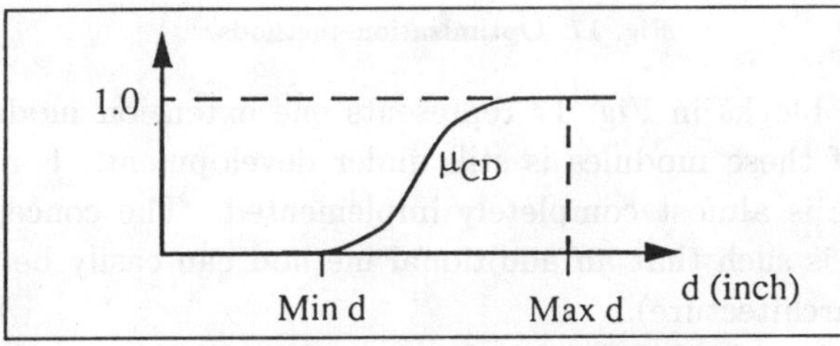


Fig. 18. Fuzzy set "comfortable-distance".

Using this measure for the distance of two neighboring rows, we are able to introduce an overall measure for one class of the airplane taking into account that row-distances could be different.

(2) The possible number of seats in a class depends of course on the "distance of rows" and of the other layout decisions (lavatories, ... ). It could be also considered a fuzzy goal as follows:

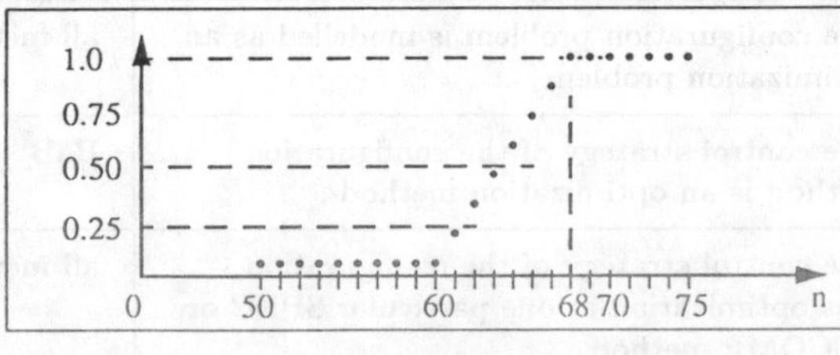


Fig. 19. Fuzzy set "large-seat-number" related to a business class.

There is a fuzzy set "Large-Seat-Number" = LSN defined over the set of integers, context dependent (on the airplane-type, class etc.) defined by a membership-function $\mu_{\text{LSN}}(n)$.

(3) Another fuzzy goal is defined by the delivery-time for determined cabin interior components which might be differ from e.g. 9 months to two years.

We introduce the fuzzy set "Fast-Delivery-Time" = FDT for a determined cabin interior component, e.g. for a galley by a membership-function $\mu_{\text{FDT}}(t)$:

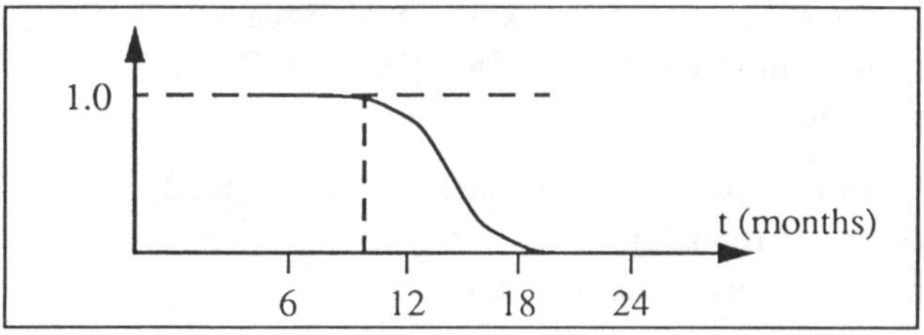


Fig. 20. Fuzzy set "fast-delivery-time".

Each cabin interior component has a so-called "lead-time" that describes the time needed from order to delivery depending on the supplier of that component.

That means the "Fast-Delivery-Time" fuzzy sets must be carefully defined for each interior component.

(4) The price for a cabin layout is defined by the sum of the prices of all its cabin interior components. The price for a single component differs depending on its place in the cabin. But, instead of a crisp goal for the whole layout, we can ask for a "low-price-layout," using a respective fuzzy set, to define what "low-price" means.

Using these four Fuzzy-Goals, we have several possibilities to model "optimality" of a cabin layout, see Ref. 41. We need an aggregation operator to define an overall criterion (top-goal). Of course, there are several possibilities. We use compensatory operators to compensate "Comfortable-Distance" (passengers appreciate this goal) with "Large-Seat-Number" (airlines goal) and "Low-Price-Layout" with "Fast-Delivery."

Therefore, for a specified cabin layout we are able to compute a degree of overall-satisfiability, and, if we are able to generate several layouts, we can find a ranking where the top layout is the optimal one according to the finite set of layout alternatives.

In 6.1. we will briefly show how we can generate such a finite set of alternatives using KONWERK's "Branch and Bound" module and how we rank the alternatives using the Fuzzy-MADM-module of KONWERK.

## 5.2. *Preliminary design of future space launch systems*

For this example we show as an example of integration of imprecision into the model a goal hierarchy which is a basis of a Fuzzy-AHP.[3] To define the hierarchy we consider[19]:

- *top goal*: Optimal Space Transportation System (STS)
- *first-level goals*: (subgoals)
  - Transportation Performance (TP)
  - Economy (E)
  - Environmental Compatibility (EC)
  - Availability (A)
  - Safety (S)
  - Access to Space (AS)
  - etc.

Exemplarily, environmental compatibility has been chosen for a more detailed consideration.

- *second-level goals* (*criteria*) e.g. Environmental Compatibility:
  - Impact of exhaust gases
  - noise duration
  - noise intensity
  - space debris
  - impact of reentering or exploding or
  - crashing parts (parts)
  - etc.

Now, the goals, specified by names only, will be characterized as fuzzy goals. For example the "noise intensity" is modeled as a linguistic variable and the respective fuzzy goal (e.g. noise intensity = low $\vee$ medium) can be derived from this linguistic variable.

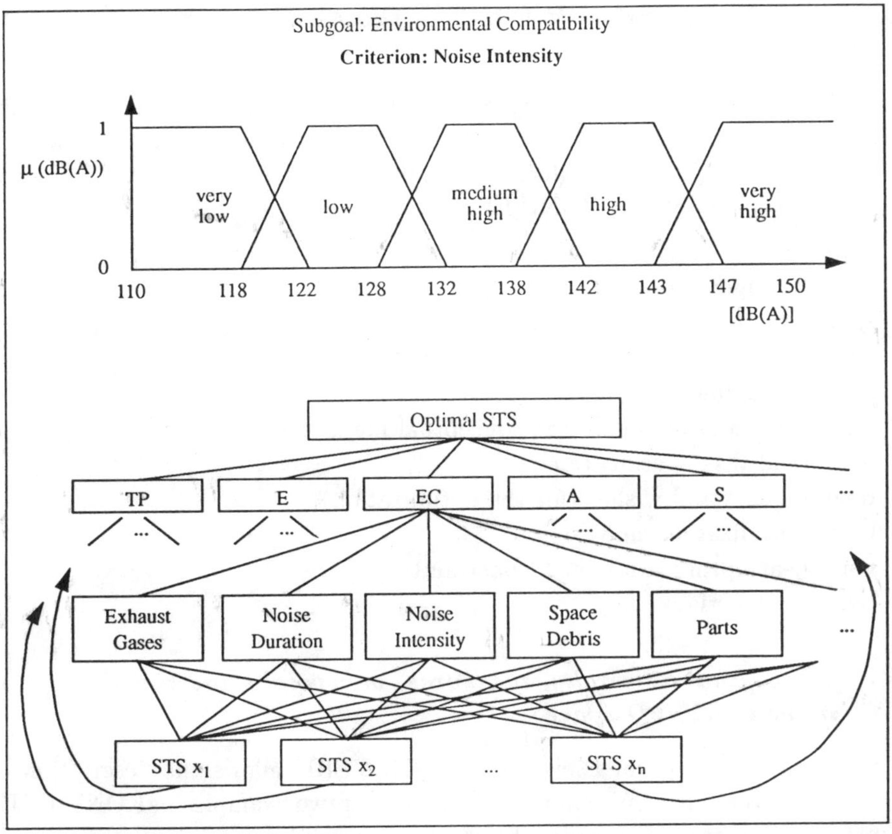


Fig. 21. Linguistic variable "noise intensity" and part of the goal hierarchy.

Figure 21 shows the modeling of the criterion "noise intensity" as a linguistic variable and part of the hierarchy of goals.

The Fuzzy-MCDM to evaluate and to select "optimal" space transportation systems for a defined mission or scenario is still under development.

The knowledge base is implemented using KONWERK and an algorithm which generates a set of alternatives of SLS which are capable of realizing the mission is implemented.

But there are problems related to the aggregation operators and to the selection of an approach "Fuzzy-AHP" which are still unsolved.

Therefore, we will go into the details of the implementation for our first example "Cabin Layout" only.

# 6. Application of "KONWERK" to the Aircraft Cabin Layout Problem

This section gives a short survey of the XKL (see 3.1.) implementation with KONWERK. Small examples for the representation of different types of knowledge demonstrate the capabilities of the modules of KONWERK. XKL needs the following modules of KONWERK:

- support-and basic modules
- spatial properties
- spatial relations
- measures
- formula module
- viewpoints and mixins in the conceptual hierarchy
- functional task specification
- requirements and wishes (for the optimization)
- basic optimization module
- non-linear optimization with constraints
- repair instructions

## 6.1. *Conceptual hierarchy, measures, viewpoints and multiple inheritance, 2D-space*

The module Conceptual Hierarchy of KONWERK allows the description of objects in a very efficient manner. The following two examples in KONWERK-syntax demonstrate the possibilities of the modules *Conceptual Hierarchy*, *Measures*, *Viewpoints & Multiple Inheritance and 2D-Space* for the representation of objects.

```
(def-do :name cabin-interior-component
        :superconcept (domain-object rectangle)
        :parameter (ref-nr [integer 2500000 2599999])
                   (x-range [float 404 inch 2408 inch])
                   (y-range [float -105 inch 105 inch])
                   (length [float 1inch 200 inch])
                   (width [float 1inch 200 inch])))
```

The representation of the object *cabin-interior-component* demonstrates the is-a hierarchy of the basic module *Conceptual Hierarchy.* By the use of the extension module *Views & Multiple Inheritance* it is possible to inherit properties not only from the *object domain-object* but also from *rectangle* which is defined in the extension module *2D-Space.* The extension module *Measures* allows the usage of the measure inch in the slot length.

```
(def-do :name galley
        :superconcept cabin-interior-component
        :relations ([galley-in (a passenger-class)])
        :parameter ((ref-nr [integer 2531000 2533999])
                    (door {1 2 4})
                    (trolleys {0 2 3 4 5 6 7 8 9 10})
                    (half-size-trolleys {0 1 2 3 4 5})
                    (meals [integer 28 140]
                    (type {longitudinal transversal})
                    [height {full half}
                            (default 'full)]))
```

The object *galley* inherits all properties of *cabin-interior-component.* In order to specialize the *cabin-interior-component* to a *galley* some of its slots are restricted (e.g. *ref-nr*) and some new slots are added.

### 6.2. *Constraints, 2D-space, formulas*

Restrictions between objects are expressed in a special constraint language. The following example demonstrates the use of the basic module Constraints and the extension module Formula.

```
[def-conceptual-constraint: name max-number-of-seats-between-door-2-and-3
 :variable-pattern-pairs ((?s :all
                             :concept bench
                             :parameter [(xpos [963.41 1562])]))
 :formulas ["sum (?s.capacity) < 155"]].
```

This constraint describes a certification rule that limits the number of seats between the doors 2 and 3 to 155. This certification rule is obeyed by identifying all benches between the doors (defined by their $x$-position) and adding their capacities (a bench may contain 1, 2, or 3 seats). Thus it is impossible to place more than 155 seats between the two doors.

The following conceptual constraint ensures the customers demand for a certain seat pitch (see Fig. 4). It does not matter whether a seat has to be placed in the first class with a seat pitch of 40 inches or in the tourist class with a seat pitch of 30 inches. This single constraint works for all seats in all classes. It uses a constraint relation that is defined in the extension module *2D-Space*.

```
(def-conceptual-constraint :name seat-pitch
 :variable-pattern-pairs ((?s1: concept seat)
                          (?s2 :concept seat
                          :parameter [(tg-left-neighbor ?s1)]})
          :constraint-calls { [to-difference-frontside ?s1 ?s2 (?s2 pitch)]))
```

### 6.3. *Control*

The following strategy for the placement of cabin attendant seats gives an impression how the basic module Control is used in XKL.

```
(def-strategy: name place-cas
 : strategy-class depth-first-strategy
 : priority 15
 : configuration-step-focus focus-cas
 : agenda-criteria (prefer-specialize
                    prefer-xpos
                    prefer-parametrization)
 : proceeding (compute take-default ask-user))
```

The strategy uses depth-first search; has a priority that determines whether it will be chosen when several strategies can be applied. It uses a focus to filter the agenda. The filtered configuration steps are processed in an order that determines that as a first step the cabin attendant seat should be specialized. The system should determine the $x$-position of the cabin attendant seat as a second step before all the other parameters are determined. The system proceeds by first trying to compute a value then looking for a default value and as a last resort asking the user for a value.

One might see, that it is possible and useful to fuzzify strategies like the one considered above in order to generate several different solutions.

In this section we gave a short description of the conventional 'configuration' approach. We described the knowledge representation and gave an impression of the proceeding. A configuration is build step by step. The system ensures consistency for each step but does not consider any optimization criteria. But, the knowledge-base with the conceptual constraints briefly touched above is also needed for optimization. Now we would like to introduce methods that enable us to 'optimize' our solution.

## 6.4. *An example for the use of the optimization module in XKL*

Let us consider a 2-class-layout with a business and a tourist class. In our example (see Table 3) the customer wants exactly 68 seats in the business class and as many seats as possible in the tourist class.

Table 3. Requirements for a cabin layout.

| Customer's requirements for | Business class | Tourist class |
|---|---|---|
| Passengers (seats) | 68 | max |
| Seat pitch | 35 | 30 |
| Seats abreast | 6 | 9 |
| Passengers per lavatory | 35 | 45 |
| Passengers per cabin attendant | 25 | 50 |
| Meals per passenger | 2 | 1 |

Thus we have two optimization tasks:

- minimization of the needed space for the business class
- maximization of the seats in the tourist class

The problem of optimizing the whole cabin is quite hard for our system. Nearly 300 objects have to be considered. The complexity of this problem is reduced by decomposing the cabin into its classes. The business class in our example consists of only about 70 objects. Due to the large number of possibilities in the placement of an item (a seat can be placed on nearly 6000 positions) this is a considerable reduction. The trade-off of this decomposition approach is that the result consists of optimal and consistent classes which are composed into a consistent cabin which may not be optimal.

In the following, we consider the layout of the business class as an example only.

Our approach is the following:

*Phase 1.* We use the Branch and Bound-module of KONWERK to generate so-called "good solutions." This module was elaborated by S. Järvinen.[15]

*Phase 2.* The finite set of good solutions, that means the finite set of layouts for the business-class will be considered as set of alternatives for the Fuzzy-MADM module of KONWERK.[21] As a result, we get a ranking of the "good-layouts" generated by the Branch and Bound algorithm.

It is impossible to go into the details of the BaB-approach, but, in the following we will give a brief outline of the method.

To define, what "good solution" means, we define three crisp goals:

- minimization of the needed space $NS_{bc}$ for the business class
- minimization of the price $PIC_{bc}$ for the cabin interior components of the business class
- minimization of the delivery time $DT_{bc}$ for the components in the business class.

We propose that if the layout is determined, we would be able to measure each of the goals above by a real number denoted again by $NS_{bc}$, $PIC_{bc}$, $DT_{bc}$. Now, we introduce utility-levels as upper bounds for each of these criteria to define what "good" means. A good layout fulfils each of these constraints:

$$\left\{ \begin{array}{rl} & NS_{bc} \leq \overline{NS}_{bc} \\ \text{and} & PIC_{bc} \leq \overline{PIC}_{bc} \\ \text{and} & DT_{bc} \leq \overline{DT}_{bc} \end{array} \right\} \qquad (*)$$

where $\overline{NS}_{bc}$, $\overline{PIC}_{bc}$, $\overline{DT}_{bc}$ are the upper bounds.

The BaB algorithm realizes now a tree search strategy combined with the bounds. It is assumed (see Fig. 22) that the configuration process is a sequential process starting either with selecting one of the possible lavatories and positioning the selected one or with the respective step for galleys. It is also assumed that we are able to estimate the criteria $NS_{bc}$, $PIC_{bc}$, $DT_{bc}$ for each of such an incomplete layout what means to give bounds for the criteria. This gives us, because of the given upper bounds $\overline{NS}_{bc}$, $\overline{PIC}_{bc}$, $\overline{DT}_{bc}$, the possibility to terminate an incomplete construction if it is identified to be not a good one.

Finally, such kind of "best-first-search" combined with bounding terminates, if a given number of good solutions is reached or with a smaller set of good solutions (which might be empty). If the set is empty one needs to decrease the bounds or to relax the constraints and to consider them as fuzzy constraints.

We want to remark that each of the good-solutions generated by the BaB algorithm is a feasible solution because KONWERK's knowledge base with the conceptual constraints ensures this feasibility by Constraint-Propagation at each of the stages of the algorithm illustrated in Fig. 22.

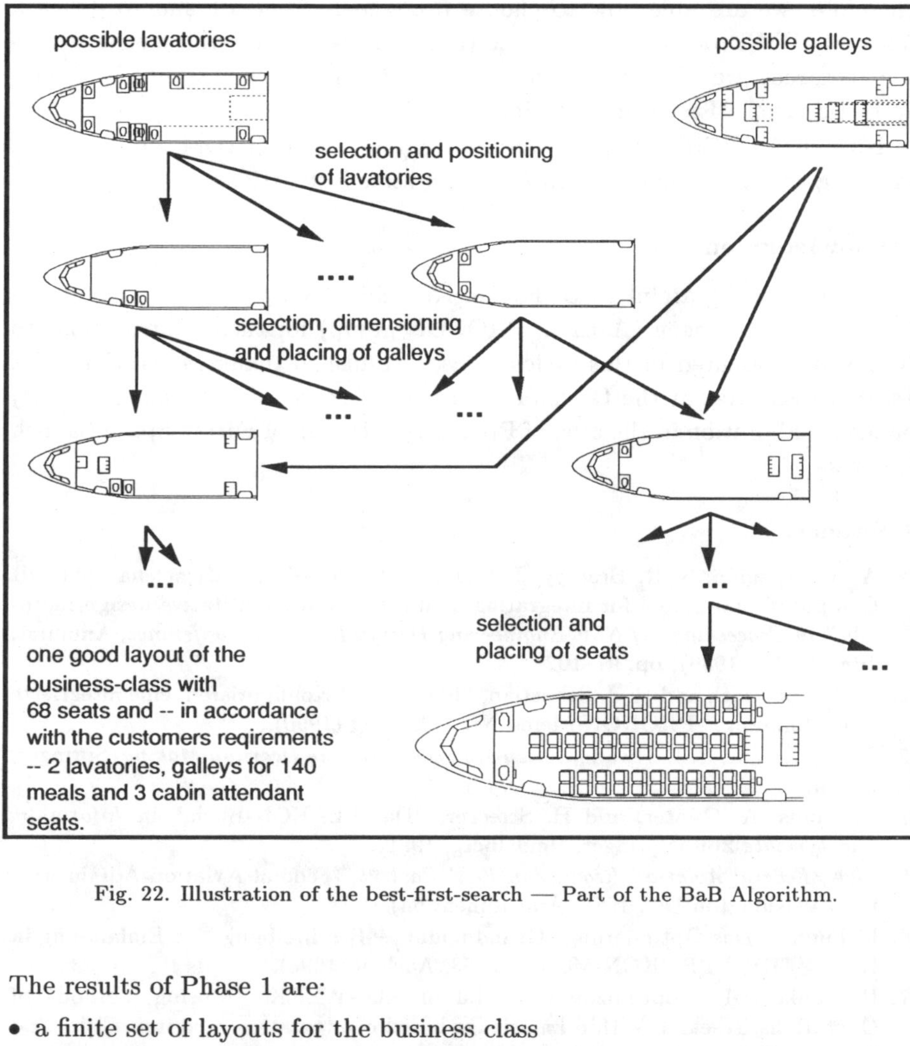


Fig. 22. Illustration of the best-first-search — Part of the BaB Algorithm.

The results of Phase 1 are:

- a finite set of layouts for the business class
- each layout is characterized by three crisp numbers $NS_{bc}$, $PIC_{bc}$, $DT_b$.

Also, for each layout alternative we can easily compute:

- the distances of seat-rows
- the number of seats

Therefore, we are able now to choose fuzzy-goals from 5.1 and to define a Fuzzy-MADM-Problem. In our case (because 68-seats for the business class was required) we choose the fuzzy goals "Comfortable-Distance," "Small-Needed-Space," "Fast-Delivery-Time," "Low-Price."

Now, KONWERK proposes an appropriate Fuzzy MADM and performs screening and/or ranking according to the users instructions.

## Acknowledgment

The authors are grateful to B. Funke, K. Müller, S. Järvinen, T. Goudarzi-Pour and M. Thärigen (Aachen, PROKON group) for their contributions to the results presented in this article. Also, we like to thank M. Wolf and his research associates at the German Aerospace Research Establishment (DLR) for the collaboration in the area of Preliminary Design of Future Space Launch Systems.

## References

1. A. M. Agogino, S. R. Bradley, J. Cagan, P. Jain, and N. Michelena, "AI/OR Computational model for integrating qualitative and quantitative design methods," in *Proceedings of NSF Engineering Design Research Conference*, Amhurst, June 11–14 (1989), pp. 97–102.
2. H.-J. Appelrath and H.-J. Sebastian, "Planen und Konfigurieren: eine integrierte Modellbildung," PROKON-Memo Nr. 2, Leipzig (1990).
3. S.-J. Chen and C.-L. Hwang, *Fuzzy Multiple Attribute Decision Making* (Springer, Berlin, 1992).
4. R. Cunis, A. Günter, and H. Strecker, "Das PLAKON-Buch," in *Informatik Fachberichte* **266** (Springer, Heidelberg, 1991).
5. FAA, *Federal Aviation Regulation FAR Part 25*, (Federal Aviation Administration, Washington DC, 1988, Amendment 64).
6. B. Funke, "Das Optimierungs-Grundmodul — Beschreibung und Einbindung in KONWERK," PROKON-Memo Nr. 53, Aachen (1994).
7. B. Funke, "The optimization modul of KONWERK — using methods of Operations Research within knowledge-based configuration of systems," in 2'nd EUFIT, Aachen, Sept. 20–23 (1994).
8. J. S. Gero, M. L. Maher, and F. Zhao, "A model for knowledge-based creative design," in *Proceedings of AAAI-88 Workshop on AI in Design*, St. Paul (1988).
9. A. Günter, H. Dörner, H. Gläser, B. Neumann, C. Posthoff, and H.-J. Sebastian, "Das Projekt PROKON: Problemspezifische Werkzeuge für die wissensbasierte Konfigurierung," PROKON-Bericht Nr. 1, Hamburg (1991).
10. A. Günter, "Modelle beim Konfigurieren," in O. Herzog, T. Christaller, and D. Schütt (Hrsg.), 17. Fachtagung für Künstliche Intelligenz "Grundlagen und

Anwendungen der Künstlichen Intelligenz" (Springer Verlag, Berlin, 1993), pp. 169–176.

11. A. Günter, "KONWERK — ein modulares Konfigurierungswerkzeug," in F. Maurer and M. M. Richter (Hrsg.), *Expertensysteme* 95, Kaiserslautern (1995).
12. A. Günter and H.-J. Sebastian, "The KONWERK-Workbench: A system of software tools to approach intelligent design support," in *Proceedings of the Second EUFIT*, Aachen (1994).
13. P. Harmon, "The market for expert systems-buildung tools," in *Intelligent Software Strategies*, **9** (2) (1993), pp. 3–9.
14. S. Järvinen, "Optimierung bei der Konfigurierung," PROKON-Memo Nr. 36, Aachen (1993).
15. S. Järvinen, "Optimierungsbasierte Konfigurierung in intelligenten Decision Support Systemen," Lizentiatarbeit, Aachen und Helsinki (1994).
16. M. Kopisch and A. Günter, "Configuration of a passenger aircraft cabin based on conceptual hierarchy, constraints and flexible control," in F. Belli and F. J. Radermacher (Hrsg.), *5th International Conference Industrial and Engineering Applications of Artificial Intelligence and Expert Systems, IEA/AIE-92* (Springer, Paderborn, 1992), pp. 421–430.
17. M. Kopisch, "Spatial relations for configuration tasks in general and for the cabin layout of the AIRBUS A340 in particular," in P. W. H. Chung, G. Lovegrove, and M. Ali (Hrsg.), *6th International Conference Industrial and Engineering Applications of Artificial Intelligence and Expert Systems*, IEA/AIE-931 (Gordon and Breach Science Publishers, Edinburgh, 1993), pp. 450–457.
18. K. Lötzerich and T. Goudarzi-Pour, "Knowledge-Based Configuration of future Space Launch Systems," in *EUFIT 94, Proceedings* (1994).
19. K. Lötzerich, H.-J. Sebastian, and T. Goudarzi-Pour, "Multi-Criteria Decision Making (MCDM) for rational evaluation and selection of future space launch Systems" (research report, 1994), preprint available.
20. S. Mittal and F. Frayman, "Towards a generic model of configuration tasks," in *Proceedings IJCAI-89* (1989), pp. 1395–1401.
21. K. Müller and M. Thärigen, "Requirement modelling imprecise knowledge and Multiple Criteria Decision Making in KONWERK — a way towards an intelligent Decision Support System," in *2 European Congress on Fuzzy and Intelligent Technologies*, Aachen, Sept. 20–23 (1994).
22. D. Navinchandra, *Exploration and Innovation in Design* (Springer Verlag, 1991).
23. K. N. Otto, "A formal representational theory for engineering design," Ph. D. Thesis (California Institute of Technology, Pasadena 1992).
24. K. Otto and V. E. Antonsson, "Modelling imprecision in product design," in *Proceeding of the 3'd Fuzzy-IEEE*, Orlando, (1994).
25. S. S. Rao *et al.*, "Multi-objective fuzzy optimization techniques for engineering design," in *Computers and Structures*, **42** (1) (1992), pp. 37–44.
26. S. S. Rao and A. K. Dhingra, "Applications of fuzzy theories to multi-objective system optimization," NASA Contractor Report 177573 (NASA Ames Research

Center, Moffett Field, CA 94035-1000), University Consortium Interchange Number CA2-223 (1991).

27. H.-J. Sebastian, "Optimierung und Unschärfe bei Konfigurierungsproblemen," PROKON-Memo Nr. 12, Aachen (1992).
28. H.-J. Sebastian, "Optimierungsbasierte Kontrollstrategien beim Konfigurieren," PROKON-Memo Nr. 49, Aachen (1993).
29. H.-J. Sebastian, "Intelligente Entscheidungsunterstützung bei der Systemkonfigurierung durch Integration von Expertensystemtechnologie mit Modellen und Methoden des Operations Research," PROKON-Memo Nr. 47, Aachen (1994).
30. H.-J. Sebastian, "Optimization-based control-strategies in expert systems for design and configuration," in *Proceedings of IFIP 7th Working Conference on Optimization-based Computer-aided Modeling and Design*, Prague (Lansa-BV, 1994).
31. H.-J. Sebastian and W. Dilger, "Knowledge-based configuration of technical systems including optimization-based reasoning," in *Proceedings of the 1st World Congress on Expert Systems* (Pergamon Press, 1991), pp. 1967–1974.
32. H.-J. Sebastian, "Intelligente Entscheidungsunterstützung bei der Systemkonfigurierung," in *Operations Research — Festschrift zum 60*, Geburtstag von H.-J. Zimmermann (Springer Verlag, 1994).
33. H.-J. Sebastian, "Modellierungsunterstützung — Ein Ansatz zur interaktiven Bestimmung eines problembezogenen Konfigurierungsmodells," PROKON-Memo Nr. 31, Aachen (1993).
34. M. Thärigen, "Realisierung der Anforderungsmodellierung in KONWERK," PROKON-Memo Nr. 50, Aachen (1994).
35. M. Thärigen, "Modellierung von Konfigurierungsaufgaben," PROKON-Memo Nr. 25, Aachen (1993).
36. C. Tong und D. Sriram (eds.), *Design Representation and Models of Routine Design* (Academic Press, 1992)
37. H. J. Zimmermann and H. J. Sebastian, "Optimization and fuzziness in problems of design and configuration," in *Proceedings of the IEEE-Conference on Fuzzy-Systems*, San Francisco (1993).
38. H.-J. Zimmermann, *Fuzzy Set Theory — and its Applications* (Kluwer, Nijhoff Pub. Boston, Dordrecht, Lancaster, 1984).
39. H.-J. Zimmermann, *Fuzzy Sets, Decision Making and Expert Systems* (Kluwer Acad. Pub. Boston, Dordrecht, Lancaster, 1989).
40. H.-J. Zimmermann and H.-J. Sebastian, "Fuzzy design — integration of fuzzy theory with knowledge-based system design," in *Proceeding of the World Congress on Computational Intelligence, Fuzzy-IEEE 94*, Orlando, Florida (1994).
41. H.-J. Zimmermann and H.-J. Sebastian, "Intelligent system design support by fuzzy-Multi-Criteria Decision Making and/or Evolutionary Algorithms," in *Proceedings Fuzzy-IEEE 95*, Yokohama (1995), to appear.

# FUZZY APPLICATIONS FOR AUTOMOBILES

HIROSHI TAKAHASHI
*Nissan Motor Co., Ltd.*
*1, Natsushima-cho,*
*Yokosuka, 237, Japan*

## 1. Introduction

When considering an optimal driver/vehicle relationship, attention is focused on their interaction. The process by which a driver operates a vehicle, decides on operations, and sets the vehicle in action can be regarded as a closed-loop control system. The automobile is a very familiar object for such closed-loop studies. In principle, anyone can operate an automobile. Understanding the driver/vehicle interaction often involves important matters. Many studies[1,2] have employed the automobile to examine human/machine interactions or human/machine interfaces. For example, a heads-up display (HUD) installed in a vehicle superimposes the oncoming landscape on the windshield and indicates such things as vehicle speed and some alarms.[3] Information about driving conditions is easily transmitted to the driver, who does not have to move his eyes a great deal. This is a good example of improving the driver/vehicle interface on the vehicle side of the system.

Transmission of information from the driver side to the vehicle side is also feasible. Takahashi *et al.*[4] found heuristically that there are intentional differences in the moving average of the amount of accelerator opening and the variance of accelerator opening when a driver travel on a winding uphill road and a straight flat highway. They described these differences qualitatively as fuzzy reasoning rules. They also processed the measured data statistically to generate membership functions, recognized the above two driving environments, and proposed a system to control the automatic transmission so as to improve drivability in each environment. In this section, some applications to improve the driver/vehicle interface from the driver side of the system are introduced.

Table 1. History of fuzzy applications.

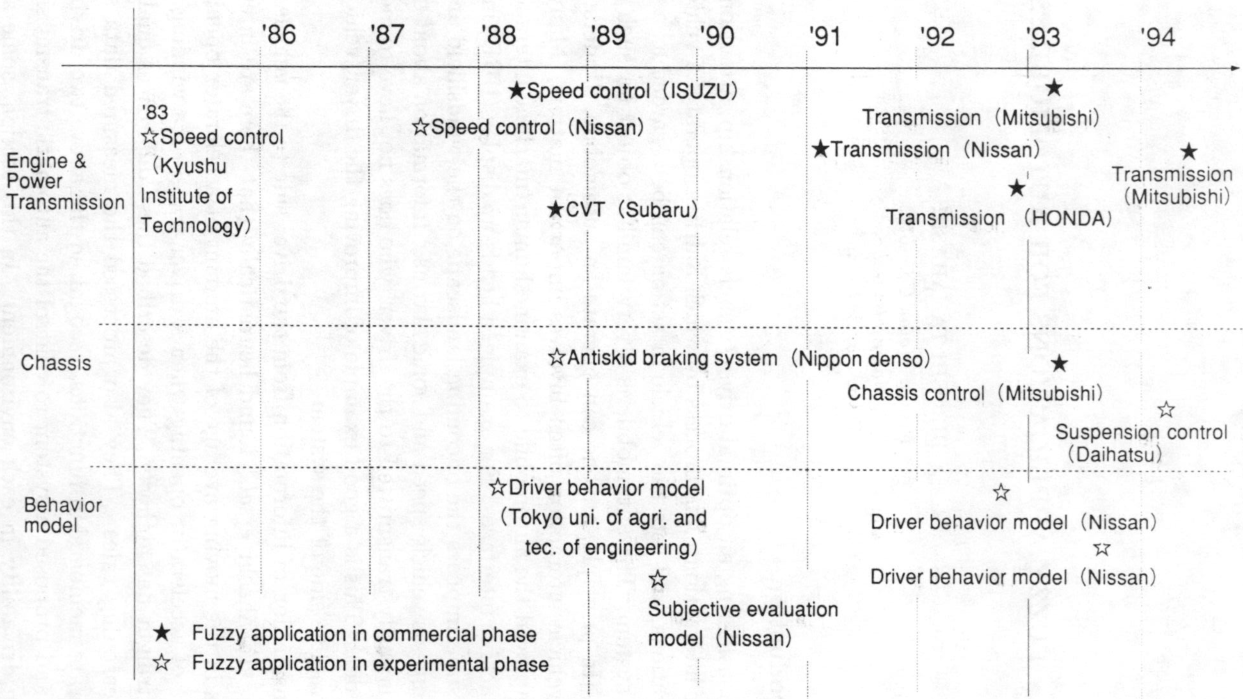

### 1.1. *Technical survey of fuzzy applications in automobiles*

Many applications of fuzzy system have been announced since 1983. The history of developing fuzzy applications are shown in Table 1.

In 1983, Maeda and Murakami developed the fuzzy speed controller using indirect fuzzy reasoning method.[5] This auto-cruise system was the feasibility study for applying fuzzy theory to real automobiles. After that, Takahashi *et al.*,[6] Uriubara *et al.*[7] applied fuzzy logic to speed control system. These systems provided drivers with good drivability. On the other hand, applications for CVT (Continuous Variable Transmission)[8] and ABS (Anti-skid Braking System)[9] were presented.

The above-mentioned applications were the first stage of fuzzy applications. The second stage was in automatic transmission technologies.

Sakai *et al.*[10] developed the shift scheduling method of automatic transmission using fuzzy logic. Their paper described how the layers of knowledge base have been extracted and constructed by applying fuzzy control to the shift scheduling of automatic transmission and presented the field test results. Their system have been applied to mass-produced automobiles.

Magoshi *et al.*[11] developed the total fuzzy control system including automatic transmission, ABS, suspension and traction control systems. Using fuzzy reasoning, the controller predicts a certain value that is related to the friction factor of driving road. Kondo *et al.*[12] have developed the adaptive shift scheduling system using fuzzy logic and neural networks. The timing of shifting into lower gear is adaptive for each driver's preference by fuzzy and neural network adaptation.

In these days, fuzzy logic is supposed to be a very popular technology in the automobile technical fields. I intend to introduce my applications in Secs. 2 and 3.

## 2. Application I — Reasoning Driving Environments[13]

It is hypothesized that if a vehicle could sense the driving environment and the driver's intentions, it could vary the control parameters and control strategies for the engine, transmission, brakes, and other vehicle systems to match the situation, and maintain good drivability at all times. The driving environment here refers to various types of roads and traffic conditions such as expressways, mountainous roads and traffic congestion. The driver's intentions mean the driving operations that the driver wants to execute, including acceleration and gear shifts. This particular case presents such a sensing system. Under the concept described here, the driver is regarded as a sensor for the vehicle. The operations performed by the driver are analyzed in order to discern the driving environment and the driver's intentions. The present research is intended to provide fundamental information for the creation of "intelligent" vehicles in the future.

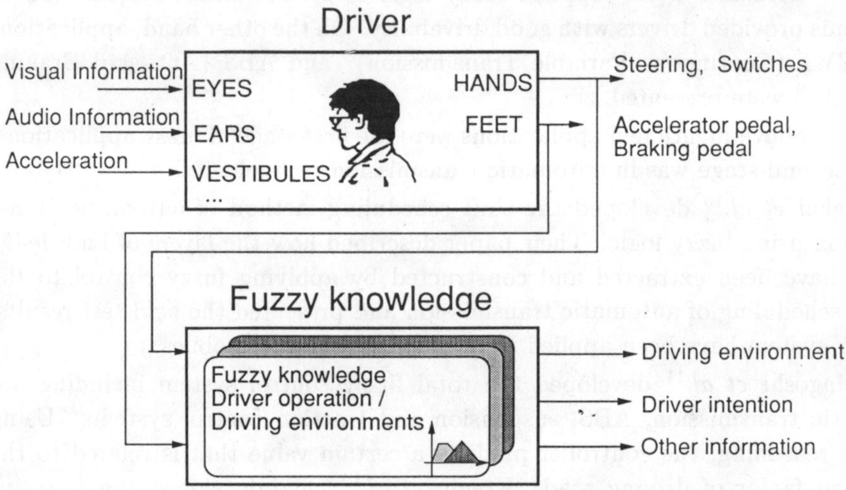


Fig. 1. Fuzzy knowledge-based human sensor.

## 2.1. *Fuzzy knowledge-based human sensor*

As illustrated in Fig. 1, while driving, the driver can be seen as functioning as a sensor that discerns the driving environment for the vehicle. The driver receives visual, auditory and acceleration information through his eyes, ears and vestibule canals. Based on these information, the driver recognizes the driving situation and judges what steering maneuvers or pedal operations should be performed. As seen from the vehicle's perspective, the driver's role is that of a "human sensor." By analyzing the "outputs" of a human sensor, it should be possible to predict the information detected by the sensor. This suggests that the driving environment and the driver's intentions might be predicted by analyzing the operations executed by the driver, such as pedal inputs and steering maneuvers. This human sensor concept is thought to be a useful approach in automotive engineering because it can enable a vehicle to obtain information on the driving environment without requiring the addition of any extra hardware such as video camera or radar. The use of this concept should make it possible to improve vehicle controllability at low cost. This section presents an example of the application of the human sensor concept to automatic transmission control.

### 2.2. *Generation of fuzzy knowledge for reasoning*

As the first step, the relationship between the driving environment and the operations performed by the driver must be identified. Conventional approaches have tried to identify the characteristics of driving operations on the basis of mathematical analysis. However, it is difficult to treat data on driving operations mathematically because the data are not ergodic. The characteristics of driving operation data will also change as a person becomes more skilled at driving. Therefore, fuzzy logic is adopted here to treat vague data on driving operations in order to clarify the relationship between the driving environment and the operations executed by the driver. The routine used to obtain this relationship consists of the following five steps:

[Step 1] *Categorization of data*

Data on driving operations are not expressed in terms of crisp variables but fuzzy variables. In order to treat driving operation data as fuzzy variables, the data distribution is found. For example, data are obtained on accelerator inputs (i.e., the pedal angle) for a few minutes and stored in memory. Subsequently, a histogram of the data is generated like that shown in Fig. 2. Based on this distribution, the data can be divided into several categories, which we will call "large," "medium," and "small." These categories represent the fuzzy variables. Assuming that these categories are sets, the data histogram can be divided into three categories. Every category has the same probability of being assigned the data sets. Histograms, like the one shown in Fig. 2, are obtained for driving tests conducted on various types of roads.

The boundary value of each category is calculated with Eq. (1) such that each one has the same probability of being assigned driving operation data. All the outputs of the human sensor are then assigned to one of the categories, which are labeled with the fuzzy variables as "large," "medium" or "small." If there are five categories, then the value of $t_i$ is determined such that it will satisfy Eq. (1).

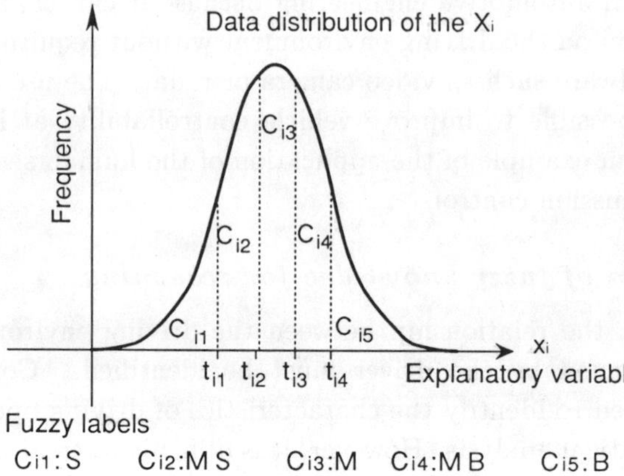


Fig. 2. Histogram of the data.

$$\int_{-\infty}^{t_{i1}} f_i(x_i)dx_i = \int_{t_{i1}}^{t_{i2}} f_i(x_i)dx_i = \int_{t_{i2}}^{t_{i3}} f_i(x_i)dx_i$$
$$= \int_{t_{i3}}^{t_{i4}} f_i(x_i)dx_i = \int_{t_{i4}}^{\infty} f_i(x_i)dx_i \quad (1)$$

where $x_i$ is the value of explanatory variable $i$ representing the output of a human sensor, $f_i(x_i)$ is the probability density function of the data set $x_i$, and $t_{ij}$ is the boundary value between category $j$ and category $j+1$. All of the driving operation data obtained for the different driving environments, such as driving on a mountainous road or a flat road, are assigned to a suitable category.

[Step 2] *Generation of fuzzy data*

The $x_i$ data obtained at a sampling interval of 0.1 to around 1 sec are assigned to a suitable fuzzy category on the basis of the boundary values. The total number of operation data values is denoted as $q$. For example, $q$ might be 100 or 1,000 values.

Calculations are then made using Eq. (2).

$$F_{ik} = [W(x_k)]/q \quad (2)$$

where $F_{ik}$ expresses the ratio of the number of values assigned to a category in relation to the total data population, $q$. $F_{ik}$ is called the category score.

$W(x_i)$ is the number of value assigned to category $k$. The patterns of $F_{ik}$ show the tendencies of a person's driving characteristics.

[Step 3] *Definition of criterion variables*

Attention was paid in this research to the relationship between the driving operation data, i.e., the explanatory variables; and the driving environment, i.e., the criterion variables. As mentioned in the previous step, the data on driving operations are assigned to categories, which means that they are expressed in the form of category scores. The next step in this study therefore was to quantify the criterion variables representing different driving environments. A camera was used to photograph the test roads while data were being obtained on the driving operations executed by each subject. Following the driving tests, subjects were asked to examine the photographs and give estimations of whether the photographs showed a mountainous road (driving environment $A$) or a flat city road (driving environment $B$). The subjects scored their estimations by assigning numerical values between 0 and 1. A value of 0 meant that they were absolutely sure a photograph did not show driving environment $A$, for example; a value of 1 meant they were absolutely sure it did. Therefore, if a subject positively believed that a photograph showed driving environment $A$, the person would assign a score of, say, 1 or 0.9 as the value of the criterion variable. On the other, the person would assign a small value of, say, 0 or 0.1 to the photograph of driving environment $B$. Many pairs of photographs of various driving environments were evaluated in this manner.

[Step 4] *Fuzzy mathematical quantification*

Table 2 is an example of the mathematical values obtained using the foregoing procedures. In this case, we want to identify the relationship between the explanatory variables, representing the fuzzy variables, and the criterion variables that were estimated in [Step 2]. A fuzzy data analysis technique was used to find this relationship. Specifically, the calculation procedure of fuzzy mathematical quantification theory no. 2 was applied to the explanatory variables and criterion variables in Table 2. This calculation procedure was proposed by Watada (1982) and is shown in Eq. (3).[14] The coefficient $C_{ik}$ is calculated by solving a characteristic equation and obtaining the eigen vectors.

$$y = \sum_{i=1}^{H} \sum_{k=1}^{l_i} C_{ik} F_{ik} \tag{3}$$

Table 2. Category scores.

| | Velocity | | | Accelerator pedal | | | S D of accelerator | | | Road condition | |
|---|---|---|---|---|---|---|---|---|---|---|---|
| | S | M | B | S | M | B | S | M | B | Mount. road | Flat road |
| 1 | 0.3 | 0.4 | 0.3 | 0.4 | 0.4 | 0.2 | 0.5 | 0.5 | 0 | 0.5 | 0.5 |
| 2 | 0.6 | 0.4 | 0 | 0.2 | 0.5 | 0.3 | 0.1 | 0.5 | 0.4 | 0.8 | 0.2 |
| 3 | 0.3 | 0.5 | 0.2 | 0.4 | 0.6 | 0 | 0.5 | 0.4 | 0.1 | 0.3 | 0.7 |
| 4 | 0 | 0.7 | 0.3 | 0.1 | 0.6 | 0.3 | 0 | 0.5 | 0.5 | 0.7 | 0.2 |
| 5 | 0.3 | 0.7 | 0 | 0.5 | 0.4 | 0.1 | 0.7 | 0.3 | 0 | 0 | 0.8 |
| 6 | 0.1 | 0.2 | 0.7 | 0.4 | 0.6 | 0 | 0.7 | 0.3 | 0 | 0.3 | 0.8 |
| 7 | 0.1 | 0.8 | 0.1 | 0.4 | 0.5 | 0.1 | 0.6 | 0.4 | 0 | 0.6 | 0.3 |
| 8 | 0.1 | 0.1 | 0.8 | 0.5 | 0.5 | 0 | 0.7 | 0.3 | 0 | 0.1 | 0.9 |
| 9 | 0.4 | 0.6 | 0 | 0.2 | 0.5 | 0.3 | 0.2 | 0.4 | 0.4 | 0.9 | 0 |
| 10 | 0 | 0 | 1 | 0.8 | 0.2 | 0 | 0.9 | 0.1 | 0 | 0 | 1 |

where $C_{ik}$ is a coefficient called the "category weight." The value of $F_{ik}$ expresses the category score of category $k$ for the explanatory variable $i$. $y$ is an index indicating the relationship between the explanatory variables and criterion variables.

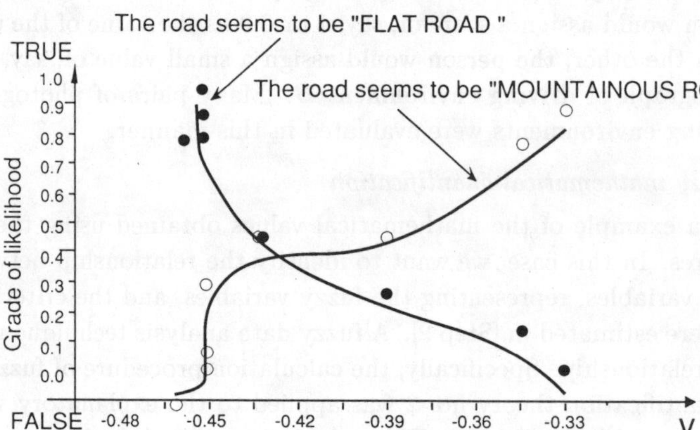


Fig. 3. Membership function to reason driving environments.

[Step 5] *Generation of membership functions*

In the functional plane shown in Fig. 3, the $x$-axis indicates the value of the index $y$, and the $y$-axis shows the value of the criterion variables. The column of numbers on the far left side of Table 2 indicates the number of the test road (driving environment). The category scores listed for the explanatory variables under each driving environment in the table can be substituted into Eq. (3) to find the respective value of $y$. The points obtained from the values of $y$ as the $x$ coordinate and the values of the criterion variables as the $y$ coordinate are then plotted to form curves like those shown in Fig. 3. These curves represent the relationship between the explanatory variables and the criterion variables. Let us consider a situation where a person is driving on an unfamiliar road. Data obtained on the driving operations performed by the driver in this situation are assigned to categories on the basis of the boundary values and data histogram, and the category scores are obtained as a result. The scores are substituted into Eq. (3) to calculate the values of $y$. Then using the function curves in Fig. 3, the values of the criterion variables are found in relation to the index $y$. Estimations are then made of the driving environment. If the values obtained from the function curves are large, it is inferred that the driving environment seems to be a mountainous road. Conversely, if the values obtained from the curves are small, it is inferred that the driving environment seems to be a flat road.

## 2.3. *Application for automatic transmission*

To test the operation of this human sensor system, it was applied to the shift control of an automatic transmission. As mentioned earlier, the underlying assumption of this research is that if the vehicle can recognize that the driving environment is, say, a mountainous road, it will vary the gear shift characteristics accordingly to provide good drivability on that type of road. For example, when driving on an uphill incline, the transmission should down-shift to a lower gear to provide more power. An experimental vehicle was mounted with the human sensor system and driven in various driving environments. Figure 4 compares the gear shift changes that were made on a winding uphill expressway with and without the system. The results indicate that the use of the human intelligent system provides intelligent transmission control.

## 2.4. *Conclusion of application I*

A new human sensor concept has been proposed in this section. This concept incorporates a fuzzy logic approach to dealing with vague driving operation data based on the use of multivariate analysis. The human sensor system

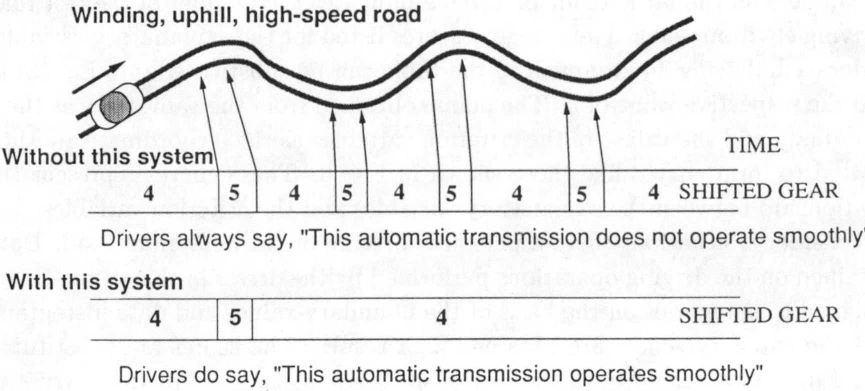


Fig. 4. Results of shift modulation.

has been tested in an experimental vehicle, and the results verified that it has the potential to be a useful tool in creating intelligent vehicles. However there are a lot of problems to apply this system for commercial products. For instance, what happens if the automatic transmission shifts into lower gear on icy roads without operating transmission lever? We are working on solving these problems for applying this concept for the commercial products.

## 3. Application II — Reasoning Subjective Evaluation[15]

A new approach to intelligent vehicle control is proposed in which the driver's subjective evaluation of drivability is fed back to modulate the control strategy parameters. It is hypothesized that a capability to infer the driver's vehicle control preferences and perception of operating and driving ease would make it possible to vary the characteristics of the engine, transmission and other vehicle subsystems to provide the type of drivability desired by the driver. The problem is there is no external quantitative yardstick for measuring a driver's subjective evaluation of the qualities constituting good drivability.

This case presents a study of an intelligent vehicle control system which incorporates a model of the driver's subjective evaluation of drivability. It was decided to use fuzzy logic to describe the subjective evaluation model because a fuzzy reasoning technique has been reported to be effective for predicting subjective evaluations. This application is meant to be one step toward the design of an intelligent vehicle capable of operating under certain specified driving conditions.

### 3.1. *Concept of subjective evaluation model feedback control*

Normal driving conditions would not be conducive to the creation of an intelligent feedback control system based on subjective evaluation of drivability. Therefore, it was necessary to define a certain driving task to be performed within a limited environment.[16,17] The driving task in this case was to maintain a given headway to a preceding vehicle.

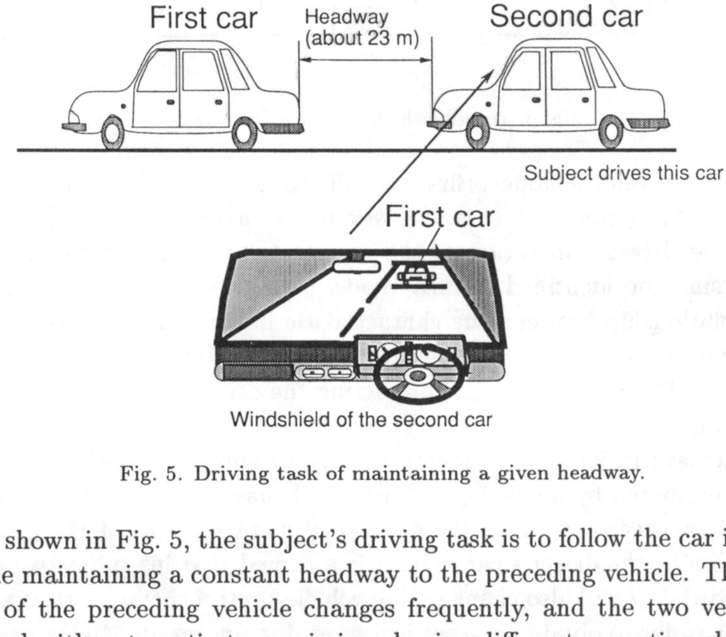


Fig. 5. Driving task of maintaining a given headway.

As shown in Fig. 5, the subject's driving task is to follow the car in front of it while maintaining a constant headway to the preceding vehicle. The driving speed of the preceding vehicle changes frequently, and the two vehicles are mounted with automatic transmissions having different operating characteristics. The driver of the second vehicle is asked to adjust his driving speed so as to center the image of the preceding car within the boundaries indicated on his windshield, as shown in the figure. The exact distance between the two cars is calculated by integrating their relative velocity.

As illustrated in Fig. 6, the input of the second car block of the driver–vehicle closed loop model is the headway error between the command distance and the actual distance to the preceding vehicle. The output of the model of second car block is the driver's operation of the accelerator.

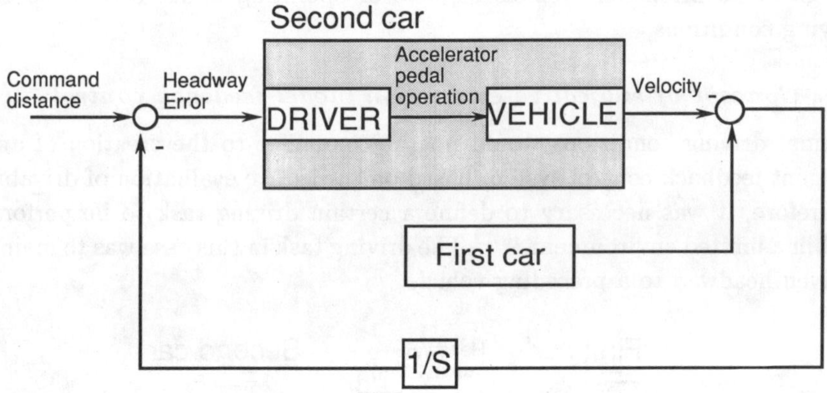


Fig. 6. Driver–vehicle model for this task.

This driver–vehicle model is first identified as an auto-regressive and moving average (ARMA) model. A characteristic index that explains the relationship between the driver's operation of the accelerator and headway error is then derived using the identified ARMA model. Knowledge rules are formulated for the relationship between the characteristic index and the driver's subjective evaluation at the time he operates the accelerator. Those rules and the ARMA model provide a basis for predicting the driver's subjective evaluation of drivability.

In other words, with this system, the driver's subjective evaluation of drivability is predicted by analyzing the driver's behavior with respect to the operation of the accelerator. If the operating characteristics of the automatic transmission in the driver's car vary, it is assumed that his subjective evaluation of drivability will also change. The vehicle control characteristics must be changed in order to obtain good subjective evaluation results. In the case of an automatic transmission, this can be accomplished by changing the gear shift characteristics. This provides the basis for a feedback vehicle control system based on a subjective evaluation model.

### 3.2. *Experimental results*

Experiments were conducted to obtain knowledge on the relationship between the ARMA model and characteristic index on the one hand and the driver's subjective evaluation of drivability on the other. Two test vehicles were used as the following vehicle. The automatic transmission of one vehicle was locked up, and set in the second gear position, and that of the other vehicle was locked up and set in the fourth gear position (overdrive). The frequency response of gear shift characteristics of the two transmissions in relation to accelerator inputs is shown in Fig. 7. This Bode diagram shows the transfer function of the two automatic transmissions in terms of the vehicle velocity vs. the throttle valve opening.

In general, the vehicle with the automatic transmission set in second gear shows sharper response than the vehicle with the transmission set in fourth gear. This can be seen from the larger gain and smaller phase delay of the former vehicle.

Table 3. Subjective evaluation of subjects.

| DRIVER | 4th gear | 2nd gear |
|---|---|---|
| Subject A (Male, Age 25) | 3 | 5 |
| Subject B (Male, Age 31) | 3 | 4 |
| Subject C (Male, Age 30) | 3 | 3 |

**Subjective evaluation scale**

1: Much poorer than the performance of the automatic transmission in 4th gear
3:The same as the performance of the automatic transmission in 4th gear
5:Much better than the performance of the automatic transmission in 4th gear

The subjective evaluation results are shown in Table 3. It is clear that the drivers had different subjective evaluations of the different power train characteristics. Two of the three subjective said that the car with the transmission set in second gear had much better drivability. Time histories were recorded of the headway error and the driver's operation of the accelerator. Table 4 gives the coefficients of the identified ARMA model.

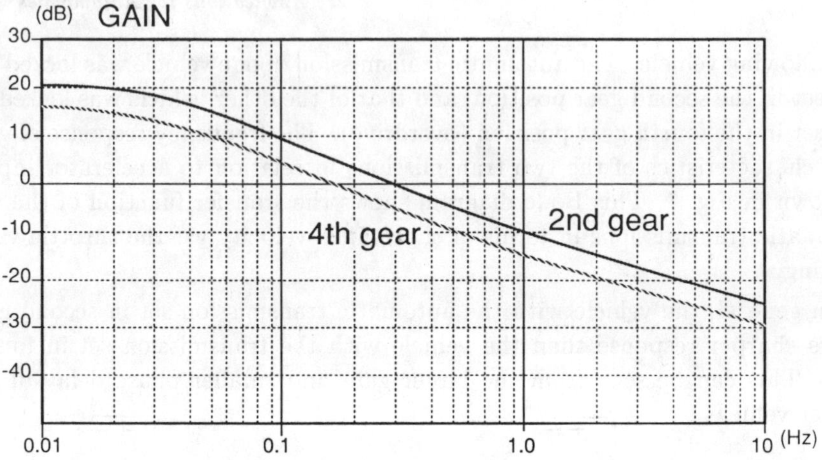


(a)

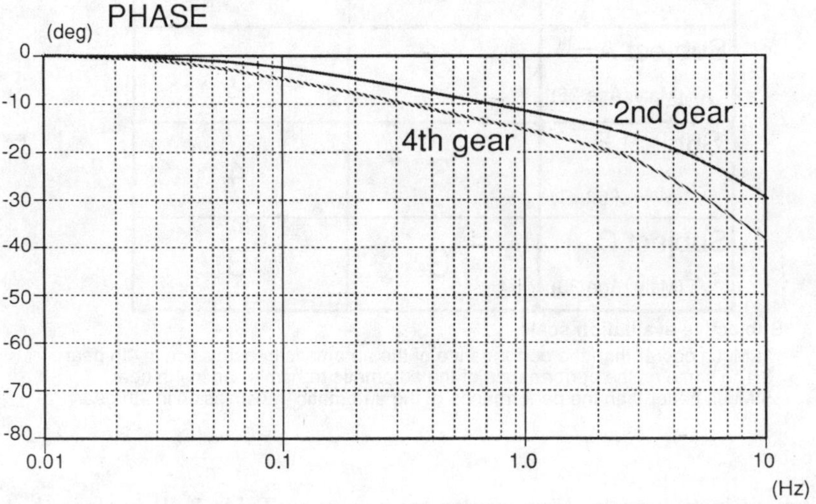


(b)

**Fig. 7. Bode diagram of two automatic transmission (a) Gain (b) Phase.**

Table 4. Coefficients of identified ARMA model.

| | Hold 4th gear | | | Hold 2nd gear | | |
|---|---|---|---|---|---|---|
| | Subject A | Subject B | Subject C | Subject A | Subject B | Subject C |
| MA -terms | | | | | | |
| α0 | 7.23 | 9.61 | 6.29 | 1.37 | 5.78 | 5.55 |
| α1 | -11.68 | -5.89 | -8.47 | -0.58 | -2.64 | -7.41 |
| α2 | 0.13 | 1.42 | 3.65 | -0.17 | -2.56 | -3.74 |
| α3 | 2.65 | -1.85 | -0.14 | 0 | 2.45 | 6.87 |
| α4 | -0.28 | -0.1 | 2.92 | 0.08 | -0.02 | 0.92 |
| α5 | 3.18 | -1.75 | -1.14 | 0.529 | 2.69 | 3.93 |
| α6 | 1.59 | -0.2 | 2.42 | 0.22 | 1.09 | -1.72 |
| α7 | 1.51 | 0.8 | -1.25 | 0.34 | 0.56 | 1.24 |
| α8 | -2.27 | 2.26 | -6.07 | 0.44 | 0.09 | -2.63 |
| α9 | 4.41 | 1.29 | 1.64 | -0.24 | -0.26 | -0.41 |
| α10 | -1.4 | -0.55 | 1.52 | 0.26 | 0.14 | -0.13 |
| AR -terms | | | | | | |
| β1 | -0.09 | -1.24 | -1.29 | -0.79 | -0.97 | -0.98 |
| β2 | -0.07 | 0.11 | 0.08 | -0.33 | -0.17 | -0.18 |
| β3 | 0.15 | 0.18 | 0.27 | 0.04 | 0.25 | 0.16 |
| β4 | -0.02 | -0.03 | -0.06 | -0.14 | -0.09 | -0.03 |
| β5 | -0.04 | -0.02 | 0 | 0.06 | 0.03 | 0.09 |
| β6 | 0.09 | 0.03 | 0.02 | 0.12 | 0.03 | 0.01 |
| β7 | 0.01 | 0.05 | 0.02 | 0.09 | 0.03 | 0 |
| β8 | -0.05 | -0.08 | -0.04 | -0.01 | -0.01 | 0 |
| β9 | 0.06 | 0.01 | 0.01 | 0 | 0.01 | 0.02 |
| β10 | 0.03 | -0.01 | -0.01 | 0 | -0.01 | -0.01 |

Table 5. Coefficients of identified ARMA model (Explanation of Table 4).

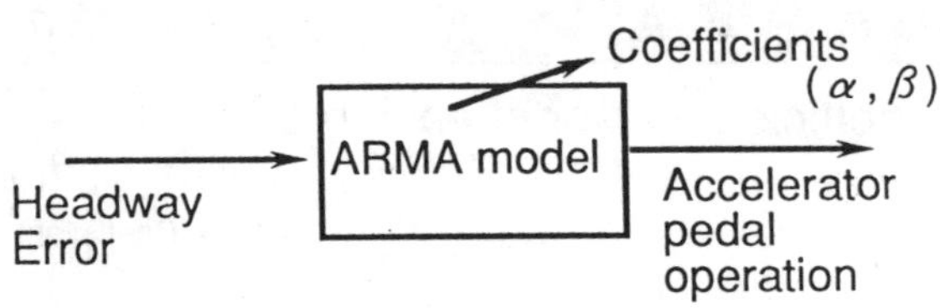


Moving average term for Headway

$$H(z) = \frac{\alpha_1 Z^{10} + \alpha_2 Z^9 + \alpha_3 Z^8 + \ldots + \alpha_{11}}{Z^{10} - \beta_1 Z^9 - \beta_2 Z^8 - \beta_3 Z^7 - \ldots - \beta_{10}}$$

Auto regressive term for Accelerator operation

## 3.3. *Generation of fuzzy reasoning knowledge and membership function*

The coefficient vector, of the identified ARMA model is expressed by

$$\theta = [\alpha + 1, \alpha_2, \ldots, \beta_1, \beta_2, \ldots], \quad (4)$$

where $i$ is coefficient of the auto-regressive term of the ARMA model and $i$ is the coefficient of the moving average term of the model. This coefficient matrix characterizes the relationship between the driver's behavior in following the vehicle ahead and the subjective evaluation of drivability. Let us consider the coefficients in terms of their physical significance. If a coefficient has a large value, it indicates that the data (i.e., headway variance and throttle valve opening) at that time have a large influence on the driver's behavior. In other words, the values of the coefficients can be regarded as being weighted values that determine how the driver operates the accelerator at that moment.

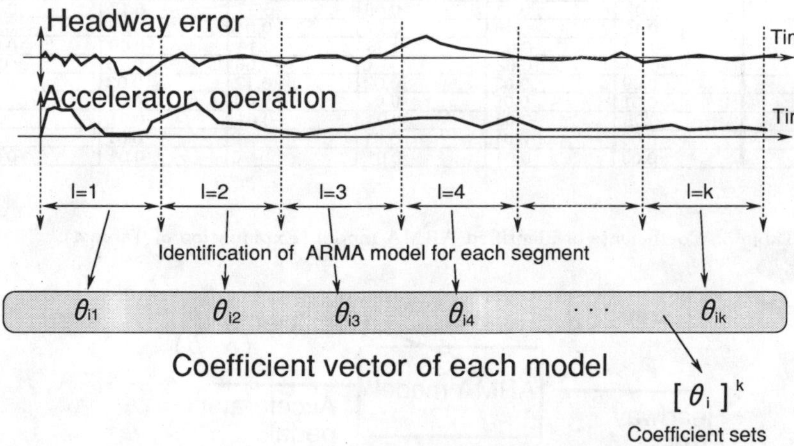


Fig. 8. ARMA model identification for data segments.

However, it is difficult to treat data on driving operations mathematically because the data are not ergodic as I mentioned at application case I. The characteristics of driving operation data will also change as a person becomes more skilled at driving. Consequently, it would not be accurate to try and express the characteristics of a person's driving behavior using only one identified ARMA model. Yet it would not be so unreasonable to assume that data obtained in many short segments would be ergodic. As shown in Fig. 8, the time histories of the headway error and accelerator operation can be divided into a number of segments and an ARMA model identified for each segment. We will let $(\theta_i)$ denote the set of coefficients of the ARMA model identified for each segment. Then, this coefficient set can be regarded as an index expressing the characteristics of a person's driving behavior, as indicated in Eq. (5).

$$[\theta_i]k = [\theta_1, \theta_2, \theta_3, \ldots, \theta_k]. \tag{5}$$

When all of the segments are viewed as a whole, it can be assumed that the same-order coefficients of the auto-regressive term and moving average term will have a Gaussian distribution. As a result, the fuzzy reasoning rules and membership functions for inferring the driver's subjective evaluation of drivability from $(\theta_i)$ can be obtained as shown in Fig. 9.

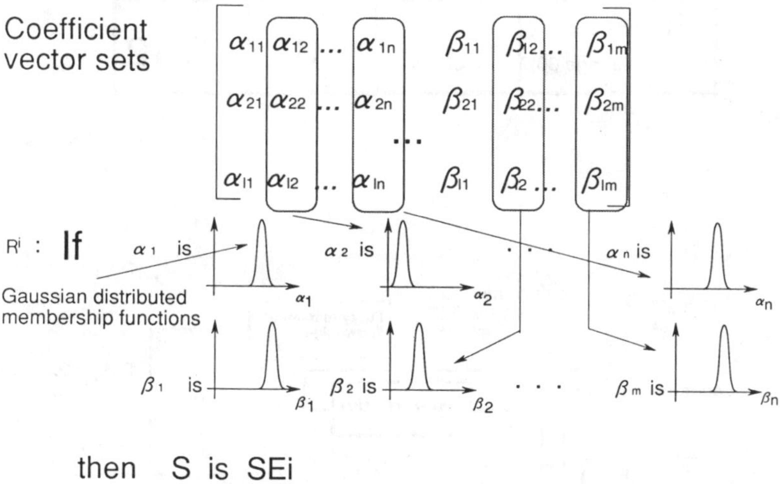


Fig. 9. Fuzzy rules and membership functions.

As indicated in Eq. (6), when (i) is given and a coefficient distribution set $\{\theta\}$ is obtained which is not contained in the rules, the system will infer from the relationship with (i) a subjective evaluation result of SE' in relation to $\{\theta\}$.

$$\begin{array}{ll} \text{(i)} & [\theta_i]k \longrightarrow \text{SE} \\ \text{(ii)} & \{\Theta\} \\ \hline \text{(iii)} & \qquad \text{SE} \end{array} \tag{6}$$

Table 6. Comparison of fuzzy reasoning vs. drivers' evaluation.

| DRIVER | 4th gear | 3rd gear | |
|---|---|---|---|
| | | Driver | Fuzzy reasoning |
| Subject A (male, age 25) | 3 | 5 | 4.82 |
| Subject B (male, age 31) | 3 | 4 | 4.23 |
| Subject C (male, age 30) | 3 | 3 | 3.12 |

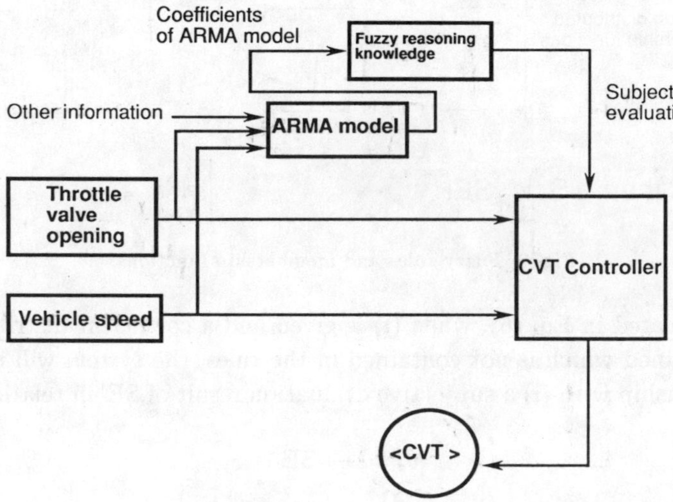


Fig. 10. System configuration of CVT.

### 3.4. *Feasibility of subjective evaluation-based feedback control*

To assess the inference accuracy of the subjective evaluation model, the control system was mounted in an experimental vehicle. The automatic transmission of the vehicle was locked up and set in third gear, representing an unfamiliar vehicle characteristic. After conducting driving tests, the values obtained with the fuzzy reasoning system were compared with the drivers' subjective evaluations, as indicated in Table 6. The similarity between the two sets of values shows that the fuzzy reasoning system can predict subjective evaluations close to those made by human drivers.

Figure 10 shows the application of subjective evaluation-based feedback control to a continuously variable transmission (CVT). This system would vary the gear ratio so as to optimize the fuzzy reasoning values sent to the CVT controller. It is expected that this concept of feedback control based on a subjective evaluation model will lead to new control strategies for automotive power trains.

### 3.5. *Conclusion of application II*

This application has proposed a method for varying control characteristics so as to optimize the subjective evaluation of drivability under a specific driving environment that involved maintaining a certain headway to a preceding vehicle. Future work will focus on the development of a subjective evaluation model for used under ordinary driving conditions.

## 4. Conclusion

This section has introduced some applications of automobiles. Especially in application II, a more detailed study into the universality (generality) of the fuzzy evaluation knowledge of the driver with more vehicle and road condition diversity is an important subject for a future study. A considerable amount of research is also required to explain driving force characteristics during routine driving by an ordinary driver and the relationship of subjective evaluations about the operating characteristics drivability. Using the proposed technique as a basic approach for future study, analysis of natural motion behavior in gradually relaxed task condition is feasible. Fluctuations in driver subjective evaluation judgment due to perturbations and the relationship of task conditions and subjective evaluation judgment also need to be investigated.

In general, fuzzy approaches are very useful for dealing with drivers' operation. The interface between vehicle and driver will be the one of the best targets of applying fuzzy technology. The vehicle will be able to predict the drivers' intentions and preference in order to provide drivers with good drivability. We strongly believe that fuzzy theory will continue to be an effective tool in such investigations.

## References


1. M. Iguchi, "Human/Machine systems control of machines by humans", in *Information Science Lectures, Vol. 29* (Kyoritu Syuppan, 1970).
2. Y. Oguchi, "Transverse acceleration characteristic of a human/vehicle system" (Oka Shobo, 1984).
3. S. Okabayashi, "Head-up display for vehicles from the point of view of visual engineering", *Light Technology* **29**, 4, 208 (1991).
4. H. Takahashi and H. Minami, "Recognizing the driving environments of a vehicle by fuzzy reasoning", *Japanese Journal of Fuzzy Theory and Systems* **2**, 3, 438 (1990).
5. M. Maeda and S. Murakami, "Vehicle speed control using fuzzy logic controller", in *Proc. of 9th System Symposium* (1983), pp. 7–11.
6. H. Takahashi *et al.*, "Automatic speed control device using self-tuning fuzzy logic", in *Proc. of IEEE Workshop on Automotive Applications of Electronics* (1988), pp. 65–71.
7. M. Uriubara *et al.*, "Development of automatic cruise control using fuzzy control", *Journal of the Society of Automotive Engineers of Japan* **42**, 2, 224 (1988).
8. Y. Kasai *et al.*, "Electronically controlled continuous variable transmission (ECVT-II)", in *Proc. of the Society of Automotive Engineers Convergence '88* (1988), pp. 33–42.
9. K. Ohba *et al.*, "Estimation of a road surface using fuzzy logic", in *Proc. of Meeting of the Society of Automobile Technology 881* (1988), pp. 109–112.
10. I. Sakai *et al.*, "Shift scheduling method of automatic transmission vehicle using fuzzy logic", in *Proc. of IEEE Roundtable Discussion on Fuzzy and Neural Systems and Vehicle Applications*, paper #27 (1991).
11. R. Magoshi *et al.*, "Environmental estimation techniques and their application to chassis control", in *Technical Review of Mitsubishi Motors Corporation* **5** (1993), pp. 48–57.
12. K. Kondo *et al.*, "Adaptive shift scheduling strategy introduced neural network in automatic transmission", in *Proc. of the Society of Automotive Engineers of Japan*, 9436611 (1994), pp. 241–244.
13. H. Takahashi *et al.*, "Intelligent controlled automatic transmission", in *Proceedings of the 12th Triennial World Congress of the International Federation of Automatic Control* **5** (1993), pp. 91–94.

14. J. Watada, "A heuristic method of hierarchical clustering for fuzzy intransitive relations", in *Recent Development in Fuzzy Sets and Possibility Theory* (Pergamon Press Inc., 1982).
15. H. Takahashi, "A study on the subjective evaluation feedback vehicle control", in *Proceedings of the Intelligent Vehicles '93* (1993), pp. 90–94.
16. D. H. Weir *et al.*, "Measurement and interpretation of driver steering behavior and performance", *International Automotive Engineering Conference*, 730098 (1973).
17. D. T. McRuer *et al.*, "A review of quasi-linear pilot models", *IEEE Transactions on the Human Factor in Electronics*, **HFE8**, 231 (1967).

# VOICE RECOGNITION USING FUZZY PATTERN MATCHING AND ITS APPLICATIONS

JUN-ICHIROH FUJIMOTO
*Information and Communication R&D Center RICOH Company Ltd.,*
*2-3,Shin-Yokohama 3-chome,Kohoku-ku, Yokohama 222, Japan*

## 1. Introduction

It is marvelous that a computer can recognize what we speak. This ability also makes it unnecessary to have keyboard. Consider a computer that can hold a discussion. The movies have shown such stories. It could be realized soon.

For more than 20 years we have been expecting the realization of such a fantastic world. However till today there is no machine which can discuss with us, not even one that can understand what we speak. Is speech recognition a "dream"? Is it impossible to realize?

As a result of the improvement of computers, many things which were impossible to realize have been put into practice. Consequently, we could consider that computers are almighty. If a human can do a problem, a computer can do it more easily. But up till now, computers do not have better recognizing ability than humans.

In the voice recognition field, existing word recognition systems were not quite useful. Now dialog recognition is the main theme of researchers. The recognition of dialog or conversation is realized not only by acoustical analysis but also by linguistic analysis. They are very important, especially the latter where grammatical analysis seemed to be useful in dialog recognition.

Recently, it has become clear that the usual grammatical analysis has little power in dialog recognition, because the grammar of spoken language is quite different from the written one. An example is Japanese, where we have an imperfect grammar in the spoken language.[1] Now many researchers believe the most important thing is to recognize conversation by understanding the meanings of spoken sentences.

While word recognition systems have achieved high accuracy in the laboratory, the same performance could not be achieved in the field. This was the main reason why the word recognition system was useless. The reason of the low accuracy in the field was due to the noise. The system could not distinguish human voice from noise. Another reason was the microphone. Every word recognition system needed a head-set or telephone hand-set microphone which can achieve good S/N ratio for the voice signal. Many users hated to use them because of the limited applications.

The speech synthesis technique, which is often compared with voice recognition, and was less sensitive to noise, has become more widespread. From such experiences researchers found that the system robustness is more important than accuracy in a laboratory.

Recently voice recognition systems which can be used in the actual field have been developed. These are far from ideal recognition machines. We cannot deny that the present technical level is too low to meet our expectation. However, that does not mean the systems are "useless." They can help our work even if the usable conditions are limiting.

Several years ago, a large computer was needed to recognize a mere word. For practical use fewer calculations, smaller size, and lower cost are necessary. A word recognition method in which these troubles are overcome by using fuzzy theory is described here. For developing the method, fuzzy pattern matching technique was invented.

First the mechanism of utterance is described, followed by the description of voice recognition method, fuzzy pattern matching method, and a noise canceling method. Concluding, some products and their applications are mentioned.

## 2. Voice Recognition

### 2.1. *Mechanism of utterance*

Before describing voice recognition, we should consider the mechanism of voice generation in a human.

The cross-section of a human head is indicated in Fig. 1. In the case of a vowel sound, the source is the vocal chord. The vibration is transmitted through the vocal tract and radiated from the mouth as sound waves. In this case, the vocal chords vibration is independent of the vowels which we want to utter. Speech information is added by varying the shape of the vocal tract, which is similar to changing the resonant frequencies.

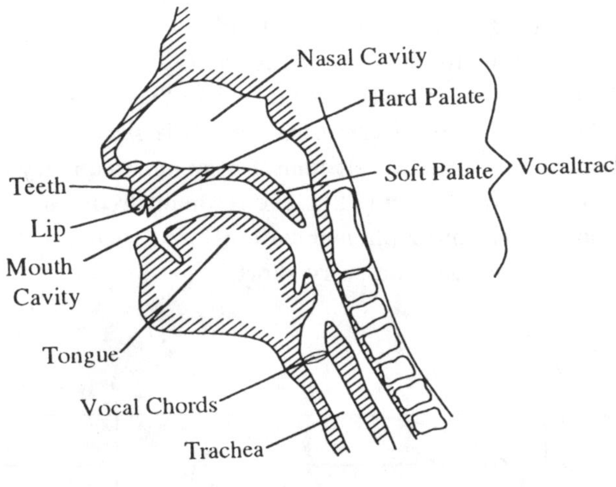


Fig. 1. Cross-section of human head.

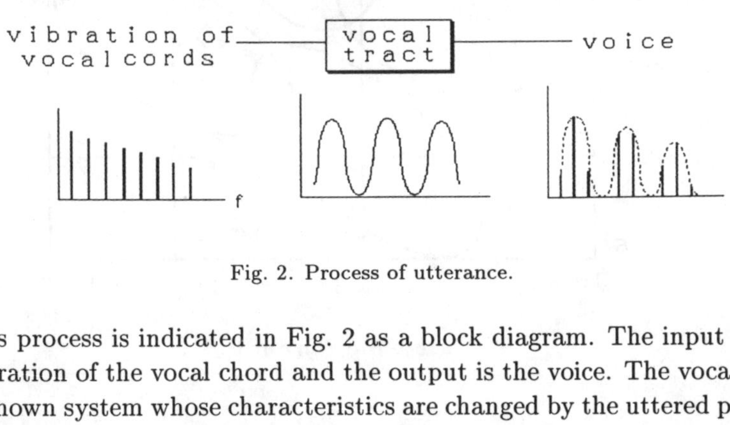


Fig. 2. Process of utterance.

This process is indicated in Fig. 2 as a block diagram. The input signal is the vibration of the vocal chord and the output is the voice. The vocal tract is an unknown system whose characteristics are changed by the uttered phoneme. The frequency characteristic of each block is also indicated in Fig. 2.

The product of the characteristics of input signal and the vocal tract gives the output voice. What is spoken can be recognized from the frequency characteristics of the envelope of the utterance. Particularly for speech recognition, the peaks of the envelope have important information because they are the resonant frequencies of the vocal tract. In the case of a consonant it is similar to that of a vowel except that a sound source is located in the vocal tract, e.g., on top of the tongue, throat, friction between tongue and teeth, etc.

The sizes of vocal chords and vocal tracts in different people are related to their different heights or weights. These differences influence the resonant frequencies. The resonant frequencies of the same phoneme are not the same among different persons. Japanese language has five vowels, /a/, /i/, /u/, /e/ and /o/. The F1-F2 map of these five vowels is shown in Fig. 3. In this figure the horizontal axis is the fundamental resonant frequency (F1) and the vertical one is the 2nd. frequency (F2). The resonant frequency of each vowel has a distribution. When many phonems are uttered within a short time, the duration of each phoneme becomes very short.

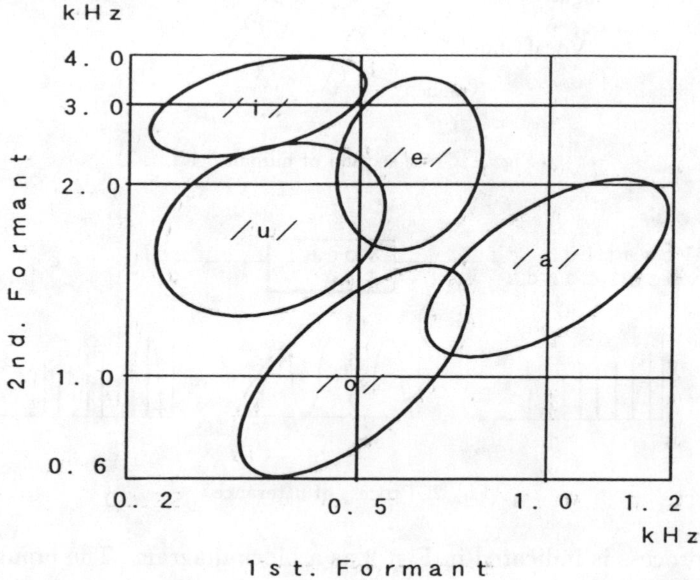


Fig. 3. F1-F2 map of Japanese vowels.

In the extreme case some phonemes are skipped in the pronounciation. The reason is as follows: For speaking a word we have to change our organ very quickly. Accordingly we have to prepare to move the organ before an utterance of a phoneme, this influences the next phoneme because of inertia of the organ. This phenomenon is known as coarticulation and it makes speech recognition difficult.

## 2.2. *Basic theory of fuzzy pattern matching*[2]

The recognition unit of a word is chosen because smaller units are adversely influenced by the coarticulation. A word is the smallest unit without considering this phenomenon. This is also the reason why word recognition systems were first developed.

Generally, voice recognition is classified into two types. One is speaker dependent and the other is speaker independent. In the former it is necessary to train the recognition machine with the operator's voice, but in the latter it can be used by many different persons.

There are two obstacles to speaker-independent spoken-word recognition in pattern matching techniques:

(1) The duration of the spoken word is not constant. The differences of speaking speed cause the fluctuation of the pattern length. Even for the same speaker speaking the same word, the duration differs.

(2) As mentioned, the resonant frequency of the vocal tract varies. The above two problems have to be solved to realize a speaker-independent recognition system.

To overcome these problems fuzzy theory is introduced in the process of pattern matching.

Recognition is performed using the technique of pattern matching. In other words, the unknown and enrolled patterns (templates) are compared and the similarities between them are calculated. Then the unknown pattern is classified into a category which gets the most similarity. Let a word name set be represented by $I = (i1, i2, \ldots, in)$, word patten set by $X = (xi1, xi2, \ldots, xin)$ and a set of membership functions by $M = (mi1, mi2, \ldots, min)$, which indicates the likelihood for each word. Here $I$ and $M$ are infinite sets but $X$ is a finite set. Assuming $y$ to be an unknown pattern ($y \subset X$), the word name $j(j \subset I)$, that is the recognition result, is determined by the following equation,

$$j = \max_{j \subset I}(Sjy) \tag{1}$$

where $Sjy$ is the similarity between $y$ and the word pattern $xj$. In this system, by holding membership functions as templetes, similarities between an unknown pattern and each word are calculated using the membership function.

TSP (time spectrum pattern) is used in pattern matching. Rather than normal TSP, local peak patterns are useful because we want to know the position of local peaks. Since binarized data is easy to treat in a computer and requires smaller amount of memory, binarized time spectrum pattern (BTSP) is adopted as a definition of a word pattern. Creating binary patterns is done by selecting the near part of the peaks on the frequency axis to be 1 and other parts to be 0. Since BTSP includes resonance frequencies of the vocal tract and their time variance, the BTSPs are effective features to treat in a system. The TSP and BTSP are indicated in Fig. 4 where the horizontal axis is the frequency and the vertical axis is the time in both patterns. Fifteen samples are analyzed on frequency and sampled every 10 ms. In Fig. 4, (a) is the normal TSP and each element is expressed in 12 bits, while (b) is the BTSP. Each element needs only 1 bit.

Word Name: one
Frame No.: 45

| | | | | | | | | | | | | | | | |
|---|---|---|---|---|---|---|---|---|---|---|---|---|---|---|---|
| 6 | 2 | 4 | 2 | 2 | 2 | 2 | 2 | 2 | 2 | 0 | 0 | 0 | 2 | 0 | 4 |
| 58 | 30 | 30 | 22 | 18 | 12 | 18 | 12 | 8 | 2 | 4 | 0 | 2 | 2 | 2 | 32 |
| 126 | 86 | 86 | 54 | 42 | 52 | 40 | 22 | 16 | 10 | 8 | 4 | 4 | 2 | 2 | 46 |
| 72 | 84 | 164 | 80 | 56 | 54 | 40 | 22 | 20 | 10 | 8 | 4 | 6 | 2 | 2 | 60 |
| 62 | 92 | 166 | 96 | 82 | 58 | 60 | 30 | 28 | 14 | 12 | 8 | 8 | 4 | 4 | 64 |
| 56 | 72 | 160 | 118 | 130 | 70 | 78 | 44 | 40 | 20 | 16 | 12 | 12 | 6 | 4 | 72 |
| 34 | 38 | 108 | 120 | 172 | 70 | 68 | 34 | 28 | 16 | 12 | 8 | 10 | 4 | 4 | 88 |
| 24 | 20 | 54 | 80 | 126 | 68 | 74 | 28 | 20 | 12 | 10 | 8 | 10 | 4 | 4 | 88 |
| 22 | 14 | 38 | 44 | 68 | 76 | 94 | 42 | 22 | 12 | 12 | 8 | 10 | 4 | 2 | 88 |
| 18 | 12 | 30 | 34 | 50 | 62 | 92 | 62 | 30 | 14 | 12 | 12 | 12 | 6 | 4 | 88 |
| 14 | 8 | 26 | 22 | 40 | 58 | 60 | 74 | 36 | 16 | 12 | 10 | 12 | 6 | 4 | 90 |
| 12 | 6 | 20 | 18 | 34 | 52 | 44 | 68 | 48 | 18 | 14 | 12 | 14 | 4 | 2 | 90 |
| 10 | 6 | 18 | 16 | 30 | 46 | 38 | 60 | 52 | 18 | 12 | 10 | 16 | 6 | 4 | 88 |
| 10 | 6 | 18 | 16 | 30 | 46 | 34 | 40 | 72 | 22 | 12 | 12 | 18 | 6 | 6 | 90 |
| 12 | 8 | 18 | 16 | 34 | 48 | 34 | 36 | 84 | 30 | 16 | 12 | 16 | 6 | 4 | 90 |
| 12 | 8 | 20 | 16 | 36 | 52 | 34 | 32 | 76 | 34 | 18 | 16 | 20 | 6 | 4 | 90 |
| 10 | 6 | 16 | 16 | 34 | 48 | 28 | 26 | 72 | 34 | 16 | 12 | 14 | 4 | 4 | 88 |
| 10 | 6 | 16 | 14 | 32 | 46 | 28 | 24 | 58 | 32 | 16 | 12 | 10 | 4 | 4 | 86 |
| 10 | 8 | 16 | 14 | 32 | 42 | 26 | 22 | 46 | 28 | 14 | 10 | 8 | 6 | 4 | 84 |
| 12 | 8 | 18 | 16 | 34 | 44 | 26 | 22 | 40 | 28 | 14 | 10 | 8 | 6 | 4 | 82 |
| 10 | 10 | 20 | 18 | 34 | 42 | 26 | 22 | 36 | 32 | 16 | 14 | 10 | 6 | 6 | 78 |
| 10 | 8 | 18 | 16 | 34 | 40 | 22 | 18 | 30 | 30 | 14 | 14 | 8 | 6 | 4 | 72 |
| 10 | 10 | 20 | 18 | 32 | 38 | 24 | 16 | 28 | 30 | 14 | 16 | 8 | 6 | 4 | 66 |
| 12 | 10 | 20 | 20 | 30 | 32 | 20 | 14 | 24 | 28 | 14 | 10 | 6 | 4 | 2 | 58 |
| 10 | 8 | 14 | 20 | 18 | 14 | 10 | 6 | 18 | 12 | 6 | 2 | 2 | 2 | 0 | 36 |
| 16 | 10 | 20 | 22 | 14 | 4 | 6 | 6 | 12 | 8 | 4 | 2 | 4 | 2 | 0 | 36 |
| 22 | 18 | 36 | 30 | 24 | 8 | 10 | 8 | 12 | 8 | 6 | 2 | 4 | 2 | 0 | 42 |
| 18 | 16 | 26 | 34 | 26 | 8 | 8 | 6 | 12 | 6 | 4 | 2 | 2 | 2 | 2 | 38 |
| 20 | 14 | 32 | 34 | 26 | 8 | 8 | 6 | 10 | 6 | 4 | 2 | 2 | 2 | 0 | 38 |
| 18 | 14 | 28 | 32 | 24 | 6 | 8 | 6 | 10 | 4 | 4 | 2 | 2 | 2 | 0 | 36 |
| 18 | 14 | 24 | 26 | 20 | 4 | 6 | 4 | 10 | 4 | 4 | 2 | 0 | 0 | 0 | 38 |
| 20 | 12 | 24 | 20 | 16 | 4 | 4 | 2 | 6 | 2 | 2 | 0 | 0 | 0 | 0 | 36 |
| 22 | 14 | 18 | 16 | 12 | 4 | 6 | 2 | 6 | 2 | 2 | 0 | 0 | 2 | 0 | 38 |
| 24 | 12 | 16 | 12 | 10 | 4 | 4 | 4 | 6 | 0 | 0 | 0 | 0 | 2 | 0 | 36 |
| 22 | 14 | 16 | 12 | 8 | 2 | 4 | 4 | 6 | 2 | 2 | 0 | 0 | 0 | 0 | 34 |
| 22 | 10 | 16 | 12 | 10 | 2 | 4 | 2 | 6 | 0 | 2 | 0 | 0 | 0 | 0 | 32 |
| 16 | 8 | 14 | 12 | 8 | 2 | 4 | 2 | 6 | 2 | 2 | 0 | 0 | 2 | 0 | 34 |
| 14 | 6 | 10 | 6 | 4 | 2 | 2 | 2 | 4 | 0 | 0 | 0 | 0 | 0 | 0 | 30 |
| 12 | 4 | 6 | 2 | 2 | 0 | 2 | 2 | 2 | 0 | 0 | 0 | 0 | 0 | 0 | 28 |
| 12 | 4 | 4 | 2 | 4 | 0 | 2 | 0 | 2 | 0 | 0 | 0 | 2 | 2 | 0 | 24 |
| 12 | 2 | 6 | 2 | 4 | 0 | 4 | 2 | 6 | 0 | 0 | 0 | 2 | 2 | 0 | 22 |
| 10 | 4 | 6 | 2 | 4 | 0 | 2 | 2 | 4 | 0 | 0 | 0 | 2 | 0 | 0 | 20 |
| 8 | 4 | 4 | 0 | 2 | 0 | 2 | 2 | 4 | 0 | 0 | 0 | 0 | 0 | 0 | 18 |
| 6 | 2 | 2 | 0 | 2 | 0 | 2 | 2 | 4 | 0 | 0 | 0 | 0 | 0 | 0 | 12 |
| 4 | 0 | 2 | 2 | 2 | 0 | 2 | 2 | 6 | 0 | 0 | 0 | 0 | 2 | 0 | 8 |

(a)

Word Name: one
Frame No.: 45

| | | | | | | | | | | | | | | | |
|---|---|---|---|---|---|---|---|---|---|---|---|---|---|---|---|
| 1 | 0 | 0 | 0 | 0 | 0 | 0 | 0 | 0 | 0 | 0 | 0 | 0 | 0 | 0 | 1 |
| 1 | 0 | 0 | 0 | 0 | 0 | 1 | 1 | 0 | 0 | 0 | 0 | 0 | 0 | 0 | 1 |
| 1 | 0 | 1 | 0 | 0 | 1 | 1 | 0 | 1 | 0 | 0 | 0 | 0 | 0 | 0 | 1 |
| 0 | 0 | 1 | 0 | 0 | 0 | 0 | 0 | 0 | 0 | 0 | 0 | 1 | 0 | 0 | 1 |
| 0 | 0 | 1 | 0 | 0 | 0 | 1 | 0 | 0 | 0 | 1 | 0 | 1 | 0 | 0 | 1 |
| 0 | 0 | 1 | 1 | 1 | 0 | 1 | 0 | 1 | 0 | 0 | 0 | 1 | 0 | 0 | 1 |
| 0 | 0 | 0 | 1 | 1 | 0 | 0 | 0 | 0 | 0 | 0 | 0 | 1 | 0 | 0 | 1 |
| 0 | 0 | 0 | 1 | 1 | 1 | 1 | 0 | 0 | 0 | 0 | 0 | 1 | 0 | 0 | 1 |
| 0 | 0 | 0 | 0 | 1 | 1 | 1 | 0 | 0 | 0 | 1 | 0 | 1 | 0 | 0 | 1 |
| 0 | 0 | 0 | 0 | 0 | 1 | 1 | 1 | 0 | 0 | 0 | 0 | 1 | 0 | 0 | 1 |
| 0 | 0 | 1 | 0 | 0 | 1 | 1 | 1 | 1 | 0 | 0 | 0 | 1 | 0 | 0 | 1 |
| 0 | 0 | 1 | 0 | 0 | 1 | 1 | 1 | 1 | 0 | 0 | 0 | 1 | 0 | 0 | 1 |
| 0 | 0 | 1 | 0 | 0 | 1 | 1 | 1 | 1 | 0 | 0 | 0 | 1 | 0 | 0 | 1 |
| 0 | 0 | 1 | 0 | 0 | 1 | 0 | 0 | 1 | 0 | 0 | 0 | 1 | 0 | 0 | 1 |
| 0 | 0 | 0 | 0 | 0 | 1 | 0 | 0 | 1 | 0 | 0 | 0 | 1 | 0 | 0 | 1 |
| 0 | 0 | 1 | 0 | 0 | 1 | 0 | 0 | 1 | 0 | 0 | 0 | 1 | 0 | 0 | 1 |
| 0 | 0 | 1 | 0 | 0 | 1 | 0 | 0 | 1 | 1 | 0 | 0 | 0 | 0 | 0 | 1 |
| 0 | 0 | 1 | 0 | 0 | 1 | 0 | 0 | 1 | 1 | 0 | 0 | 0 | 0 | 0 | 1 |
| 0 | 0 | 0 | 0 | 1 | 1 | 0 | 0 | 1 | 1 | 0 | 0 | 0 | 0 | 0 | 1 |
| 0 | 0 | 1 | 0 | 1 | 1 | 0 | 0 | 1 | 1 | 0 | 0 | 0 | 0 | 0 | 1 |
| 0 | 0 | 1 | 0 | 1 | 1 | 0 | 0 | 1 | 1 | 0 | 0 | 0 | 0 | 0 | 1 |
| 0 | 0 | 1 | 0 | 1 | 1 | 0 | 0 | 1 | 1 | 0 | 0 | 0 | 0 | 0 | 1 |
| 0 | 0 | 1 | 0 | 1 | 1 | 0 | 0 | 1 | 1 | 0 | 1 | 0 | 0 | 0 | 1 |
| 0 | 0 | 0 | 0 | 1 | 1 | 0 | 0 | 1 | 1 | 0 | 0 | 0 | 0 | 0 | 1 |
| 0 | 0 | 0 | 1 | 1 | 1 | 0 | 0 | 1 | 1 | 0 | 0 | 0 | 0 | 0 | 1 |
| 1 | 0 | 1 | 1 | 1 | 0 | 0 | 0 | 1 | 1 | 0 | 0 | 0 | 0 | 0 | 1 |
| 1 | 0 | 1 | 1 | 1 | 0 | 0 | 0 | 1 | 1 | 0 | 0 | 0 | 0 | 0 | 1 |
| 0 | 0 | 1 | 1 | 1 | 0 | 0 | 0 | 1 | 0 | 0 | 0 | 0 | 0 | 0 | 1 |
| 1 | 0 | 1 | 1 | 1 | 0 | 0 | 0 | 1 | 0 | 0 | 0 | 0 | 0 | 0 | 1 |
| 1 | 0 | 1 | 1 | 1 | 0 | 0 | 0 | 1 | 0 | 0 | 0 | 0 | 0 | 0 | 1 |
| 1 | 0 | 1 | 1 | 1 | 0 | 0 | 0 | 1 | 0 | 0 | 0 | 0 | 0 | 0 | 1 |
| 1 | 0 | 1 | 1 | 1 | 0 | 1 | 0 | 1 | 0 | 0 | 0 | 0 | 0 | 0 | 1 |
| 1 | 0 | 1 | 1 | 1 | 0 | 0 | 0 | 1 | 0 | 0 | 0 | 0 | 0 | 0 | 1 |
| 1 | 0 | 1 | 1 | 0 | 0 | 0 | 0 | 1 | 0 | 0 | 0 | 0 | 0 | 0 | 1 |
| 1 | 0 | 1 | 1 | 1 | 0 | 0 | 0 | 1 | 0 | 0 | 0 | 0 | 0 | 0 | 1 |
| 1 | 0 | 1 | 1 | 1 | 0 | 0 | 0 | 1 | 0 | 0 | 0 | 0 | 0 | 0 | 1 |
| 1 | 0 | 1 | 0 | 0 | 0 | 0 | 0 | 0 | 0 | 0 | 0 | 0 | 0 | 0 | 1 |
| 1 | 0 | 1 | 0 | 0 | 0 | 0 | 0 | 0 | 0 | 0 | 0 | 0 | 0 | 0 | 1 |
| 1 | 0 | 0 | 0 | 0 | 0 | 0 | 0 | 0 | 0 | 0 | 0 | 0 | 0 | 0 | 1 |
| 1 | 0 | 1 | 0 | 0 | 0 | 0 | 0 | 1 | 0 | 0 | 0 | 0 | 0 | 0 | 1 |
| 1 | 0 | 1 | 0 | 0 | 0 | 0 | 0 | 0 | 0 | 0 | 0 | 0 | 0 | 0 | 1 |
| 1 | 0 | 0 | 0 | 0 | 0 | 0 | 0 | 0 | 0 | 0 | 0 | 0 | 0 | 0 | 1 |
| 1 | 0 | 0 | 0 | 0 | 0 | 0 | 0 | 0 | 0 | 0 | 0 | 0 | 0 | 0 | 1 |
| 0 | 0 | 0 | 0 | 0 | 0 | 0 | 0 | 1 | 0 | 0 | 0 | 0 | 0 | 0 | 0 |

(b)

Fig. 4. (a) TSP and (b) BTSP of word "one".

A general membership function which expresses normal fuzziness, for example hot, hard, middle age, etc., has one peak. In the described case, a membership function is utilized for classification of a word pattern which consists of time-varying resonant frequencies. A vowel does not always have one peak, but in many cases has multiple peaks. The resonances of vowels are called formants. Some formants are indicated in Fig. 3. Consequently, the membership function has multiple peaks. This fact causes trouble in that the degree of each peak in the membership function is not able to reach 1.0. The first and the second formants of a vowel $j$ and the membership function of the same vowel which shows the localization of formants are indicated in Figs. 5(a) and (b).

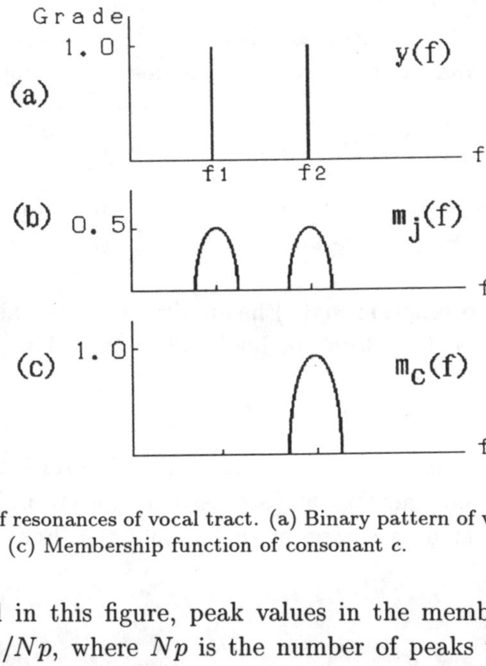


Fig. 5. Examples of resonances of vocal tract. (a) Binary pattern of vowel $j$ (b) Membership function of vowel $j$ (c) Membership function of consonant $c$.

As indicated in this figure, peak values in the membership function are normalized as $1/Np$, where $Np$ is the number of peaks that appears in the analyzed frequency band.

When all the formants correspond to the peaks of a membership function, the degree of membership should be 1. Then the degree D of membership is defined as follows:

$$Djy = \int mj(f) \wedge y(f)df \tag{2}$$

where $f$ is frequency, $y(f)$ indicates location of resonant frequencies $f1,\ f2$.

$$y(f) = \mathbb{1}(f - f1) + \mathbb{1}(f - f2) \tag{3}$$

In Eq. (3), $\mathbb{1}(f)$ presents

$$\mathbb{1}(f) = 1 \quad \text{for} \quad f = 1 \qquad\qquad (4)$$
$$\qquad 0 \qquad\qquad f \neq 0\,.$$

For a more general treatment, resonant frequencies both of vowels and consonants are represented as "local peaks." Some phonemes have different numbers of local peaks on the frequency axis. Many consonants have one local peak. The membership function of such a consonant is illustrated in Fig. 5(c) as $mc(f)$. However the degree of membership between $y(f)$, indicated in Fig. 5(a), and $mc(f)$, indicated in (c), becomes 1 from Eq. (2) although they are different phonemes. To overcome this impropriety, degree $Dz$ of anti-membership is defined as

$$D_{zjy} = \int m_{zj}(f) \wedge y(f) df\,. \qquad (5)$$

Equation (5) means if one of the local peaks correspond to zero of the membership function $mj(f)$, the location of local peaks will show a different phoneme from $j$.

If the membership function $mj(f)$ has multiple peaks, the well known relation $mzj(f) = 1 - mj(f)$ does not hold. Then the relation is modified as follows:

$$mzj(f) = 1/Np - mj(f)\,. \qquad (6)$$

Equation (6) is not important. In the algorithm described here, the degrees of membership and anti-membership are used to classify unknown patterns. To facilitate treatment by a computer these equations (Eqs. (2) and (5)) are modified.

## 2.3. *Equipment*

The membership function $mj(f,t)$ of word pattern $xj$ indicates the degree of likelihood of the word pattern $xj$. The function $mj(f,t)$ has to know the characteristics of the fluctuations among many speakers. Furthermore, it is desirable for membership functions to be constructed in the actual machine. Then, each membership function is constructed by superposition of many BT-SPs for each word. The BTSPs and the membership function of a word are indicated in Fig. 6. At first two BTSPs are superimposed and corresponding elements in each pattern are added. Next the third BTSP is superimposed on the added pattern by the same procedure. Repeating this procedure, we get

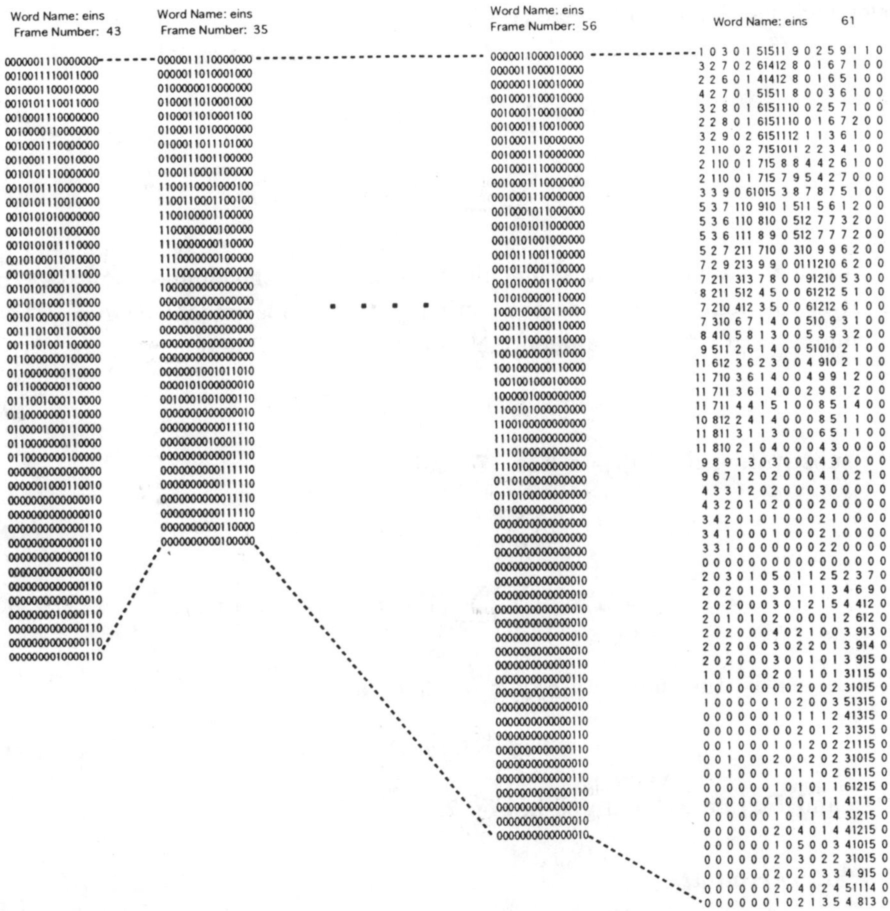


Fig. 6. Construction process of membership function of word "eins(one)".

the membership function shown on the right side in Fig. 6. This membership function has to store the information of the frequency and time fluctuation. The fluctuation of frequency has no influence on the size of the frequency axis, however fluctuation in time influences patttern length. Pattern lengths are made to be coincident by a linear expansion technique, which requires few calculations, before superimposing BTSPs. The membership function shown in Fig. 6 consists of fifteen BTSPs. Usually, since four bits are used for each element, the maxmum value of the element is 15.

For efficient processing, Eqs. (2) and (5) are modified as follows:

$$Djy = \sum_t \sum_f mj(f,t) \wedge y(f,t) \tag{7}$$

$$Dzjy = \sum_t \sum_f mzj(f,t) \wedge y(f,t)\,. \tag{8}$$

In an actual machine it is difficult to normalize by the number of peaks $Np$, so it is convienient to define the similarity to be independent on $Np$. Then assuming $m'j(f,t)$ to be a denormalized membership function,

$$m'j(f,t) = mj(f,t)/Np\,. \tag{9}$$

By substituting Eq. (9) into Eqs. (7) and (8)

$$Djy = (1/Np) \sum_t \sum_f m'j(f,t) \wedge y(f,t) \tag{10}$$

$$Dzjy = (1/Np) \sum_t \sum_f (1 - m'j(f,t)) \wedge y(f,t)\,. \tag{11}$$

Consequently, the similarity $Sjy$ is redefined as

$$Sjy = Djy/Dzjy\,. \tag{12}$$

$Djy$ and $Dzjy$ are illustrated in Fig. 7.

## 2.4. *Experiments*

The scheme of the recognition system is shown in Fig. 8. The voice signal is converted to frequency domain of the range of 250 Hz to 6250 Hz. After this processing, fifteen points are derived from the frequency analyzed signal every 10 ms. In the recognition process the unknown voice utterance is converted to a BTSP. The BTSP is superimposed on each membership function template in a dictionary and the similarity is calculated. In the last step the word name for the pattern that gets the highest similarity to the unknown BTSP is output as the recognition result.

A typical set of 120 Japanese command words for office automation equipment was selected for the recognition experiment. Spoken utterance of these words by many people were recorded on audio tape. Reproduced voices are input to a computer via an $A/D$ convertor. In the first step, only the voice part is segmented and separated from the surrounding sound using a segmentation program. Next, the segmented data is added to the recognition algorithm program.

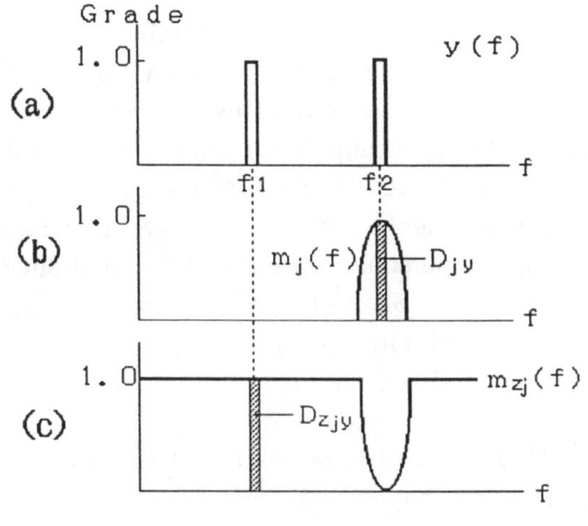


Fig. 7. Relation of $D_{jy}$ and $D_{zjy}$.

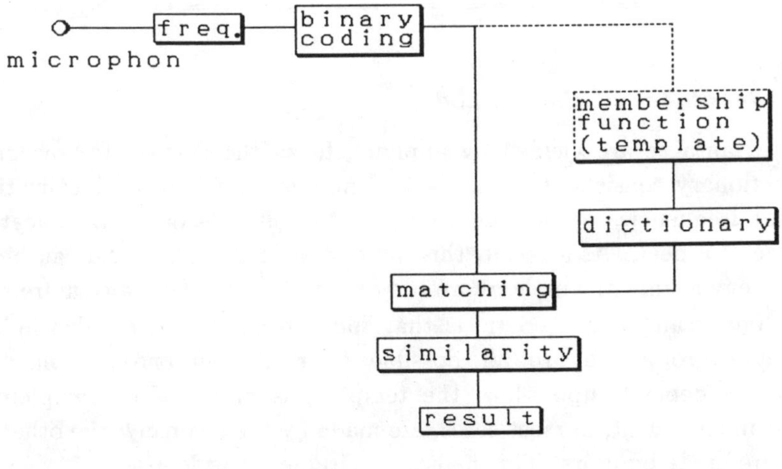


Fig. 8. Recognition scheme. (Dot line indicates speaker dependent.)

In the speaker-independent case the average recognition rate, 93.2% was achieved for 80 persons (40 females and 40 males).

Two foreign languages were tested for recognition rates employing this algorithm. One was English and the other was German. 92.8% and 93.7% were achieved respectively when enrolling 120 words.

In the speaker-dependent case, 98.5% is achieved under the same conditions as that of the speaker-independent's.

If the users of this recognition system can be limited to a few, then speaker-independent templates are not as efficient. In such a case, a higher recognition rate is possible to achieve by means of training the system where all members speak all enrolled words. In this group fluctuations among a limited number of people are accounted for in membership functions. This usage is considered a compromise between speaker-dependence and speaker-independence. Consider a group that consists of ten persons. This method was applied to 4 Japanese groups. The results were 97.1%, 95.1%, 95.0% and 96.5%. This usage is considered suitable for a small office group.

These results are indicated in Table 1.

Table 1. Recognition rate for three languages.
Enrolling 120 words.

| | Japanese | English | German |
|---|---|---|---|
| Speaker-independent | 93.2 | 92.8 | 93.7 |
| Groupe use | 95.9 | | |

## 2.5. *Advantages of the method*

Some advantages are yielded by applying fuzzy theory to voice recogniton. The dictionary consists of membership functions that indicate localization of resonant frequencies of the vocal tract. One bit is enough to indicate the presence of a peak. As a result, this method needs less data and can be used in noisy environment. Furthermore the calculations for recognition are small.

An important characteristic is that speaker-dependent, speaker-independent, and group use are all possible to realize in one system. The application depends upon how the template is made. The templetes for speaker-independent, in most cases, are made by the company; the other templates are made by users. This makes it suitable for wide usage.

An example of the merits of mixing the use of speaker dependence and independence is described. Three speaker-independent templates allow an easy enroll system to be available in a toy. The templates of three words, "YES", "NO" and "REPEAT", should been enrolled before shipment.

*Procedure*

When a user turns on the main switch of the toy, which is the voice synthesis machine, they are greeted with "Thank you for buying this great toy."

"Please train this machine with your own voice."

"First, please say the command TURN TO THE RIGHT for turning the machine to the right after the beep sound."

"Are you ready?"

"If you are ready, please say YES."

"If you want to quit enrollment, please say NO."

"If you cannot understand, please say REPEAT, then I will explain in more detail."

(user) "Yes."

(machine) "Your voice is a little bit soft. I cannot hear you completely".

"Please repeat more loudly."

(user) "YES"

(machine) "Ok." "Please follow me."

"TURN TO THE RIGHT"

(user) "TURN TO THE RIGHT"

........

........

At last if the the three word "YES," "NO" and "REPEAT" are enrolled by the user, this machine becomes a complete speaker-dependent machine.

## 3. Noise Reduction

A noise reduction method, which was developed for practical use, is described in this section.

As mentioned in the introduction, noise is the biggest enemy for voice recognition. If we want to use speech recognition in the field, this problem has to be solved.

The recognition method described here has good tolerance for noise. Only the distributions of the local peaks are used as the feature for recognition. This has the advantage of picking up voice in noisy conditions. However it is not sufficient on its own.

It has been known that training under the same noise conditions as when the machine was in the case of speaker-dependent, actual use was the best way to improve recognition in noisy environments. It is of course true but not realistic except in special cases. In most cases noise changes and often it is difficult to train under the same conditions as when in actual use, such as in a driving car, in a market, or in a factory, etc. In these cases one needs to train in a runnning car, in a crowded market or an operating factory. Furthermore, the noise level can still change.

The most widely known method is SSM (Spectrum Subtraction Method).[3] In this method the noise spectrum $N'(f)$ is detected before using the system. When the system get a voice input signal $X(f)$, the noise spectrum is substracted from the input signal.

$$S'(f) = X(f) - N'(f)\,. \tag{13}$$

The signal $X(f)$ is the addition of voice signal $S(f)$ and noise one $N(f)$, then

$$S'(f) = S(f) + N(f) - N'(f)\,. \tag{14}$$

If $N(f)$ approaches $N'(f)$ then the method operates optimally. Since all data is treated in the spectrum domain in this method, the calculations are simple. It is suitable for practical use.

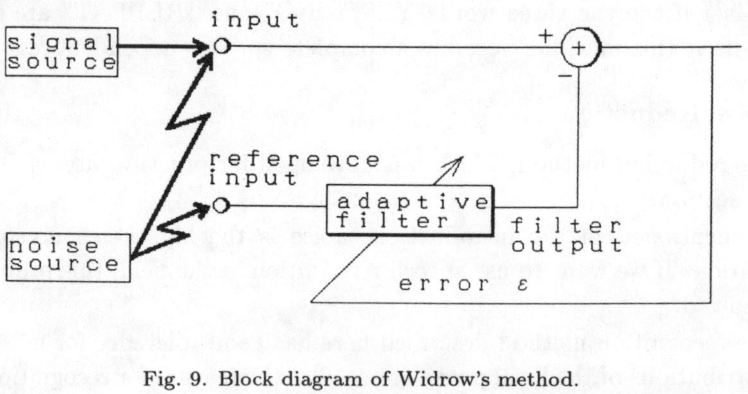


Fig. 9. Block diagram of Widrow's method.

It is rare that the relation $N(f)$ is nearly equal to $N'(f)$ for a long time. When the actual noise changes from the training noise spectrum, the method not only loses the noise canceling ability but also has bad influences on the detected signal.

As a countermeasure to this, adaptive training was developed. The noise spectrum is detected whenever the voice signal is absent.

Typically, noise is always changing in a dynamic manner. This is a limitation of the method. As a result, this method is not suitable for a dynamic noise condition.

A famous adaptive noise reduction is proposed by Widrow.[4] In this method the signal and noise are considered to have no correlation. A block diagram is shown in Fig. 9.

By minimizing the square of the error $\varepsilon$,

$$\begin{aligned}\varepsilon &= x(t) - n'(t) \\ &= [s(t) + n(t)] - n'(t)\,,\end{aligned} \tag{15}$$

the $\varepsilon$ approaches to desired voice signal, where

$$x(t) = s(t) + n(t)\,. \tag{16}$$

He proposed LMS algorithm for this calculation. For a voice recognition system the calculation is not small. Furthermore our system uses a frequency domain rather than time domain signal. Our method[5,6] is a modification of Widrow's in frequency domain. We also use a main and reference microphone. The block diagram is shown in Fig. 10. Fourier transforms of $x(t)$ and $n'(t)$ are denoted as $X(f)$ and $N'(f)$, respectively. Assuming the signal from the

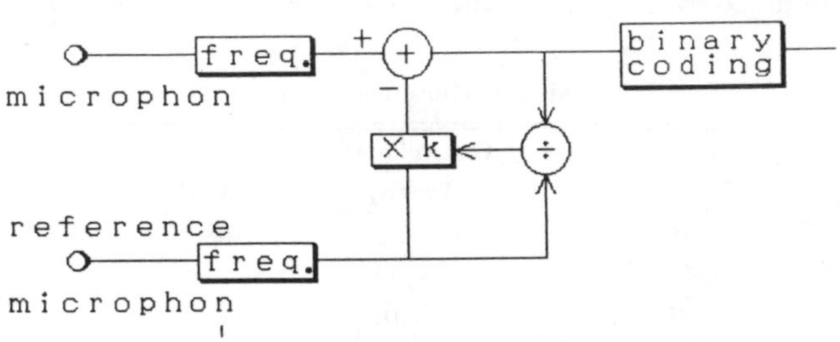


Fig. 10. Block diagram of noise reduction.

main microphone is $X(f)$ and that from the reference microphone is $N'(f)$ in the frequency domain, the output signal will be:

$$E = X(f) - k(f) \cdot N'(f), \tag{17}$$

where $k(f)$ is a constant. Before any utterances are emitted, if the constant $k$ is adjusted to the $E$ to be 0, we can get the required signal in actual usage. In the method the reference microphone always monitors the noise such that noise reduction can be achieved even if the noise changes. Few further calculations are necessary. These are the main advantages for actual use.

A limitation of this method is related to motion of the noise source, which are changes of noise loudness between the main and reference microphones.

The performance of this method was checked in cars traveling on highways. Firstly the speaker-dependent method was checked.

The training was done in a stopped car which was almost noiseless, then recognition was tested in a running car during traveling. The main microphone was set on the sun visor of the front window, the other microphone was located at the back of the seat. The main microphone was about 25 cm away from the speaker's mouth. Two experiments were done. One using voice data recorded in a speeding car with 100 km/h on a highway and open windows. The other was done using voice data recorded in speeding car on a city street. The noise conditions are indicated in Table 2. The uttered words are 60 Japanese city names. By taking the average results of three males, 97.8% and 93.9% are obtained on the city street and the highway, respectively. When the car stereo was turned on, the rate went down to 71.1%.

In contrast, a common word recognition system, on the market at that time, gave almost no results. We could train the system in the quiet conditions, but in many cases it gave no result if we uttered a word, or gave unexpected results caused by the noise even without the music from the car stereo.

Table 2. Noise conditions.

| Condition | Car velocity (km/h) | Noise level (dBA) |
|---|---|---|
| Stop | 0.0 | 58 |
| City street | $0 \sim 50$ | $58 \sim 65$ |
| Highway | $80 \sim 100$ | $70 \sim 73$ |
| With music | $80 \sim 100$ | $73 \sim 80$ |

Before that, it was difficult to realize a noise-robust speaker-independent system. We tried to check a speaker-independent method in the same condition. We collected the voice data of 15 persons in both moving and stopped cars. The voice data of the 3 persons who were already used for speaker-dependent recognition was combined with 12 others to make speaker-independent templetes. The average results for the three persons are also indicated in Table 3. If the standard pattern could be tuned up using more data, better results could be obtained.

Table 3. Recognition rate in noise (60 words).

| Condition | Speaker-dependent | Speaker-independent |
|---|---|---|
| Stop | 98.3 % | 87.2% |
| City street | 97.8 | 81.1 |
| Highway | 93.9 | 73.3 |
| With music | 71.1 | |

For commercial products, improvements were done on the following: On the microphone, many thresholds, capacity of word number, etc. Consequently, the required accuracy for actual use was obtained.

As in Eq. (15), the minimization of Eq. (17) gives accurate noise reduction in conditions where noise changes dynamically. For minimizing error, the fuzzy control is useful and is currently being studied.

## 4. Applications

### 4.1. *Products*

The first speaker-independent system, model RV100I, which used the fuzzy pattern matching is shown in Fig. 11. This product, which had an RS232C interface and could recognize 120 words, has been on the market since 1985. The size, 235 × 440 × 70 mm, was the smallest but it did not have noise reduction. The price of the system was ¥500,000, fantastically low at that time. After that we studied a more user-acceptable system. As a result, the following products were brought into the market.

Two types of boards and two types of LSI sets are available. In these products both speaker-dependent and speaker-independent methods can be used. The user can choose according to their requirement. The external view of the small LSI(RF5A128) is shown in Fig. 12.

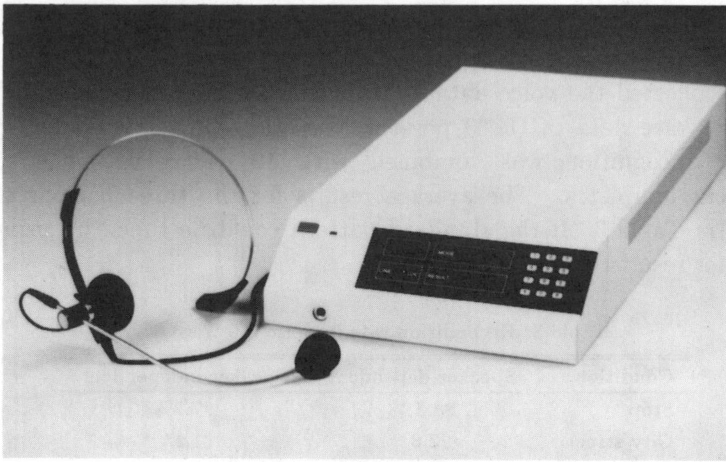

Fig. 11. RV100I — The first product of voice recognition using fuzzy pattern matching.

Fig. 12. Voice recognition LSI using fuzzy pattern matching.

The first LSI is small and cheap with a capacity to recognize 13 words. The other is a set of 3 LSIs which can accept 120 words.

One of the boards has the same specification as the LSI set. The other board is PC-VOICE RECOGNIZER. This will be described in the next section.

## 4.2. *Applications*

Some applications using the fuzzy pattern matching are introduced here.

*PC-VOICE RECOGNIZER*

A diagram of this is shown in Fig. 13. This product needs no software when used for voice recognition. The recognition software is residented in the computer memory. The computer has a conversion table between recognition result

Fig. 13. PC-VOICE RECOGNIZER.

and codes of the keyboard. If one speaks "return," PC-VOICE RECOGNIZER will recognize the word, and the result is converted into the appropriate keyboard code. Consequently, the computer treats the result in the same way as the command input from the keyboard. A big advantage is that many software packages, which were not made specially for voice recognition, can be operated by voice command. Many software packages on the market were checked to determine whether the commands from PC-VOICE RECOGNIZER were accepted or not. There were very few packages that would not accept voice input.

As an example, the famous software LOTUS 1-2-3 will be shown here. The first part of the procedure is to make the PC-VOICE RECOGNIZER software inherent, the second is selecting the conversion table of LOTUS, followed by voice training. If "LOTUS" is said, the program starts without modification to the software.

For example, when one wants to sum the numerical table in the program using the keyboard, five or six keys must be hit after hitting the "ESC" key. In the same program, if PC-VOICE RECOGNIZER is used, at first the command "SUM" is uttered, following it the cursor is moved from the start point to the end point saying only "FROM HERE" and "TO HERE," then we can get the required sum. If a pie chart is required, only the voice command "PIE CHART" should be uttered. Again five or six keys must be hit if the keyboard is used.

Many word processing software accept our PC-VOICE RECOGNIZER other than LOTUS 1-2-3. Using this software, it is possible to enter commands by voice, and it also interests the user. For instance if "MY HOME" is uttered, then the address will appear on the display, or when "I'M TIRED" is uttered, the program will end.

*Applications of the LSI*

Applications of the LSI products are introduced here. One use of voice recognition is in applications for handicapped people. Two such applications are shown here: One is a bed and the other is a wheel chair.

There are patients who cannot move themselves. This bed is for such people who want to move their bed to watch TV or to do something else. This can be done without help. If they say "HEAD UP," the head part of the bed goes up little by little. "HEAD DOWN," "LEG UP" or "LEG DOWN" are also applicable. The commands, "MORE UP" or "MORE DOWN" are also acceptable. Furthermore it has some external switches which are convienient for switching on or off connected equipment. If one says "LIGHT ON/OFF" or "TV ON/OFF," the equipment will be adjusted accordingly. The bed is shown in Fig. 14.

A voice-controlled wheelchair was made with a similar application. One who has trouble using hands, can control the chair by voice, e.g., "GO," "TURN TO RIGHT," or "STOP."

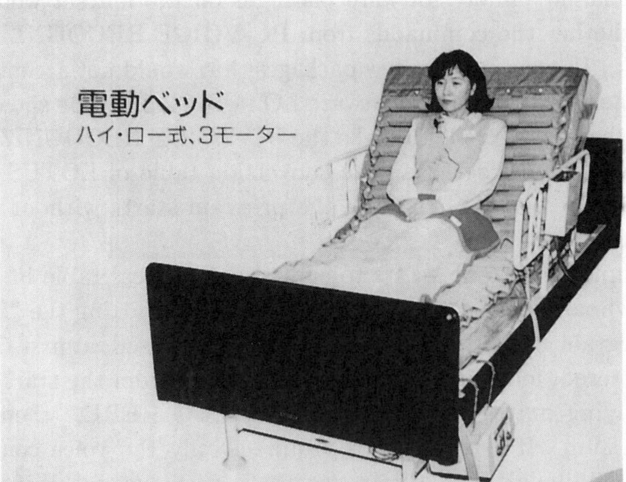


Fig. 14. Voice controllable bed.

## 5. Conclusion

This voice recognition system is the first voice recognition product which adopted the fuzzy theory. It was then introduced at the International Convention of Fuzzy Sets Association in 1987.[7]

By using the fuzzy pattern matching algorithm, it became easy to realize speaker-independent recognition. It became easy to make a dictionary of reduced data and size. That in itself is a big advantage. The next important characteristic is the mixed use of speaker-dependent, speaker-independent and group use in one system. Differences depend upon how the templates are made. This merit makes the usage wider.

After studying usable systems, the noise reduction method was developed. That made it possible to use the system in a noisy environment.

The described system is only for word recognition. It is the first step for our hope to realize voice recognition. Although our dream is still far away, from a different viewpoint, however, it has become possible to use voice recognition even if it has limited ability in word recognition.

While fuzzy theory has been a common technique and has many applications, it has only just begun to be widely used in voice recognition. The fuzzy vector quantization is an example of this technique.[8] There are many other possible applications for the theory. It is hoped that fuzzy theory will enhance the development of voice recognition.

## Acknowledgment

The author thanks all members of RICOH Voice Recognition teams in R&D Group and Electric Device Division. The author also wishes to thank Mr. J. F. Cullen of RICOH California Research Center, who was also a member of the team, for his help in revising the English of this manuscript.

## References

1. T. Muroi, *Proc. Acoust. Soc. Jap.* **3**, 1 (1993).
2. J. Fujimoto, T. Nakatani, and M. Yoneyama, *Fuzzy Sets and Systems* **32**, 181 (1989).
3. S. F. Boll, *IEEE Trans.* **ASSP-27**, 2 (1979).
4. B. Widrow *et al.*, *Proc. IEEE* **63**, 1692 (1975).
5. S. Nakamura *et al.*, *Proc. IECE* (1989), p. A-1-10.
6. T. Ariyoshi, M. Matsushita, and J. Fujimoto, *Proc. Acoust. Soc. Jap.* **3**, 10 (1990).
7. J. Fujimoto *et al.*, *IFSA Congress* (1987), p. 819.
8. H. P. Tseng, M. Sabin, and E. Lee, *Proc. ICASSP* (1987), p. 641.

# INTELLIGENT HOME APPLIANCES USING FUZZY TECHNOLOGY

NOBORU WAKAMI, HIROYOSHI NOMURA and SHOICHI ARAKI

*Central Research Laboratories,*
*Matsushita Electric Industrial Co. Ltd.,*
*3-4 Hikari-dai Seika Souraku, Kyoto 619-02, Japan*

## 1. Introduction

Fuzzy logic has been applied to many kinds of home appliances in Japan since 1990. The fuzzy logic enables the home appliances to be intelligent and easy to operate by dealing with qualitative knowledge of experts.

In this chapter, we explain about the following four topics:

a. Application to a washing machine.
An application of fuzzy inference to a washing machine is introduced. This fuzzy washing machine went on sale in 1990. It triggered a flood of applications of fuzzy inference to many kinds of home appliances in Japan.

b. Application to an air-conditioning system.
An air-conditioning system with a pyroelectric infrared rays detector can identify the number of occupants in a room from thermal images. A fuzzy clustering algorithm is applied to distinguish occupants from the background.

c. Self-tuning method of fuzzy rules : "*neuro-fuzzy*".
The fuzzy inference enables home appliances to be intelligent easily. However, when the number of input variables and fuzzy rules increase, it is difficult to design the fuzzy rules. In order to overcome this problem, some methods were developed to tune fuzzy rules by a learning function of neural networks. In this section, a self-tuning method of fuzzy rules is introduced. This method is called "*neuro-fuzzy*" techonology.

d. Application of the *neuro-fuzzy* technology.
As mentioned above, by means of the *neuro-fuzzy* technology, we can design fuzzy rules automatically. In this section, as an example of *neuro-fuzzy* application, we introduce a kerosene fan heater. The kerosene fan heater estimates an index of thermal comfort (PMV value) by fuzzy inference and controls amount of kerosene combustion, direction of wind, etc. according to the estimated PMV value.

## 2. Application to a Washing Machine

The first home appliance[1] in the Japanese market that adopted the application of the fuzzy inference was a fully automatic washing machine. This washing machine went on sale in 1990. It triggered a flood of applications of fuzzy inference to many kinds of home appliances (vacuum cleaner, rice cooker, air-conditioner, microwave oven, etc.) in Japan. (see Fig. 1)

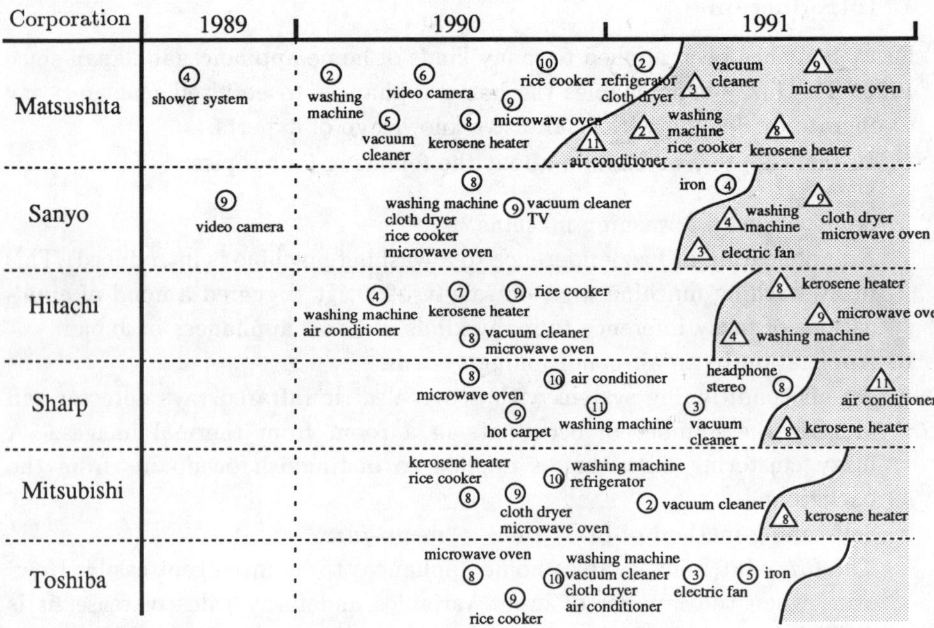


Fig. 1. Home appliances using fuzzy logic and *neuro-fuzzy* technologies.

In order to wash clothes skillfully and easily, we applied the fuzzy inference to the washing machine based on the basic principle of laundry: "When dirt has been removed, washing is stopped".

Using the fuzzy inference, the optimum washing time is determined from outputs of a wash sensor. The wash sensor can measure the dirtiness of the laundry by detecting the transmittance of the wash water with an optical sensor. The mechanism of the wash sensor and the functions realized by the fuzzy inference are outlined in the following:

## 2.1. *Mechanism of the wash sensor*

Figure 2 shows the composition of the washing machine. The wash sensor is installed near the drain valve. Figure 3 shows the mechanism of the wash sensor in detail. The wash sensor consists of an infrared LED (light emitting diode) and a photo transistor. The light beam generated from the infrared LED, passes through the wash water in the pipe and enters the photo transistor. The photo transistor outputs a voltage in proportion to the intensity of the light.

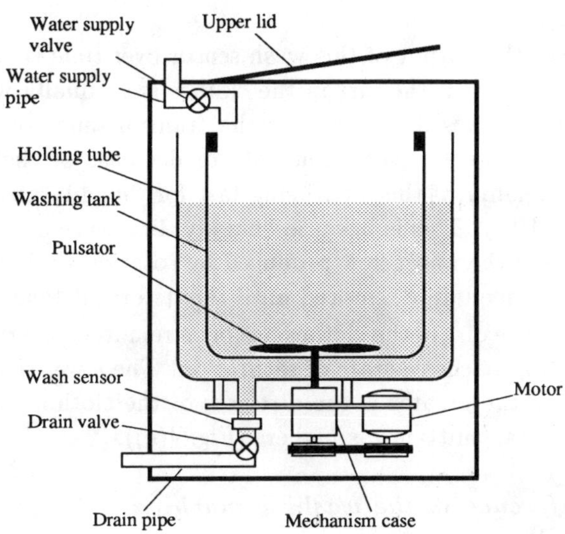


Fig. 2. Composition of a washing machine.

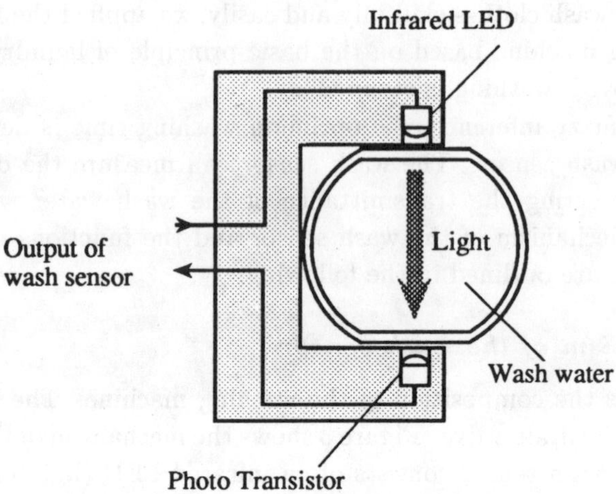


Fig. 3. Wash sensor.

The voltage is the output of the wash sensor, which shows the transmittance of the wash water.

The change of the output of this wash sensor over time is shown in Fig. 4. When washing is started, the dirt in the clothes is gradually washed out and the wash water becomes dirty, causing the transmittance of wash water to decrease as shown in Fig. 4(a). The rate of decrease of the transmittance depends on the quality of the dirt, being fast for "muddy dirt" and slow for "oily dirt" (Fig. 4(b)). This is because "muddy dirt" is removed easily by the mechanical force of the water flow produced by rotation of the pulsator, while "oily dirt" is not adequately removed until the detergent takes effect.

When most of the dirt in the clothes has been removed, the transmittance of the wash water approaches a state of saturation. The lower the transmittance at the saturation, the greater is the dirtiness of the clothing; the dirtiness is lighter when the transmittance is higher. (Fig. 4(c)).

## 2.2. *Fuzzy inference in the washing machine*

It is difficult to obtain the optimum relation between dirtiness and washing time experimentally, because there are numerous kinds of dirtiness (wash sensor output patterns). Collecting detailed experimental data for all such types is nearly impossible. In addition, since the washing mechanism in the washing

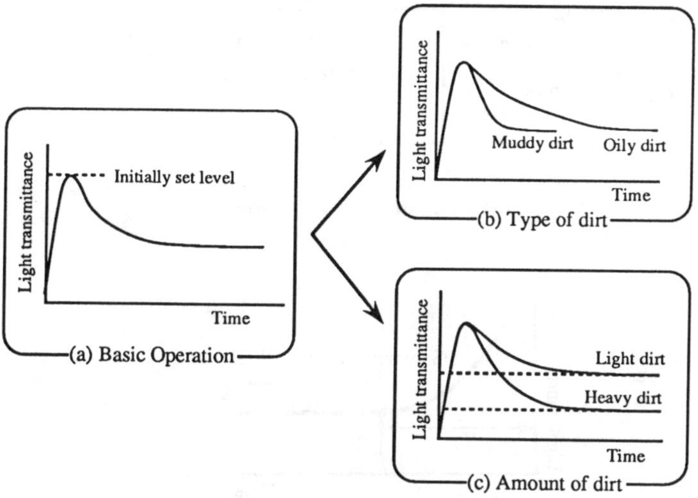


Fig. 4. Change of the output of wash sensor.

machine is not touched on sufficiently and the relation between dirtiness and wash time is not strictly linear, it is difficult to obtain an unique numerical formula relating them. Therefore, in order to determine the wash time from the output of the wash sensor, we employed the fuzzy inference. Using the fuzzy inference, the relation between the dirtiness and the wash time is expressed by linguistic rules, and the nonlinear property of the relation is approximated by an interpolation function of the fuzzy inference.

Figure 5(a) shows a diagram of the fuzzy inference in the washing machine and Fig. 5(b) explains input/output variables used in the fuzzy inference. From the time the transmittance takes to reach saturation, $T_s$, (which is related to the type of dirt), and the output level at the saturation, $V_s$, (which is related to the amount of dirt), the remaining wash time $W_t$ is determined by the fuzzy inference. Some fuzzy rules to determine the wash time are shown below.

- If the transmittance $V_s$ is low and the saturation time $T_s$ is long, the wash time $W_t$ is very long ($T4$).

(a) Diagram of fuzzy inference.

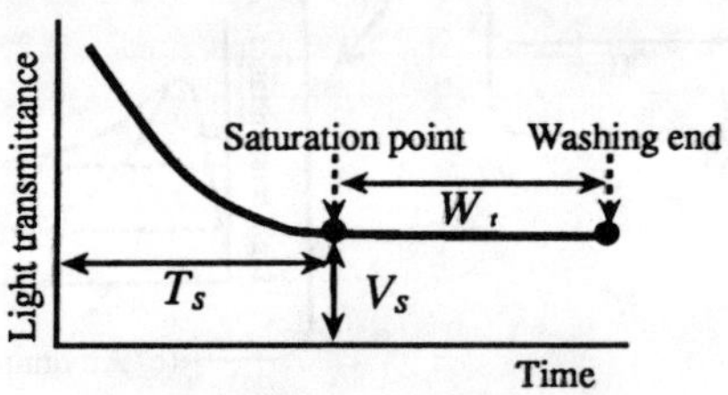


(b) Input/output variables.

Fig. 5. Fuzzy inference in washing machine.

- If the transmittance $V_s$ is high and the saturation time $T_s$ is short, the wash time $W_t$ is very short ($T3$).

Table 1 shows the rule table, where $T_1 \sim T_6$ are wash times. Figure 6 shows the membership functions using in the fuzzy rules. In this case, we employed a simplified fuzzy inference as a fuzzy inference.

Table 1. Rule table.

| $T_s$ \ $V_s$ | Low | Middle | High |
|---|---|---|---|
| Short | $T_1$ | $T_2$ | $T_3$ |
| Long | $T_4$ | $T_5$ | $T_6$ |

In the simplified fuzzy inference, the consequent part of the fuzzy rule is expressed by a real number. Therefore, the simplified fuzzy inference enables one to simplify the inference operations and the tuning of the membership

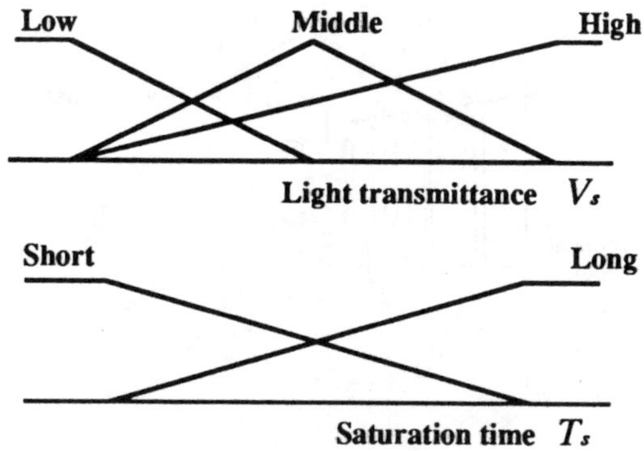


Fig. 6. Membership functions.

function, and to reduce the amount of required memory. In this case, the creation of rules and the tuning of membership functions were carried out based on skilled launderers' know-how.

By using the fuzzy inference to the fully automatic washing machine, the non-linear relation between the degree of dirtiness and the wash time can be expressed by rules, and it makes it possible to determine the optimum wash time based on the amount and types of dirt detected by the wash sensor.

As a result of the fuzzy inference, excessive washing or inadequate washing can be prevented, and saving of both energy and time can be realized.

## 3. Application to an Air-Conditioning System

Recently image processing by fuzzy logic has been actively applied to video equipments (televisions, video tape recorders, ... ). Particularly in video cameras, many functions (auto-focus control, auto-iris control, electronic image stabilizer etc.) have been improved by fuzzy logic.[2,3] These image processing technologies have began to be applied to air-conditioning systems. In this section, we explain an air-conditioning system with "eyes."

In order to achieve a more comfortable thermal environment, we have to detect not only physical environmental conditions (air temperature, humidity, etc.) but also the information about the occupants in a room (the number, positions, etc.). A pyroelectric infrared ray detector is developed to take a thermal image inside the room.[4] We introduce a segmentation algorithm of a

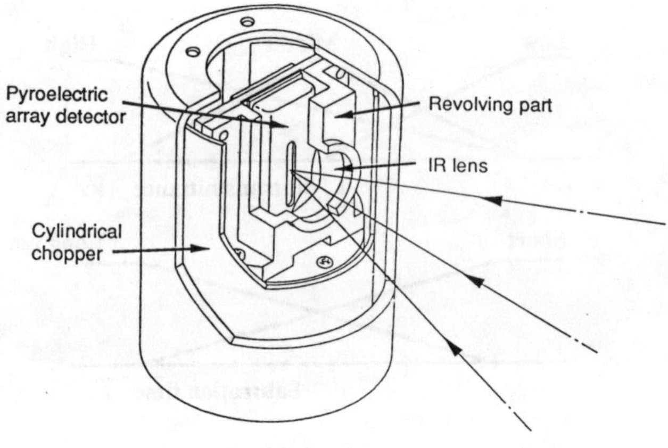


Fig. 7. Infrared rays detector.

thermal image to identify the number and positions of occupants using fuzzy logic.[5]

### 3.1. *Structure of the pyroelectric infrared rays detector*

A schematic structure of the developed pyroelectric infrared rays detector is shown in Fig. 7. The sensor consists of a one-dimensional linear pyroelectric array detector, a horizontal scanning mechanical body, a silicon infrared rays (IR) lens and a mechanical chopper. The array detector which has 8 elements and the IR lens are united to be a revolving part scanning horizontally. The part is located in the center of a cylindrical chopper. The chopper has 3 windows, so that it opens and closes at 60 degree intervals. It opens and closes three times for one rotation. The revolving part and the chopper are linearly rotated at a constant ratio. A two-dimensional thermal image is produced by a horizontal scanning of the vertical array detector. An example of thermal image and its corresponding visual image are shown in Figs. 8(a) and 8(b), respectively. The thermal image is composed of $8 \times 20$ elements (see Fig. 8(c)). Each element has a value of temperature expressed by 8 bit.

### 3.2. *Identifying the number of occupants*

A segmentation algorithm to identify the number of occupants is described here. The algorithm consists of three procedures: Removal of background, identification of each occupant, and region growing. The principle of our approach is based on the feature of thermal images that the temperature of occupants is higher than that of the background and each occupant has a local temperature peak in an image. We explain our algorithm using the image shown in Fig. 8(a).

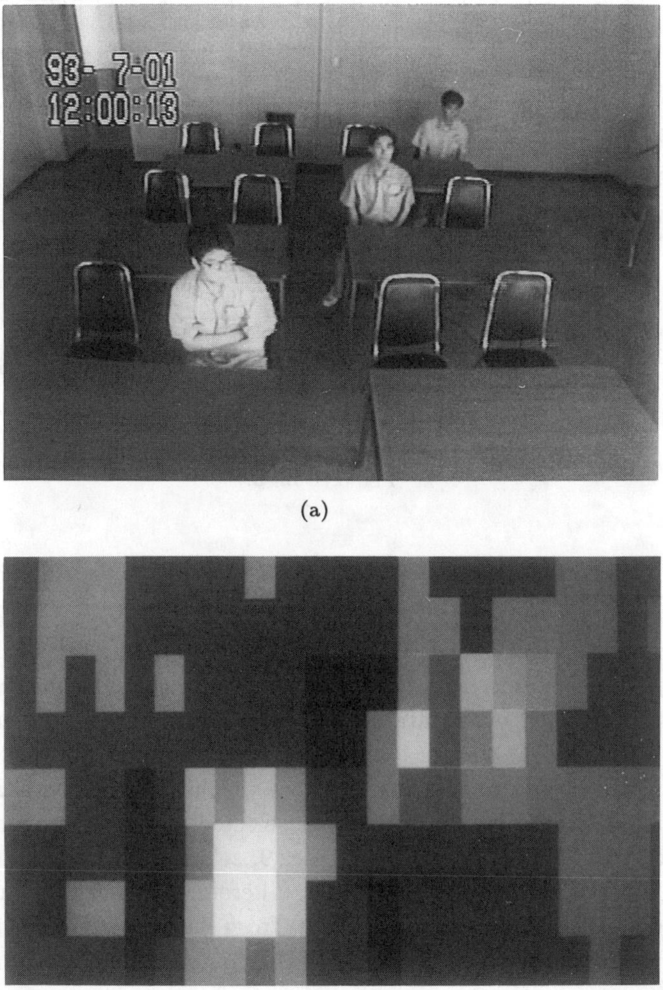


(a)

(b)

Fig. 8. An example of image.

search step

| element number | 1 | 2 | 3 | 4 | 5 | 6 | 7 | 8 | 9 | 10 | 11 | 12 | 13 | 14 | 15 | 16 | 17 | 18 | 19 | 20 |
|---|---|---|---|---|---|---|---|---|---|---|---|---|---|---|---|---|---|---|---|---|
| 1 | 81 | 84 | 82 | 64 | 68 | 74 | 75 | 80 | 74 | 72 | 73 | 69 | 85 | 75 | 76 | 77 | 77 | 80 | 83 | 75 |
| 2 | 78 | 82 | 82 | 71 | 73 | 71 | 63 | 73 | 72 | 62 | 72 | 66 | 78 | 78 | 69 | 83 | 86 | 84 | 83 | 76 |
| 3 | 80 | 72 | 82 | 74 | 82 | 73 | 70 | 71 | 74 | 62 | 55 | 59 | 96 | 101 | 111 | 132 | 118 | 84 | 71 | 68 |
| 4 | 71 | 75 | 77 | 65 | 71 | 75 | 72 | 72 | 66 | 42 | 38 | 80 | 164 | 162 | 131 | 113 | 86 | 74 | 68 | 66 |
| 5 | 78 | 76 | 69 | 55 | 63 | 120 | 151 | 135 | 93 | 66 | 66 | 79 | 107 | 108 | 98 | 93 | 91 | 81 | 82 | 80 |
| 6 | 53 | 69 | 77 | 32 | 50 | 149 | 215 | 207 | 133 | 84 | 73 | 70 | 70 | 67 | 63 | 65 | 73 | 78 | 84 | 78 |
| 7 | 54 | 78 | 87 | 51 | 47 | 121 | 186 | 197 | 138 | 77 | 71 | 58 | 66 | 70 | 72 | 65 | 69 | 82 | 89 | 85 |
| 8 | 62 | 72 | 76 | 57 | 65 | 79 | 92 | 102 | 86 | 64 | 65 | 56 | 64 | 67 | 64 | 72 | 73 | 76 | 82 | 73 |

(c)

Fig. 8 (*Cont'd.*)

### 3.2.1. *Removing background using the fuzzy c-means algorithm*

A fuzzy *c*-means algorithm (FCM) is applied to remove the background, or to distinguish occupants from the background in a thermal image. Clustering of FCM is performed on a data set of temperature of $8 \times 20$ elements of an image (see Fig. 8(c)). Figure 9 shows an example of clustering result. The elements which belong to the cluster whose weighted mean temperature is the highest with high membership values are selected as the component of regions representing occupants. As shown in Fig. 9, selected elements depend on the threshold level of the membership value. Therefore, we set the threshold level at 0.5 to roughly distinguish the occupants from the background at first. Figure 10 shows the initial selected elements as the components of occupants.

### 3.2.2. *Identifying the number of occupants*

In order to distinguish each occupant, local temperature peaks in an image are located from among the elements selected as occupants. A particular element

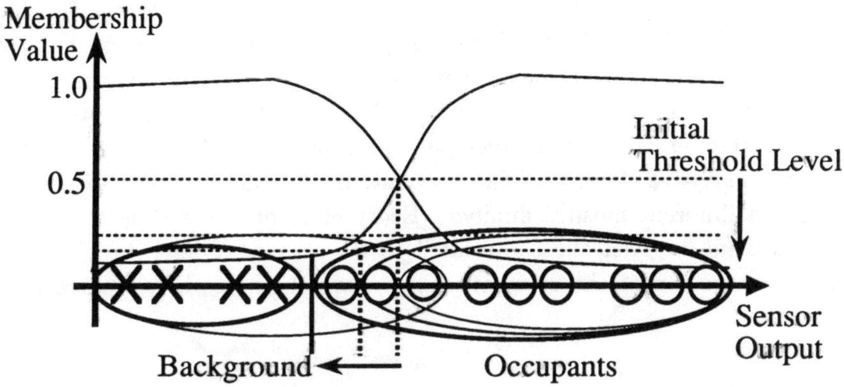


Fig. 9. An example of clustering result.

search step

| element number | 1 | 2 | 3 | 4 | 5 | 6 | 7 | 8 | 9 | 10 | 11 | 12 | 13 | 14 | 15 | 16 | 17 | 18 | 19 | 20 |
|---|---|---|---|---|---|---|---|---|---|---|---|---|---|---|---|---|---|---|---|---|
| 1 | 81 | 84 | 82 | 64 | 68 | 74 | 75 | 80 | 74 | 72 | 73 | 69 | 85 | 75 | 76 | 77 | 77 | 80 | 83 | 75 |
| 2 | 78 | 82 | 82 | 71 | 73 | 71 | 63 | 73 | 72 | 62 | 72 | 66 | 78 | 78 | 69 | 83 | 86 | 84 | 83 | 76 |
| 3 | 80 | 72 | 82 | 74 | 82 | 73 | 70 | 71 | 74 | 62 | 55 | 59 | 96 | 101 | 111 | 132 | 118 | 84 | 71 | 68 |
| 4 | 71 | 75 | 77 | 65 | 71 | 75 | 72 | 72 | 66 | 42 | 38 | 80 | 164 | 162 | 131 | 113 | 86 | 74 | 68 | 66 |
| 5 | 78 | 76 | 69 | 55 | 63 | 120 | 151 | 135 | 93 | 66 | 66 | 79 | 107 | 108 | 98 | 93 | 91 | 81 | 82 | 80 |
| 6 | 53 | 69 | 77 | 32 | 50 | 149 | 215 | 207 | 133 | 84 | 73 | 70 | 70 | 67 | 63 | 65 | 73 | 78 | 84 | 78 |
| 7 | 54 | 78 | 87 | 51 | 47 | 121 | 186 | 197 | 138 | 77 | 71 | 58 | 66 | 70 | 72 | 65 | 69 | 82 | 89 | 85 |
| 8 | 62 | 72 | 76 | 57 | 65 | 79 | 92 | 102 | 86 | 64 | 65 | 56 | 64 | 67 | 64 | 72 | 73 | 76 | 82 | 73 |

Fig. 10. Initial selected elements.

will search for an element which has the highest temperature among its eight neighboring elements. If the temperature of the searched element is higher than the original one, the searched one is taken to be the parent of the original one. The result of peak location is shown in Fig. 11. An arrow linked between two elements points the parent. Elements with no parent are the local peaks of temperature. A distinct peak surrounded by a circle is regarded as the representative element of an occupant. Each peak is given a distinct label such as a different natural number. Every element selected as a component of occupants is given the same label as its parent. Shown in Fig. 12, the region composing of the elements which have the same label represents one occupant.

search step

| element number | 1 | 2 | 3 | 4 | 5 | 6 | 7 | 8 | 9 | 10 | 11 | 12 | 13 | 14 | 15 | 16 | 17 | 18 | 19 | 20 |
|---|---|---|---|---|---|---|---|---|---|---|---|---|---|---|---|---|---|---|---|---|
| 1 | 81 | 84 | 82 | 64 | 68 | 74 | 75 | 80 | 74 | 72 | 73 | 69 | 85 | 75 | 76 | 77 | 77 | 80 | 83 | 75 |
| 2 | 78 | 82 | 82 | 71 | 73 | 71 | 63 | 73 | 72 | 62 | 72 | 66 | 78 | 78 | 69 | 83 | 86 | 84 | 83 | 76 |
| 3 | 80 | 72 | 82 | 74 | 82 | 73 | 70 | 71 | 74 | 62 | 55 | 59 | 96 | 101 | 111 | 132 | 118 | 84 | 71 | 68 |
| 4 | 71 | 75 | 77 | 65 | 71 | 75 | 72 | 72 | 66 | 42 | 38 | 80 | 164 | 162 | 131 | 113 | 86 | 74 | 68 | 66 |
| 5 | 78 | 76 | 69 | 55 | 63 | 120 | 151 | 135 | 93 | 66 | 66 | 79 | 107 | 108 | 98 | 93 | 91 | 81 | 82 | 80 |
| 6 | 53 | 69 | 77 | 32 | 50 | 149 | 215 | 207 | 133 | 84 | 73 | 70 | 70 | 67 | 63 | 65 | 73 | 78 | 84 | 78 |
| 7 | 54 | 78 | 87 | 51 | 47 | 121 | 186 | 197 | 138 | 77 | 71 | 58 | 66 | 70 | 72 | 65 | 69 | 82 | 89 | 85 |
| 8 | 62 | 72 | 76 | 57 | 65 | 79 | 92 | 102 | 86 | 64 | 65 | 56 | 64 | 67 | 64 | 72 | 73 | 76 | 82 | 73 |

Fig. 11. Peak location.

### 3.2.3. *Region-growing algorithm*

The region-growing algorithm settles the regions of occupants in an image. It is performed by lowering the threshold level of the membership value of

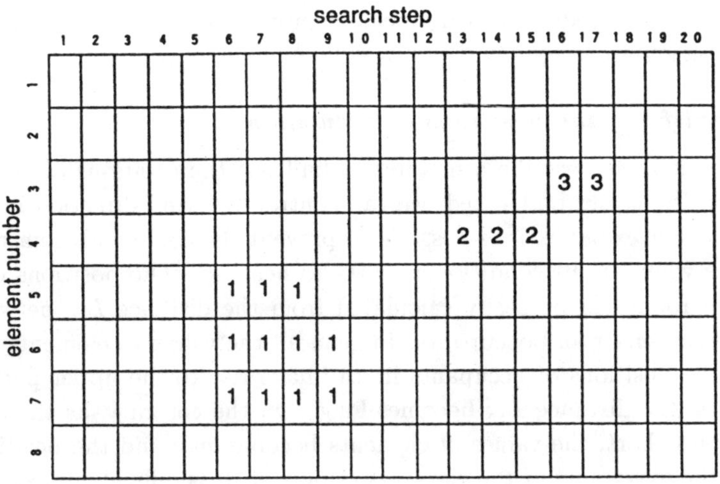


Fig. 12. Initial label.

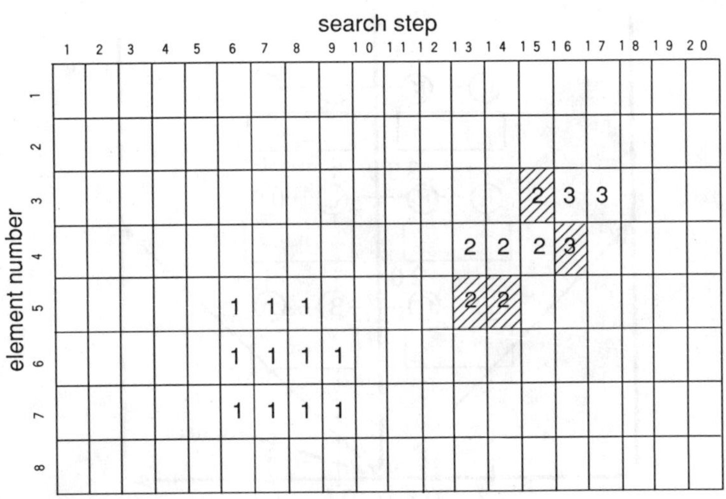


Fig. 13. Result of region-growing.

elements to be regarded as occupants under the restriction that the number of occupants determined by locating peaks is not changed. Figure 13 shows the result of region-growing.

### 3.3. *Identifying the positions of occupants*

In this section, we describe a method to find out the positions of occupants. If their positions can be located, we can control the wind direction in various situations. For example, it is possible to prevent the cool wind from reaching occupants when an air-conditioner starts to heat up. The positions of occupants in a room can be easily calculated from the distance $L_{so}$ between the infrared rays sensor and occupants. In general, the values of elements become low and the positions of occupants in an image rise to the upper part of the elements as the distance $L_{so}$ becomes long. On the contrary, as the distance $L_{so}$ becomes short, the values of elements become high and the positions descend to the lower part of the elements. However, it is difficult to express this relationship in a simple model. Therefore, we tried to identify the relationship between the distance $L_{so}$, the output of local temperature peak and its element number using fuzzy rules.

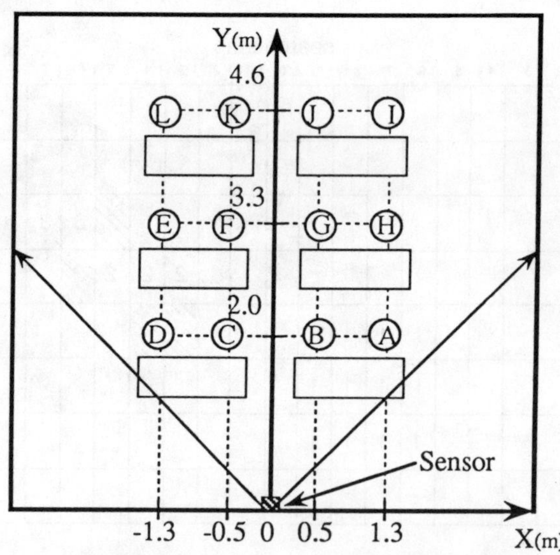


Fig. 14. Room layout for experiment.

Table 2. Reasoning result of distance.

| Position | Peak temp. | Element no. | Reasoning result | Real distance |
|---|---|---|---|---|
| A | 162 | 6 | 2.58 | 2.63 |
| A | 188 | 6 | 2.52 | 2.63 |
| B | 203 | 6 | 2.35 | 2.34 |
| B | 193 | 6 | 2.48 | 2.34 |
| C | 207 | 6 | 2.36 | 2.34 |
| C | 201 | 6 | 2.41 | 2.34 |
| D | 209 | 6 | 2.34 | 2.63 |
| D | 184 | 6 | 2.56 | 2.63 |
| E | 173 | 4 | 3.57 | 3.72 |
| E | 161 | 4 | 3.68 | 3.72 |
| F | 175 | 4 | 3.55 | 3.52 |
| F | 176 | 4 | 3.55 | 3.52 |
| G | 168 | 4 | 3.61 | 3.52 |
| G | 161 | 4 | 3.68 | 3.52 |
| H | 167 | 4 | 3.62 | 3.72 |
| H | 167 | 4 | 3.62 | 3.72 |
| I | 129 | 3 | 4.86 | 4.91 |
| I | 131 | 3 | 4.84 | 4.91 |
| J | 128 | 3 | 4.87 | 4.76 |
| J | 138 | 3 | 4.78 | 7.76 |
| K | 131 | 3 | 4.84 | 4.76 |
| K | 125 | 3 | 4.89 | 4.76 |
| L | 136 | 3 | 4.80 | 4.91 |
| L | 121 | 3 | 4.92 | 4.91 |

Figure 14 shows a room layout for the experiment. The sensor is set at 2.3 m at above the floor and its decline angle is 28 degree. We generated nine fuzzy rules for the estimation of the distance $L_{so}$.[6] Reasoning results are shown in Table 2. The average error between the estimated value and real value is less than $0.1m$. These results show that the fuzzy inference can estimate the distance $L_{so}$ with desirable accuracy.

## 4. Self-Tuning Method of Fuzzy Rules

As mentioned above, the fuzzy inference enables home appliances to be intelligent easily. However, when the numbers of input variables and rules increase, it is difficult to design fuzzy rules. In conventional fuzzy systems, human operators have to determine the rules and the membership functions by trial and error, and waste much time to optimize the rules.

In order to overcome this problem, we developed some methods to design fuzzy rules automatically applying a learning function of neural networks.[6,7,10,11] In this section, we introduce a self-tuning method by a delta rule. This method is called "neuro-fuzzy" technology.

### 4.1. *Self-tuning method by a delta rule*

We explain the self-tuning method,[9,10] i.e., a method to tune the fuzzy rules by a delta rule. The delta rule is a kind of learning algorithm of a neural network.[8]

This self-tuning method has the following rule format.

$$\text{Rule } i\text{: If } x_1 \text{ is } A_{i1} \text{ and } \cdots \text{ and } x_m \text{ is } A_{im} \text{ then } y \text{ is } w_i \tag{1}$$

where $i$ is a rule number ($i = 1, \ldots, n$), $x_1, \ldots, x_m$ are input variables, $A_{ij}$ is a fuzzy set, and $w_i$ is a real number. This is the simplified fuzzy inference rule. The membership function in the antecedent part has a triangular form as shown in Fig. 15.

The output $y$ of the fuzzy inference is derived from the following equations.

$$\mu_i = \prod_{j=1}^{m} \mu_{A_{ij}}(x_j) \tag{2}$$

$$y = \frac{\sum_{i=1}^{n} \mu_i \cdot w_i}{\sum_{i=1}^{n} \mu_i} \tag{3}$$

where $\mu_{A_{ij}}$ is a membership function of the fuzzy set $A_{ij}$, and $\mu_i$ is a membership value of $i$th rule.

Figure 16 shows a block diagram of this self-tuning method. The simplified fuzzy inference can be expressed by the three-layered network structure, which have processing units (○), fuzzy sets $A_{ij}$, and real numbers $w_i$ (□). The processing unit of the output layer is to derive the weighted average in accordance with Eq. (3), and the processing units of the hidden layer are to calculate the membership value $\mu_i$ using Eq. (2). In this method, the tuning

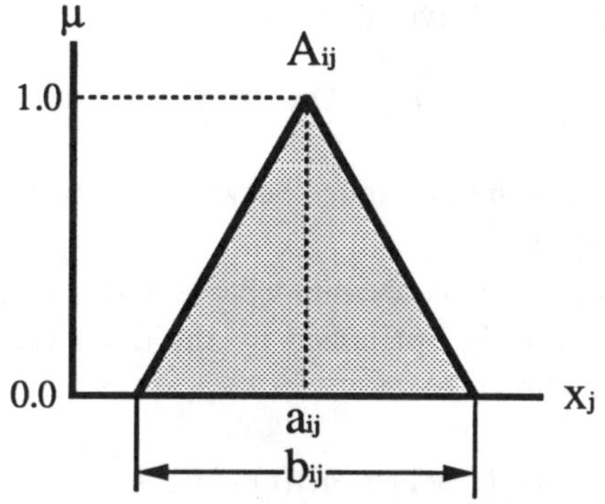


Fig. 15. Membership function.

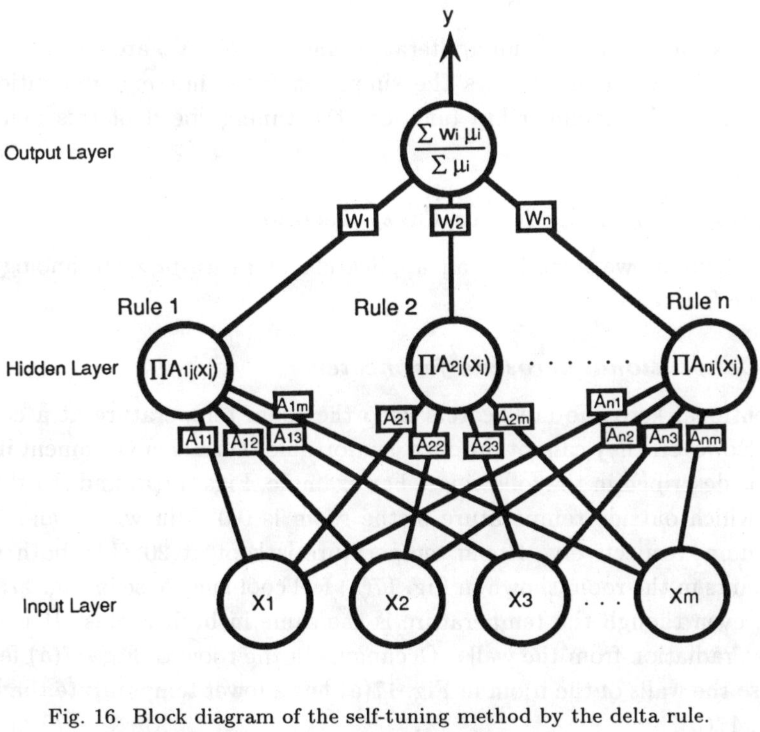


Fig. 16. Block diagram of the self-tuning method by the delta rule.

parameters expressing by (□) are iteratively adjusted by the delta rule to minimize the following objective function $E$.

$$E = \frac{1}{2}(y' - y)^2 . \tag{4}$$

The objective function is the difference between the output $y$ of fuzzy inference and the desirable output $y'$.

The parameters $w_i$ of the consequent parts and the $a_{ij}, b_{ij}$ of the membership functions are iteratively optimized by the delta rule. The learning rules of the parameters are shown below.

$$a_{ij}(t+1) = a_{ij}(t) - K_a \cdot \frac{\partial E}{\partial a_{ij}} \tag{5}$$

$$b_{ij}(t+1) = b_{ij}(t) - K_b \cdot \frac{\partial E}{\partial b_{ij}} \tag{6}$$

$$w_i(t+1) = w_i(t) - K_w \cdot \frac{\partial E}{\partial w_i} \tag{7}$$

where $t$ is the number of tuning iteration and $K_a, K_b, K_w$ are constants.

Since this method employs the simplified fuzzy inference operation and simple triangular membership function, the tuning speed of this method is fast.

## 5. Application of the Neuro-Fuzzy Technology

In this section, we introduce an application of *neuro-fuzzy* technology to a kerosene fan heater.

### 5.1. *Conventional kerosene fan heater*

Conventional kerosene fan heaters keep the room temperature at a constant level. However, they cannot produce comfortable thermal environment in some cases as described in the followings: For example, Figs. 17(a) and (b) show the room which outside temperature of the room is 0.0°C in winter and 10.0°C in autumn, respectively. Room temperature is kept at 20°C in both rooms. Occupants in the room shown in Fig. 17(a) feel cool and those in Fig. 17(b) feel warm, even though the temperature is the same in both rooms. It is caused by heat radiation from the walls. Occupants in the room of Fig. 17(a) feel cool because the walls of the room in Fig. 17(a) has a lower temperature than those in Fig. 17(b).

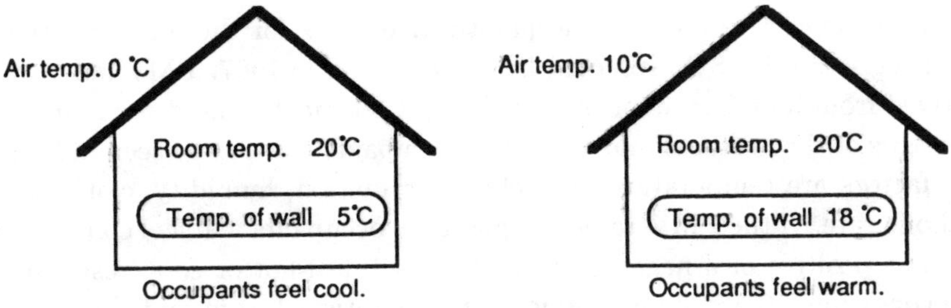


(a) In winter (b) In autumn

Fig. 17. Thermal sensation in different seasons.

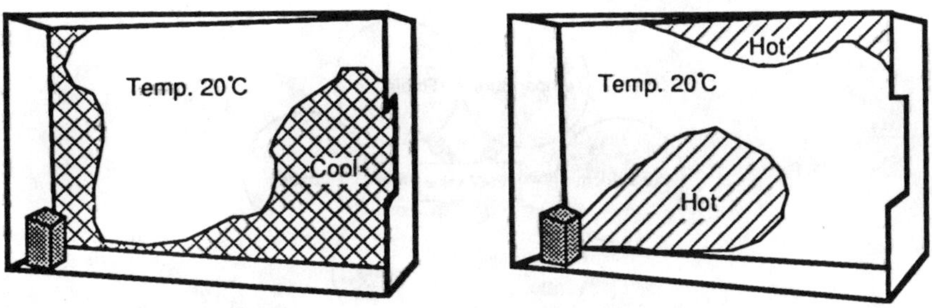


(a) In the night (b) In the day

Fig. 18. Thermal sensation at different times in a day.

Figures 18(a) and (b) show another example. Season and room temperature are assumed to be the same in both rooms but the time is in the night and day respectively. Occupants feel cool in the meshed region shown in Fig. 18(a) since the temperature of the wall is low and no sunlight is in the room in the night. On the contrary, occupants feel hot in the obliquely lined region shown in Fig. 18(b) because the wall temperature is high and the sun shines into the room in the day.

In such cases, conventional fan heater could not always produce a comfortable thermal environment for occupants without considering some factors influencing the thermal sensation of the occupants, such as radiation heat.

### 5.2. *PMV index*

PMV (Predicted Mean Vote) is a quantitative index for thermal sensation of human registered in ISO presented by P .O. Fanger in 1967. PMV index value is derived from four factors of the thermal environment around occupants and two factors of the personal information of occupants. As can be seen in Fig. 19, PMV factors are temperature, air velocity, radiation, humidity, clothing and metabolic rate. If the PMV value is equal to 0, occupants will feel comfortable. If it has a positive or a negative value, they will feel hot or cool, respectively. If the condition $-0.5 < (\text{PMV value}) < 0.5$, is satisfied, it is said that at least 80% of the occupants feel comfortable. For this reason, fan heaters should be controlled to satisfy this condition. However, it is too difficult for a fan heater which is apart from the occupants to detect PMV factors. Then, we try to estimate the PMV value from the information which can be detected by built-in sensors in a fan heater.

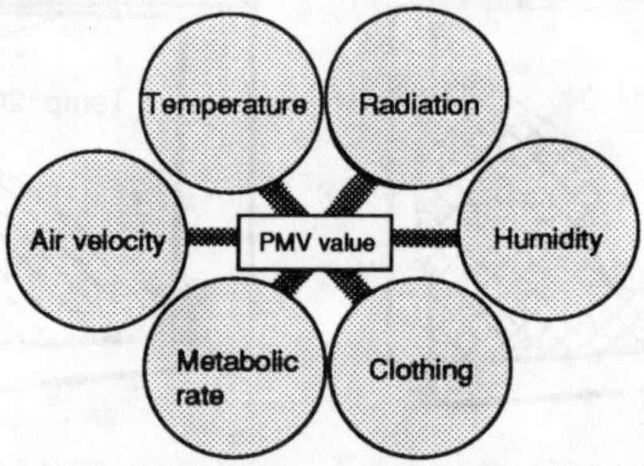


Fig. 19. PMV factors.

### 5.3. *Relationship between the PMV factors and sensor information of a fan heater*

A fan heater is equipped with some sensors detecting the temperature, radiation and humidity in a room. The amount of kerosene combustion, louver angle and the amount of wind can be controlled by a fan heater. It is well-known to engineers of a fan heater that information detected by a fan heater is mutually related to the PMV factors. Figure 20 shows their relationship. The clothing and metabolic rate of occupants can be regarded as a constant value because fan heaters are utilized only in winter.

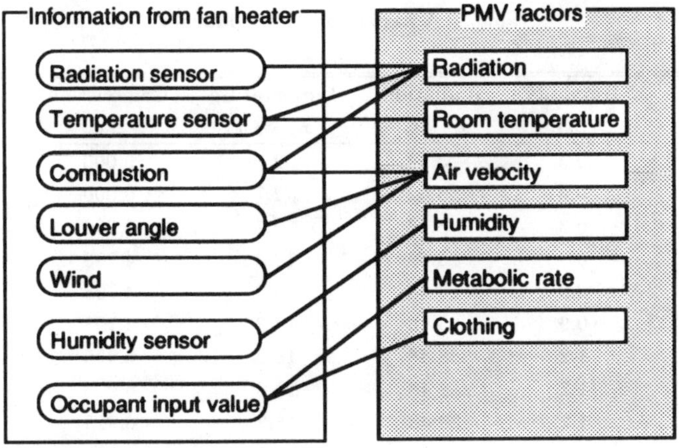


Fig. 20. Relationship between PMV factors and sensor information

However, there are some cases that a scantily-clad or frequently moving occupant is in the room. To deal with such cases, we consider a set point of temperature input by the occupants. It reflects their clothing and activities. Its value is converted into the occupant input value representing his clothing and activities. The louver angle and the amount of wind are not considered because they are controlled according to the amount of kerosene combustion. In accordance with above consideration, we estimate the PMV value by fuzzy reasoning from the occupant input value, temperature, radiation, humidity and amount of kerosene combustion. *Neuro-fuzzy* technology is applied to identify the relationship between the PMV value and the detected information by a fan heater in fuzzy rules.

### 5.4. *Construction of fuzzy rules for reasoning PMV value*

We applied *neuro-fuzzy* technology to construct fuzzy rules for reasoning PMV value from the five inputs ($x_0$, $x_1$, $x_2$, $x_3$, $x_4$) shown below.

$x_0$ : occupant input value

$x_1$ : output value of humidity sensor

$x_2$ : output value of radiation sensor

$x_3$ : output value of temperature sensor

$x_4$ : amount of kerosene combustion

Table 3. A part of observed data set.

| Input value $x_0$ | Humidity $x_1$[Kpa] | Radiation $x_2$ [°C] | Temperature $x_3$ [°C] | Combustion $x_4$ | PMV $y^r$ |
|---|---|---|---|---|---|
| 2.0 | 0.3 | 3.42 | 2.7 | 0.000 | −2.7070 |
| 1.0 | 0.3 | 3.42 | 2.7 | 0.000 | −5.6250 |
| 0.1 | 0.3 | 3.42 | 2.7 | 0.000 | −14.8780 |
| 2.0 | 0.9 | 23.68 | 14.3 | 4.798 | −0.5680 |
| 1.0 | 0.9 | 23.68 | 14.3 | 4.798 | −7.4100 |
| 0.1 | 0.9 | 23.68 | 14.3 | 4.798 | −7.8070 |
| 2.0 | 1.2 | 28.48 | 18.2 | 6.925 | −0.4060 |
| 1.0 | 1.2 | 28.48 | 18.2 | 6.925 | −0.9820 |
| 0.1 | 1.2 | 28.48 | 18.2 | 6.925 | −5.3550 |
| 2.0 | 1.8 | 33.88 | 22.7 | 9.220 | 1.3060 |
| ⋮ | ⋮ | ⋮ | ⋮ | ⋮ | ⋮ |

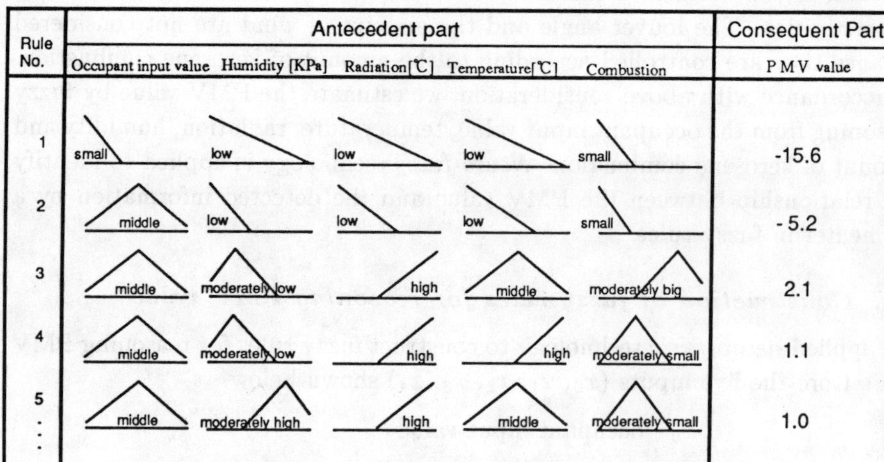


Fig. 21. A part of acquired rules and their membership functions.

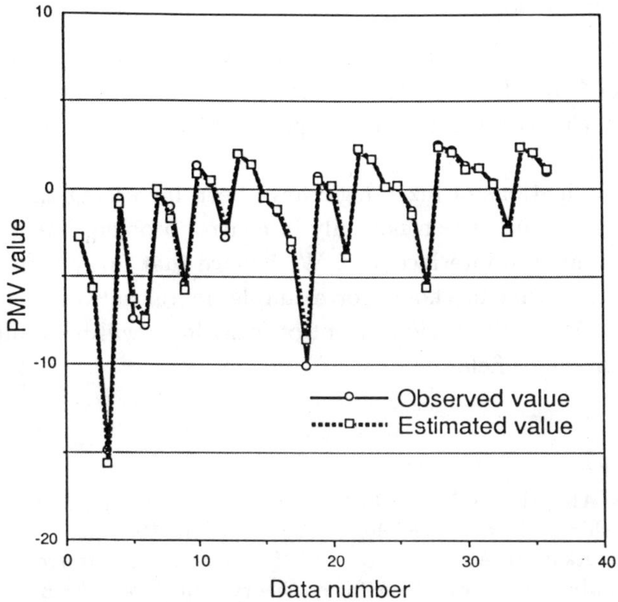


Fig. 22. Reasoning result of the PMV value.

Data sets for constructing fuzzy rules were collected in a laboratory which can produce various thermal environment. Under some representative thermal environments, input values ($x_0$, $x_1$, $x_2$, $x_3$, $x_4$) and PMV value $y^r$ were observed. A part of the observed data set is shown in Table 3. Using the observed data set, fuzzy rules were constructed by applying the *neuro-fuzzy* technology. The number of fuzzy rules was set to be 675. A part of the acquired rules and their membership functions are shown in Fig. 21. The third rule means "If occupant input value is moderate, humidity is moderately low, radiation is high, temperature is average, the amount of combustion is moderately high, and the PMV value is equal to 2.1. The rule is interpreted that the PMV value is big because radiation is high even if temperature is moderate. Acquired rules are successful in identifying the relationship between the PMV value and the detected information by a fan heater. The reasoning result of the PMV value is shown in Fig. 22. Circles "○" show the observed PMV value and squares "□" show the estimated PMV value by fuzzy reasoning. In this case, the average reasoning error is reduced to 0.051.

## 6. Conclusions

In this chapter, we explained four topics. By using fuzzy logic, all applications described in this chapter are successful in adding new functions and improving the control result compared with the conventional home appliances.

In Japan, applications of fuzzy logic have been increasing since 1990, and research is being carried out on not only in control problem but also in information retrieval, human interface, etc. We believe that fuzzy logic will evolve by fusing with the other methods (for example neural network, genetic algorithm, chaos, etc.) and the basic concept of fuzzy logic will be applied to even more products in wider fields.

## References


1. S. Kondo, S. Abe, H. Terai, M. Kiuchi, and H. Imahashi, "Fuzzy logic controlled washing machine," in *Proc. IFSA '91* (1990), pp. 97–100.
2. Y. Egusa, H. Akahori, A. Morimura, and N. Wakami, "An electronic video camera image stabilizer operated on fuzzy theory," in *Proc. IEEE Int. Conf. on Fuzzy Systems* (1992), pp. 851–858.
3. Y. Egusa, H. Akahori, A. Morimura, and N. Wakami, "An application of fuzzy set theory for an electronic video camera image stabilizer," *IEEE Transactions on Fuzzy Systems* (forthcoming).
4. N. Yoshiike, K. Arita, and K. Morinaka, "Human information sensor," in *Proc. 7th Int. Conf. on Solid-State Sensor and Actuators* (1993), pp. 1015–1018.
5. S. Araki, H. Nomura, and N. Wakami, "Segmentation of thermal images using the fuzzy c-means algorithm," in *Proc. IEEE Second Int. Conf. on Fuzzy Systems* (1993), pp. 719–724.
6. S. Araki, H. Nomura, I. Hayashi, and N. Wakami, "A self-generating method of fuzzy inference rule," in *Proc. Int. Fuzzy Engineering Symposium* (1991), pp. 1047–1058.
7. I. Hayashi, H. Nomura, and N. Wakami, "Artificial neural network driven fuzzy control and its application to the learning of inverted pendulum system," in *Proc. 3rd IFSA Congress* (1989), pp. 610–613.
8. D. E. Rumelhurt, G. E. Hinton, and R. J. Williams, "Learning internal representations by error propagation," in *Parallel Distributed Processing* (1986), pp. 318–362.
9. H. Nomura, I. Hayashi, and N. Wakami, "A self-tuning method of fuzzy reasoning by delta rule and its application to a moving obstacle avoidance," *Journal of Japan Society for Fuzzy Theory and Systems* **4**, 2 (1992) (in Japanese).

10. H. Nomura, I. Hayashi, and N. Wakami, "A learning method of fuzzy inference rules by descent method," in *Proc. IEEE Int. Conf. on Fuzzy Systems* (1992), pp. 203–210.
11. H. Nomura, I. Hayashi, and N. Wakami, "A self-tuning method of fuzzy reasoning by genetic algorithm," in *Proc. Int. Fuzzy System and Intelligent Control Conference* (1992), pp. 236–245.

# FUSION TECHNOLOGY OF FUZZY AND CHAOS THEORY, AND ITS APPLICATIONS

RYU KATAYAMA

*Hypermedia Research Center, Sanyo Electric Co., Ltd.*
*1-18-13, Hashiridani, Hirakata, Osaka, 573, Japan*
*Phone: +81-720-41-7854 Fax: +81-720-41-1413*
*E-mail: katayama@hr.hm.rd.sanyo.co.jp*

## 1. Introduction

As well as fuzzy logic and neural networks, a novel paradigm, chaos theory,[2,3] is now emerging as one of the key technology of soft computing.[1] The term "chaos" means, in the scientific field, the complicated and noise-like phenomena originated from nonlinearity involved in deterministic dynamical systems.[4] Recently, there is a growing interest to discover the law of nature hidden in these complicated phenomena, and the attempt to utilize chaos theory for engineering purpose is now quickening.[4–11,69,75] Chaos is not only an important research field in science, but also has potential to be applied in many fields such as information processing based on dynamical system theory,[12,13] control and suppression of chaotic vibrations in mechanical systems and electrical systems,[14,16] synchronization of chaos and communication using chaos,[15] nonlinear signal processing such as adaptive equalization,[17] plant control,[48] analysis and estimation of biological phenomena,[18] or even the prediction of economic market price and analysis of sociological phenomena.[19,66,75] This new interdisciplinary field is called "chaos engineering" or "engineering chaos", which is defined by K. Aihara, as generic studies on technological foundations of chaos.[5–9]

On the other hand, fuzzy theory is widely applied to ill-structured problems such as control, evaluation, and decision-making problems whose mathematical model cannot be easily obtained, by expressing the experts' knowledge and skilled operators' experience using linguistic rules. In particular, fuzzy control was put into practice in the field of plant control such as a cement kiln plant, water purification plant, automatic train operation based on predictive fuzzy control and so on,[20] where conventional control theory cannot be easily applied because of the lack of the mathematical model. Furthermore, fuzzy control has been widely applied to a number of consumer electronics products and home appliances from the end of the '80s to the beginning of the '90s, especially in Japan, Korea, Taiwan and other South-east Asian countries.[21] Concerning the acquisition of fuzzy rules and membership functions, where cut and try methods are used in the past, optimization techniques, neural networks, and neuro-fuzzy technology have been introduced to automate the tuning process of fuzzy models,[22] and their application areas have grown rapidly.

In accordance with these trends, new interdisciplinary researches to combine chaos engineering with fuzzy control, and neuro-fuzzy technology are rapidly growing.[7–11] For example, by analyzing chaotic signals observed from natural and artificial complex systems and finding out hidden deterministic rules by the use of fuzzy rules and neuro-fuzzy techniques, we can interpret and identify the deterministic laws using the linguistic rules, whose mathematical model is not easily obtained. These approaches make it possible to predict, control, and identify complex phenomena which were regarded as noise in the past.

In this chapter, we survey fusion researches of chaos and fuzzy theory, and especially focus on nonlinear prediction methods based on neuro-fuzzy technology,[10,59] which are now intensively studied.

## 2. Fusion Technology of Fuzzy and Chaos

Research on the fusion technology of fuzzy and chaos theory was initiated by Diamond.[23] In order to extend state space from real vector space to fuzzy set space and consider dynamical systems defined on the fuzzy set space, it is required to introduce some kind of topology in the fuzzy set state space. Diamond introduced Haudorff metric as this metric, and established sufficient conditions that mapping from real vector space $R^n$ to fuzzy set space is chaotic, and called it "fuzzy chaos."[23,26] On the other hand, conventional chaos or chaotic mapping defined from $R^n$ to $R^n$ is called "crisp chaos".[27] Kloeden extended Diamond's work and derived sufficient conditions that mapping defined by difference equation from fuzzy set space to fuzzy set space is chaotic.[24,25]

As a numerical experimental research, Horiuchi *et al.* considered a sequence which is obtained by iterative mapping of fuzzy sets, and classified several operators such as $t$-norm, averaging, and absolute difference by examining the dynamical behavior of the sequence.[28] Yamakawa *et al.* analyzed the transition of fuzzy information and shape of membership functions in the iterative fuzzy inference system with extension principle.[29] Grim regarded a multiple fuzzy inference system with self reference as a difference dynamical system, and examined the behavior of fuzzy information using liar's paradox.[30] He also demonstrated that strange attractors and fractal structures are observed in these iterative systems.

In conjunction with fuzzy sets and fractals, Ezawa *et al.* introduced fuzzy Koch curve and fuzzy Julia Set, where iterative mapping constants are given by triangular fuzzy numbers.[31] They also showed that the fractal dimensions for fuzzy Koch curves are given by fuzzy numbers.

There are other researches on fuzzy and chaos, where the freedom of parameters which defines the characteristic of fuzzy sets is more restrictive. Teodorescu treated "chaotic fuzzy systems," where only characteristic parameters of membership functions are chaotic.[27] He also classified these membership functions into "strong chaotic," "chaotic," and "weak chaotic" according the complexity of their behavior. Possible application areas are modeling of chaos in economy and sociological phenomena.[19,34,66,75]

Diamond's approach is a purely mathematical one where fuzzy sets are treated in points in Banach space. However, the physical meaning to consider fuzzy sets as state space vectors in dynamical systems is not clarified yet. On the other hand, it is worthwhile to consider a system where the meanings of symbols are given by fuzzy sets, and their meanings are dynamically changed based on the time evolution of dynamical systems.[12,13]

On this point of view, Sato *et al.* proposed a chaotic fuzzy associative memory system, where a chaotic memory search is applied to carrying out inference on a fuzzy associative memory system to support human creative thinking.[32]

Katai *et al.* discussed chaotic structures of fuzzy symbolic dynamics, in conjunction with the hierarchical structure of natural systems and problem solving by self-organization.[33]

It is considered that these approaches may initiate a new field such as symbol processing and knowledge information processing based on dynamical systems.

## 3. Fuzzy Control and Chaos

Concerning the OGY(Ott, Grebogi, York) method, which is famous as a control method of chaos, see the work of Ott *et al.*[14] With respect to the synchronization of chaos and communication using chaos, refer to Pecola.[15]

Since fuzzy control is in general a kind of nonlinear control, stability theory for nonlinear control systems[35] and stability theory for fuzzy control systems[36] are useful design guides to avoid chaos in fuzzy control systems. There is another study to use Lyapunov spectrum for evaluating the extent of stability in inverted pendulum controlled by fuzzy logic.[37] TDOF (two degree of freedom) control is also proposed to control a system where disturbance signal or plant dynamics are chaotic.[16] TDOF fuzzy control is also proposed.[76] However, it is not applied to the control of chaotic systems, and thus it is considered to be a interesting future research topic. All of these are concerned with design principles to avoid chaos in control systems.

Other studies concerning fuzzy control systems and chaos are nonlinear prediction for unknown disturbance and load combined with fuzzy control system,[8,48] application of fuzzy logic system to control bouncing ball, where fuzzy controller is designed by the use of genetic algorithm.[38]

## 4. Prediction of Chaotic Time Series by Self Generating Neuro-Fuzzy Model

### 4.1. *State space reconstruction by embedding*

The basic idea of nonlinear prediction is to identify a deterministic law which governs the behavior of complex systems from observed time series data. Therefore, nonlinear prediction is carried out as follows:

[1] Observation of time series data.
[2] State space reconstruction by embedding.
[3] Identification of nonlinear dynamics in the reconstructed state space.
[4] Nonlinear prediction using identified dynamics.

Let us explain the state space reconstruction of the dynamics based on the embedding. We consider a general $n$-dimensional discrete time dynamical system defined as follows:

$$\mathbf{z}(t+1) = \mathbf{F}(\mathbf{z}(t)) \tag{1}$$

where $\mathbf{z}(t) \in R^n$ is a state vector and $\mathbf{F}: R^n \to R^n$ is a $C^1$ map in the $n$-dimensional state space. In general, it is usually very difficult to observe all the elements of $\mathbf{z}(t)$ in experiments. In other words, one can usually measure only a single variable time series $y(t) \in R$ through the observation mapping $g: R^n \to R$ such that

$$y(t) = g(\mathbf{z}(t)). \tag{2}$$

In this case, it is difficult to know the explicit representation of nonlinear system dynamics as (1). In such a case, we often use a method to reconstruct the orbit of the attractor in the $m$-dimensional state space from single time series by the delay time vectors.[40] The reconstructed vector $\mathbf{X}(t) \in R^m$ is given as

$$\mathbf{X}(t) = (y(t), y(t+\tau), y(t+2\tau), \ldots, y(t+(m-1)\tau)) \tag{3}$$

where $\tau$ is a delay time (Fig. 1). Then, Takens' embedding theorem[40] represents the relationship between the original dynamical system and the dynamics of this reconstructed state vector.

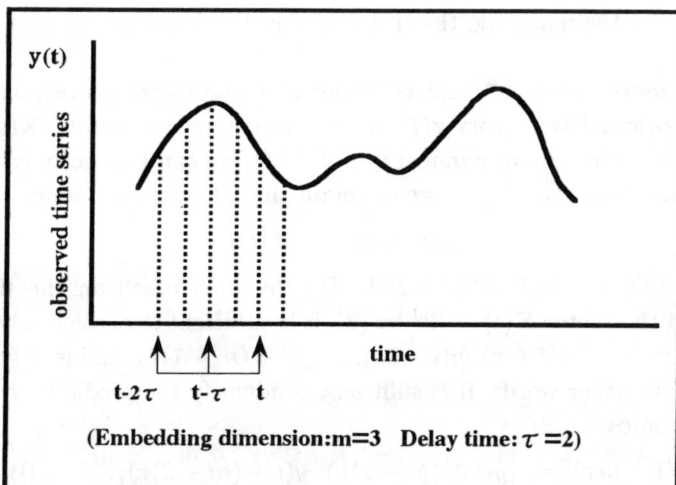


Fig. 1. Reconstruction dynamics in $m$-dimensional state space from experimental data.

### *Takens' embedding theorem*[40]

Let the attractor of the original dynamical system is a $n$-dimensional compact manifold $M^n$, and $C^2$ class map $F: M^n \to M^n$ and $g: M^n \to R$ are given.

Then, if inequality $m \geqq 2n + 1$ is satisfied, the map $\Phi : M^n \to R^m$ defined by (4) is an embedding.

$$\Phi(\mathbf{z}(t)) = \big(g(\mathbf{z}(t)),\, g(\mathbf{F}(\mathbf{z}(t))),\, g(\mathbf{F}^2(\mathbf{z}(t))), \dots, g(\mathbf{F}^{m-1}(\mathbf{z}(t)))\big) \qquad (4)$$

where $\mathbf{F}^1(\mathbf{z}(t)) \equiv \mathbf{F}(\mathbf{z}(t))$, $\mathbf{F}^{m+1}(\mathbf{z}(t)) \equiv \mathbf{F}(\mathbf{F}^m(\mathbf{z}(t)))$, $m = 1, 2, \dots$ .

This theorem is interpreted as follows: By "embed" we mean that there is a mapping $\Phi$ from the original manifold $M^n$ to the $m$-dimensional Euclidian space $R^m$ such that $\Phi^{-1}$ exists for all points in the image of the manifold, and that both $\Phi$ and $\Phi^{-1}$ are continuously differentiable.

Since the map $\Phi$ defined by (4) is an embedding, it should be noted that the topological property of the original trajectory $\mathbf{z}(t)$ of the manifold is preserved even in the transformed trajectory $\mathbf{X}(t)$ in the Euclidian state space, which are obtained by the mapping $\Phi$. In other words, the fixed points, limit cycles, and chaotic attractors on the original manifold, correspond to the fixed points, limit cycles, and chaotic attractors of the mapped image in the state space. Moreover, the stability of these trajectories are preserved, and because of the smoothness of the mapping, the ratios of the characteristic exponents are also preserved.

Furthermore, since $\Phi^{-1}$ exists, there is a one-to-one correspondence between the original trajectory $\mathbf{z}(t)$ and reconstructed trajectory $\mathbf{X}(t)$. Therefore, by identifying the dynamical law of $\mathbf{X}(t)$, we can indirectly estimate the dynamical system (1). Hence, we estimate the following dynamical system

$$\mathbf{X}(t+1) = \mathbf{f}(\mathbf{X}(t)) \qquad (5)$$

instead of (1), where $\mathbf{f} : R^m \to R^m$. By the way, based on the the way to reconstruct the vector $\mathbf{X}(t) \in R^m$ by (3), it is sufficient to predict only $y(t+m\tau)$ from $\mathbf{X}(t) = (y(t), y(t+\tau), y(t+2\tau), \dots y(t+(m-1)\tau))$ to identify $\mathbf{f} : R^m \to R^m$ in (5). In other words, it is sufficient to identify the predictor $f : R^m \to R$ in the dynamics

$$y(t+m\tau) = f\big(y(t+(m-1)\tau),\, y(t+(m-2)\tau), \dots, y(t)\big) . \qquad (6)$$

Replace "$t$" with "$t - m\tau$" in (6) and define $\mathbf{x}(t) \in R^m$ such that

$$\mathbf{x}(t) = \big(y(t-\tau), y(t-2\tau), \dots, y(t-m\tau)\big) , \qquad (7)$$

then we have

$$y(t) = f(\mathbf{x}(t)) = f\big(y(t-\tau),\, y(t-2\tau), \dots, y(t-m\tau)\big) . \qquad (8)$$

Therefore, hereafter, we consider to identify the predictor $f : R^m \to R$ in the nonlinear autoregressive model (8).

When the observed data is given, it is important to determine the delay time $\tau$ and embedding dimension $m$ in (8). To determine the suitable delay time $\tau$, several methods have been proposed.[63] In case that $\tau$ is too short, the correlation between two adjacent state space vectors gets relatively big value, and this makes the reconstructed attractor distribute along the line $y(t+\tau) = y(t)$, when embedding dimension is two. This is not preferable, since the specific shape characteristics of each attractor cannot be extracted. On the contrary, in case $\tau$ is too long for reconstructing the state space vector, the shape characteristics of each attractor cannot be extracted, either, since the deterministic relation between two adjacent state space vectors is lost because of the sensitivity of the initial condition inherent in the deterministic chaos. One of the methods to determine the suitable delay time, is to select the delay time at which the value of autocorrelation of $y(t)$ becomes zero for the first time. This method reduces the correlation between two adjacent state space vectors, while preserving the deterministic relation between the two adjacent state space vectors.

Concerning the determination of embedding dimension $m$, Grassberger and Procaccia suggested a method to estimate the correlation dimension by use of a correlation integral.[42] However, one of the problems of Grassberger's method is that it requires a large number of observed time series data as the embedding dimension increases.

On the other hand, Sugihara and May proposed a method to estimate the embedding dimension based on the nonlinear prediction.[43] In this method, identification and prediction procedures are employed to calculate the correlation coefficient between the predicted and observed data by using a reconstructd state space vector. In contrast with Grassberger's method, Sugihara's method requires fewer data values to estimate the embedding dimension.

However, in Sugihara's method, simplex method[43] is used for nonlinear prediction, which requires abruptly much computation time as the dimension increases. To cope with this problem, self-generating radial basis function network is applied to estimate the embedding dimension of the nonlinear dynamical systems including chaos.[63]

### 4.2. *Nonlinear prediction using neuro-fuzzy model*

To construct the predictor $f$ in nonlinear autoregressive model (8), there are in general two approaches such as local approximation techniques and global approximation techniques.[39] In local approximation techniques, a local predictor is generated in the suitably partitioned state space which approximates the dynamics of the reconstructed trajectory. As local approximation techniques, there are piece-wise linear prediction method,[44] method by using Gram–Shudmit orthogonalization procedure,[45] simplex method,[43] and tesselation.[46] Hybrid methods with local approximation techniques and fuzzy inference are nonparameric regressive model by using membership functions[47] and local fuzzy reconstruction method.[48]

On the other hand, in global approximation techniques, the map of recontructed dynamics is approximated by the combination of standard basis function such as multilayer neural network,[49] radial basis function network (RBFN),[50–53] and a method by fuzzy relation.[54] Since all of multilayer neural networks with sigmoidal transfer function,[55] RBFN with Gaussian units,[56,57] and fuzzy models with product-sum-gravity inference[58] are regarded as universal function approximators, these are suitable for approximating smooth nonlinear predictors.

In these nonlinear models, RBFN model can be regarded as both three-layer network[51–53] or a fuzzy model of Class $C^\infty$.[77] Hence, RBFN is also called neuro-fuzzy model.[59,77] Moreover, concerning the speed of convergence, RBFN is much faster than conventional multilayer neural networks with sigmoidal transfer function.[51]

Let $f^*$ be an approximated predictor of $f$ in the nonlinear autoregressive model defined by (8). Then $f^*$ is represented by RBFN as in (9).

$$f^* = \sum_{k=1}^{n} w_k \mu_k(\mathbf{x}), \ \mathbf{x} = \big(y(t-\tau), y(t-2\tau), \dots, y(t-m\tau)\big) \tag{9}$$

where $n$ is a number of basis functions, and $w_k$, $k = 1, \dots, n$ are coefficient of $k$th basis function. In addition,

$$\mu_k(\mathbf{x}) = \prod_{i=1}^{m} A_{ik}(y(t-i\tau)) = \prod_{i=1}^{m} A_{ik}(x_i) \tag{10}$$

where $A_{ik}$ is the radial basis function for the input component $y(t-i\tau) = x_i$ which has $a_{ik}$ as its center, and $b_{ik}$ as its width of the distribution. Various function forms for $A_{ik}$ can be taken such as (11a)–(11h).

[Gaussian]

$$A_{ik}(x_i) = \exp\left(\frac{-(x_i - a_{ik})^2}{b_{ik}}\right) \tag{11a}$$

[linear]

$$A_{ik}(x_i) = |x_i - a_{ik}| \tag{11b}$$

[cubic]

$$A_{ik}(x_i) = |x_i - a_{ik}|^3 \tag{11c}$$

[thin-plate-spline]

$$A_{ik}(x_i) = (x_i - a_{ik})^2 \log\left(|x_i - a_{ik}|\right) \tag{11d}$$

[isosceles triangle]

$$A_{ik}(x_i) = \begin{cases} 1 - \dfrac{2|x_i - a_{ik}|}{b_{ik}}, & (2|x_i - a_{ik}| < b_{ik}) \\ 0, & (2|x_i - a_{ik}| \geq b_{ik}) \end{cases} \tag{11e}$$

[multiquadric]

$$A_{ik}(x_i) = \left((x_i - a_{ik})^2 + b_{ik}^2\right)^{1/2} \tag{11f}$$

[inverse multiquadric]

$$A_{ik}(x_i) = \left((x_i - a_{ik})^2 + b_{ik}^2\right)^{-1/2} \tag{11g}$$

[RBFN with Non-integer exponents]

$$A_{ik}(x_i) = \begin{cases} 1 - \dfrac{2|x_i - a_{ik}|^{\beta_{ik}}}{b_{ik}}, & (2|x_i - a_{ik}| < b_{ik}) \\ 0, & (2|x_i - a_{ik}| \geq b_{ik}) \end{cases} \tag{11h}$$

RBFN with Non-integer exponents (11h) with the value $\beta_{ik} = 1$ corresponds to isosceles triangle (11e). In this model, $\beta_{ik}$ is considered to be a smoothing parameter, or parameter to control the curvature of the basis function.[63]

Here, we use Gaussian basis functions defined by (11a). If we regard this model as a fuzzy model, the $k$th fuzzy rule is expressed by (12),

Rule $k$:

$$\text{If } x_1 = A_{1k} \text{ and} \cdots \text{and } x_i = A_{ik} \text{ and} \cdots \text{and } x_m = A_{mk}\,, \quad \text{Then } y(t) = w_k\,, \quad k = 1, \ldots, n$$

or equivalently expressed as,

$$\text{If } y(t-1) = A_{1k} \text{ and} \cdots \text{and } y(t-i) = A_{ik} \text{ and} \cdots \text{and } y(t-m) = A_{mk}\,, \quad \text{Then } y(t) = w_k\,, \quad k = 1, \ldots, n\,. \tag{12}$$

The three-layer neural network representation of RBFN and the fuzzy model representaion are shown in Fig. 2 and Fig. 3, respectively.

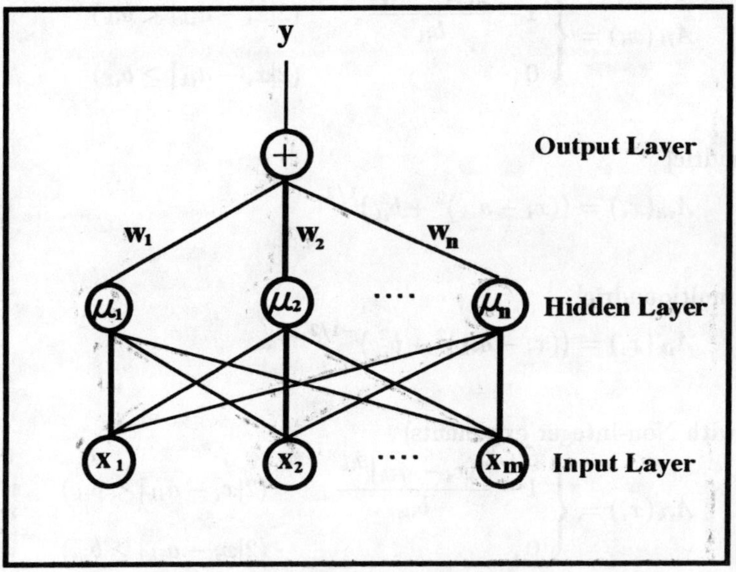


Fig. 2. Radial basis function network as three-layer neural network.

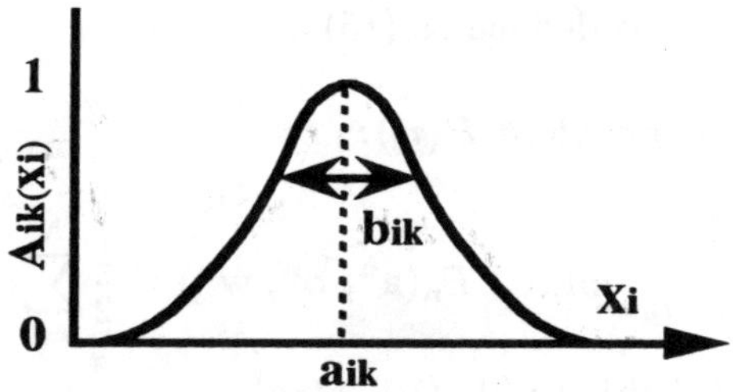


**Rule k:**

**IF x1=A1k(x1) and x2=A2k(x2) and ⋯and xm=Amk(xm)**
**THEN y=wk, k=1,2, ⋯ ,n**

**Membership Function:** $A_{ik}(X_i)=\exp(-(X_i-a_{ik})^2/b_{ik})$

**Grade:** $\mu_k(X)=\prod_{i=1}^{m} A_{ik}(X_i)\ ,\quad k=1,2,\cdots,n$

**Output:** $y=\sum_{k=1}^{n} w_k\,\mu_k(X)$

Fig. 3. Radial basis function network as fuzzy model.

Let us assume that from observed time series of $y(t)$, the $N$ training data

$$(\mathbf{x}^1, y^1), \ldots, (\mathbf{x}^p, y^p), \ldots, (\mathbf{x}^N, y^N) \tag{13}$$

are given. Let us introduce the following parameter vectors $\mathbf{a}^n \in R^{mn}$, $\mathbf{b}^n \in R^{mn}$, $\mathbf{w}^n \in R^n$ defined by (14) when the number of RBF is $n$,

$$\begin{aligned}
\mathbf{a}_k &= (a_{1k}, \ldots, a_{ik}, \ldots, a_{mk}) \in R^m, \quad k = 1, \ldots, n \\
\mathbf{a}^n &= (\mathbf{a}_1, \ldots, \mathbf{a}_k, \ldots, \mathbf{a}_n) \in R^{mn} \\
\mathbf{b}_k &= (b_{1k}, \ldots, b_{ik}, \ldots, b_{mk}) \in R^m, \quad k = 1, \ldots, n \\
\mathbf{b}^n &= (\mathbf{b}_1, \ldots, \mathbf{b}_k, \ldots, \mathbf{b}_n) \in R^{mn} \\
\mathbf{w}^n &= (w_1, \ldots, w_k, \ldots, w_n) \in R^n
\end{aligned} \tag{14}$$

Then, given the $N$ training data (13), the identification problem $P(n)$ with $n$ basis functions is defined as (15).

***Identification problem $P(n)$:***

$$\min_{(\mathbf{a}^n,\mathbf{b}^n,\mathbf{w}^n)} E_n(\mathbf{a}^n,\mathbf{b}^n,\mathbf{w}^n) = \frac{1}{2}\sum_{p=1}^{N}(y^p - {}^*y^p)^2 \tag{15}$$

subj. to (9), (10), (11a)

where $y^p$ and ${}^*y^p$ are the output data and the inference output for $p$th training data input $\mathbf{x}^p$, respectively.

Let optimal solution for Problem $P(n)$ be (${}^*\mathbf{a}^n$, ${}^*\mathbf{b}^n$, ${}^*\mathbf{w}^n$) and the specified model error value be $\varepsilon > 0$, then the overall identification problem is formulated as follows:

***Overall identification problem:***

Given the $N$ training data (13) and the specified model error $\varepsilon > 0$, obtain the minimal number $n$ of RBF and optimal solution (${}^*\mathbf{a}^n$, ${}^*\mathbf{b}^n$, ${}^*\mathbf{w}^n$) for Problem $P(n)$ which satisfies the inequality (16).

$$E_n({}^*\mathbf{a}^n, {}^*\mathbf{b}^n, {}^*\mathbf{w}^n) < \varepsilon\,. \tag{16}$$

### 4.3. *Self-generating algorithm by maximum absolute error selection method*

Several design methods for RBFN are proposed.[51,53,59] Moody *et al.* proposed a $k$-means clustering algorithm[51] by which centers of RBF are selected according to the distribution of learning data. Chen *et al.* proposed an Orthogonal Least Squares (OLS) method,[53] where the essential basis is gradually selected by the Gram–Schmidt orthogonalization procedure, from the subset regression point of view. But in these methods, relatively large number of the basis functions are required, since the tuning parameters are limited to only the coefficients of RBF. To solve this problem, Katayama *et al.* proposed a self generating algorithm by MAE (Maximum Absolute Error) selection method,[59] with relatively small number of basis functions, where not only the coefficients, but also centers and widths of RBF are tuned. Furthermore, the performance of these three methods are compared with function approximation problems.[60,79] According to this evaluation, the following results are obtained.[60,79]

(1) Since $k$-means method is an inherently two-pass type algorithm, it requires much computation time.

(2) When tuning parameter is restricted to only the coefficients of RBF, OLS method and MAE method require almost the same number of basis, but the computation time of MAE method is about half of that of OLS method. This is because the new basis selection algorithm of MAE method is much simpler than that of OLS method.

(3) MAE method with all the learning model parameters is superior to other methods concerning the number of RBF, and also computation time.

Therefore, we use MAE method for nonlinear autoregressive model identification. The MAE method consists of the following two processes.[59]

(1) Parameter tuning process with fixed number of RBF.

(2) New basis function generation process such that the center of new basis is located at the point where the maximal inference error value takes place.

Let us explain the details individually.

#### 4.3.1. *Gradients methods for parameter tuning*

In order to solve the identification problem $P(n)$, the gradients $\partial E_n/\partial a_{ik}$, $\partial E_n/\partial b_{ik}$, and $\partial E_n/\partial w_k$, $i = 1, \ldots, m$, $k = 1, \ldots, n$, are derived as (17) under the equality constraints (9), (10), (11a).

$$\frac{\partial E_n}{\partial a_{ik}} = \frac{-2}{b_{ik}} \sum_{p=1}^{N} (w_k \mu_k(\mathbf{x}^p)(y^p - {}^*y^p)(x_i^p - a_{ik})) \qquad i = 1, \ldots, m\,,\ k = 1, \ldots, n \tag{17a}$$

$$\frac{\partial E_n}{\partial b_{ik}} = \frac{-1}{b_{ik}^2} \sum_{p=1}^{N} (w_k \mu_k(\mathbf{x}^p)(y^p - {}^*y^p)(x_i^p - a_{ik})^2) \qquad i = 1, \ldots, m\,,\ k = 1, \ldots, n \tag{17b}$$

$$\frac{\partial E_n}{\partial w_k} = -\sum_{p=1}^{N} (\mu_k(\mathbf{x}^p)(y^p - {}^*y^p)) \qquad k = 1, \ldots, n \tag{17c}$$

By using these gradients (17), we can solve the identification problem $P(n)$ by appropriate gradient methods such as steepest descent method, conjugate gradient method, and Quasi-Newton method. The learning algorithm by steepest descent method is given by (18),

$$a_{ik}(h+1) = a_{ik}(h) - \alpha \cdot \partial E_n/\partial a_{ik} \quad (18a)$$

$$b_{ik}(h+1) = b_{ik}(h) - \alpha \cdot \partial E_n/\partial b_{ik} \quad (18b)$$

$$w_k(h+1) = w_k(h) - \alpha \cdot \partial E_n/\partial w_k \quad (18c)$$

$$i = 1, \ldots, m\,,\ k = 1, \ldots, n$$

where $h$ is iteration number and $\alpha$ is an optimal step size obtained by solving a one-dimensional search problem.

### 4.3.2. *Self-generating algorithm for overall identification problem*

When the value of the error function $E_n(\mathbf{a}^n, \mathbf{b}^n, \mathbf{w}^n)$ takes the minimum during the tuning process for fixed number of RBF, we generate a new basis function. Although several generating strategies may be possible, we adopt one that will generate a new basis function in such a way that its center is located at the point where the maximum of the absolute inference error occurs in the input space. The self-generating algorithm based on this idea is described as follows:

[Step 1]

Set the model error $\varepsilon$ to be satisified and stop criteria constant $\varepsilon_1$ for the parameter tuning process for the fixed number of RBF. Also, considering the case that model error stop criteria $\varepsilon$ is too small for the given problem, we set STMAX for the maximum limit of total iteration number $st$.

[Step 2]

Let $h$ be the iteration number for the fixed number of RBF, and $st$ be the total iteration number for the entire procedure. Let $h = 0$, $st = 0$.

[Step 3]

For $N$ training data (13), compute $(\mathbf{a}^n(h+1), \mathbf{b}^n(h+1), \mathbf{w}^n(h+1))$ from $(\mathbf{a}^n(h), \mathbf{b}^n(h), \mathbf{w}^n(h))$ by the updating rule (18) and the gradients (17).

[Step 4]

Let $E_n(h) \equiv E_n(\mathbf{a}^n(h), \mathbf{b}^n(h), \mathbf{w}^n(h))$ for simplicity. Then, if the inequality

$$E_n(h+1) < \varepsilon \quad (19)$$

or the condition for total iteration limit

$$\mathrm{STMAX} \leqq st \tag{20}$$

is satisfied, $n$ is the least number of RBF which satisfies the inequality (16) for the overall identification problem. In this case, let $(\mathbf{a}^n(h+1), \mathbf{b}^n(h+1), \mathbf{w}^n(h+1))$ be an optimal solution $({}^*\mathbf{a}^n, {}^*\mathbf{b}^n, {}^*\mathbf{w}^n)$ and terminate the procedure.

Otherwise, go to step 5.

[Step 5]

Let us define the decreasing rate of $E_n$ by (21).

$$D_n(h+1) \equiv \left|(E_n(h+1) - E_n(h))/En(h)\right| \tag{21}$$

If inequality (22)

$$D_n(h+1) < \varepsilon_1 \tag{22}$$

is satisfied, go to step 6 to generate a new basis function. If

$$D_n(h+1) \geqq \varepsilon_1 \tag{23}$$

holds, let $h = h+1$, $st = st+1$, and go back to step 3.

[Step 6]

Let $(\mathbf{x}^q, y^q)$ be a input/output vector such that the absolute inference error takes the maximum at this point among the $N$ training data $(\mathbf{x}^p, y^p)$, $p = 1, \ldots, N$. That is to say, find $(\mathbf{x}^q, y^q)$ which is not selected as center beforehand and satisfying (24),

$$|y^q - {}^*y^q| = \max_{1 \leqq p \leqq N} |y^p - {}^*y^p| \,. \tag{24}$$

Then, generate a $(n+1)$th new radial basis function according to (25),

$$a_{i,n+1} = x_i^q \,, \qquad i = 1, \ldots, m \tag{25a}$$

$$b_{i,n+1} = b_0 \,, \qquad i = 1, \ldots, m \tag{25b}$$

$$w_{n+1} = y^q - {}^*y^q \tag{25c}$$

where $b_0$ is a given constant for the initial width of RBF. Let $n = n+1$, $h = 0$, and go back to step 3. The initial values for other model parameters are set identical which are obtained by the just previous procedure.

### 4.4. *Numerical example*

In this section, we apply the self-generating algorithm to the nonlinear prediction of optical chaotic time series.[59] Let us consider the Ikeda map[62,39] defined by (26),

$$x(t+1) = a + b \cdot \left(x(t) \cdot \cos(z(t)) - y(t) \cdot \sin(z(t))\right) \tag{26a}$$

$$y(t+1) = b \cdot \left(x(t) \cdot \sin(z(t)) + y(t) \cdot \cos(z(t))\right) \tag{26b}$$

$$z(t) = 0.4 - 6.0/\left(1 + x(t)^2 + y(t)^2\right) \tag{26c}$$

where $x$, $y$ are the real part and imaginary part of electric field of a ring cavity respectively. The model generates relatively simple strange attractor with $a = 1.0$, $b = 0.7$, which is shown in Fig. 4.

We apply two input/two output recurrent RBFN as is illustrated in Fig. 5 to identify the chaotic time series generated by (26). The learning data are obtained by solving (26) with initial condition $x(0) = y(0) = 0$. We set $N = 100$, $\varepsilon = 0.05$, $\varepsilon_1 = 0.01$, and we assume the initial number of rules $n = 0$. By applying MAE method, the stop condition $E_n < \varepsilon$ is satisfied when $n = 10$. The learning curve (Error $E_n$ vs total iteration number $st$) is shown in Fig. 6, where "*" shows the iteration $st$ at which new basis function is generated. Figure 6 represents that from $n = 2$ to $n = 3$, $E_n$ decreases rapidly. Moreover, self-generating process of rules in input space is shown in Fig. 7. As we see in Fig. 7, new rules are successively generated in the area where absolute inference error takes the maximal value, along the trajectory on strange attractor in Fig. 4. This process can be regarded as if generated rules are breeding cells in a living organism. Furthermore, the surface defined by the learning data and identified surface are shown in Fig. 8. Thus, surface approximation is realized very well.

The original time series $x(t)$ and one step ahead of the predicted time series $x'(t)$ by the identified model are shown in Fig. 9. Thus, $x(t)$ is nicely predicted by the identified model, and this result indicates that short time prediction of the chaotic time series (26) is realized. Moreover, strange attractor generated by the pair of one-step-ahead predicted value $(x'(t), y'(t))$ is shown in Fig. 10.

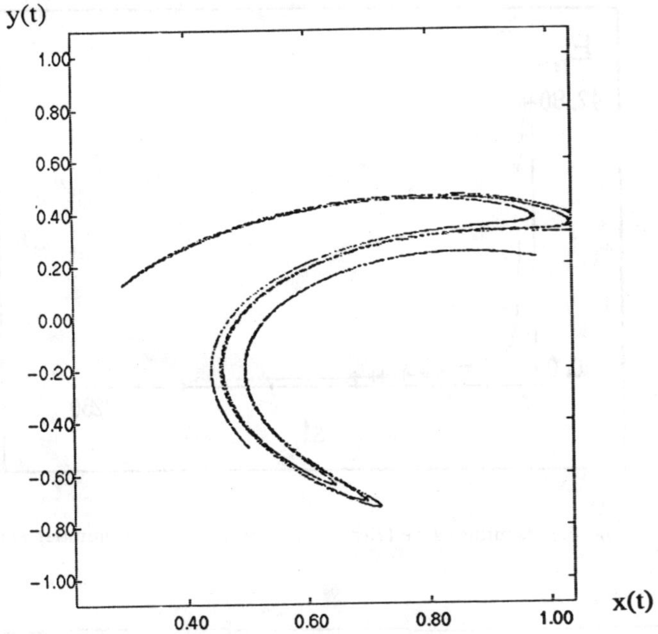


**Fig. 4. Strange attractor of Ikeda map.**

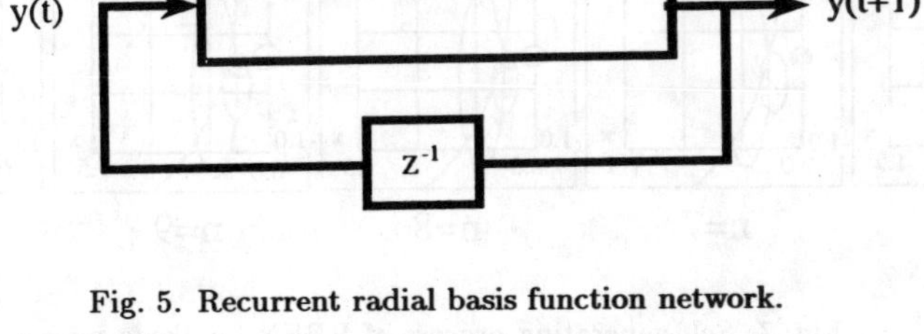


**Fig. 5. Recurrent radial basis function network.**

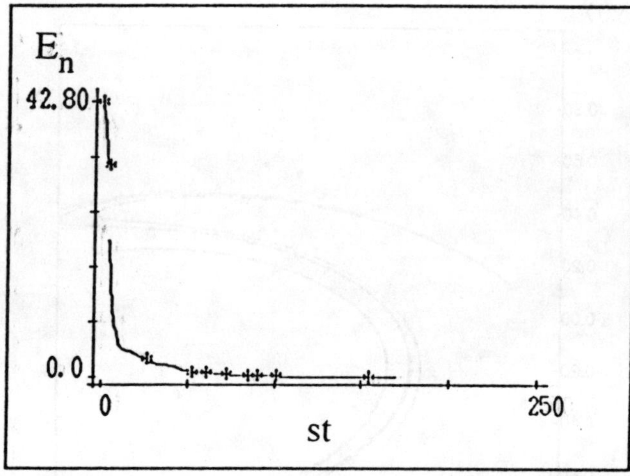


Fig. 6. Learning curve (Error $E_n$ vs total iteration number $st$).

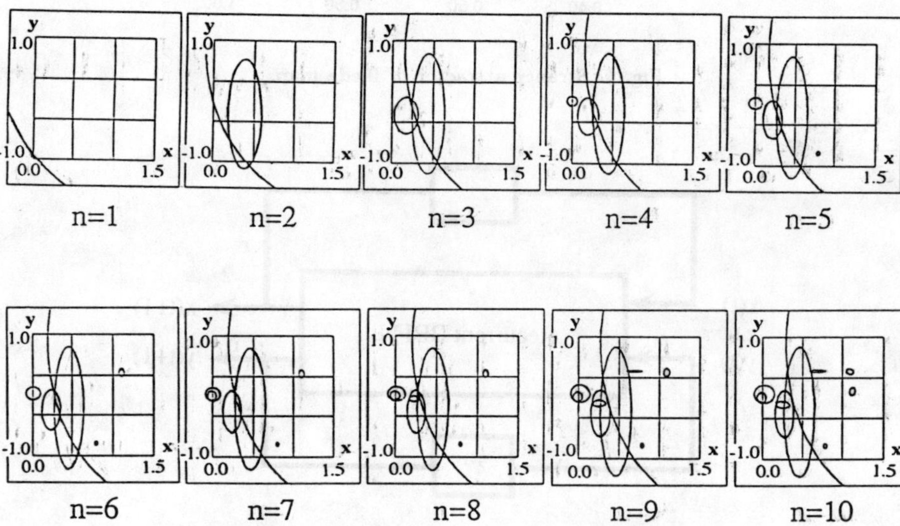


Fig. 7. Self-generating process of RBFN for Ikeda map.

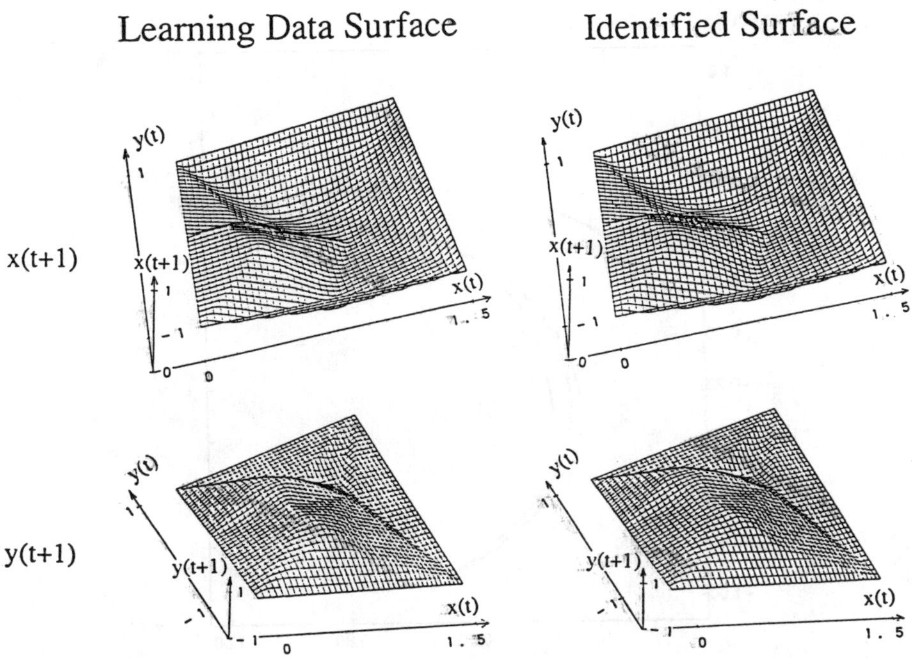


Fig. 8. Learning data surface and identified data surface for Ikeda map.

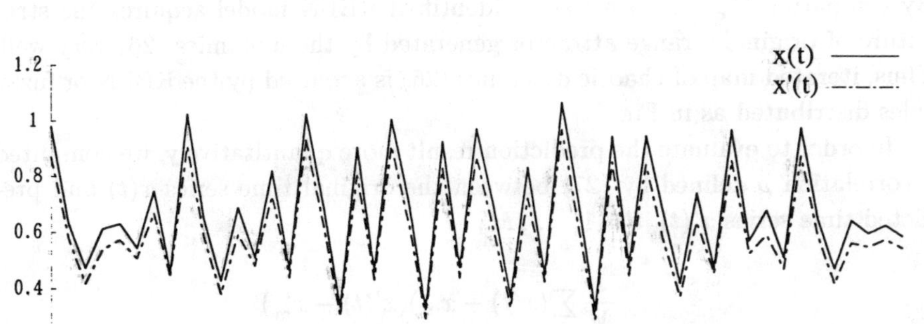


Fig. 9. Comparison of original time series $x(t)$ and the predicted time series $x'(t)$ one step ahead.

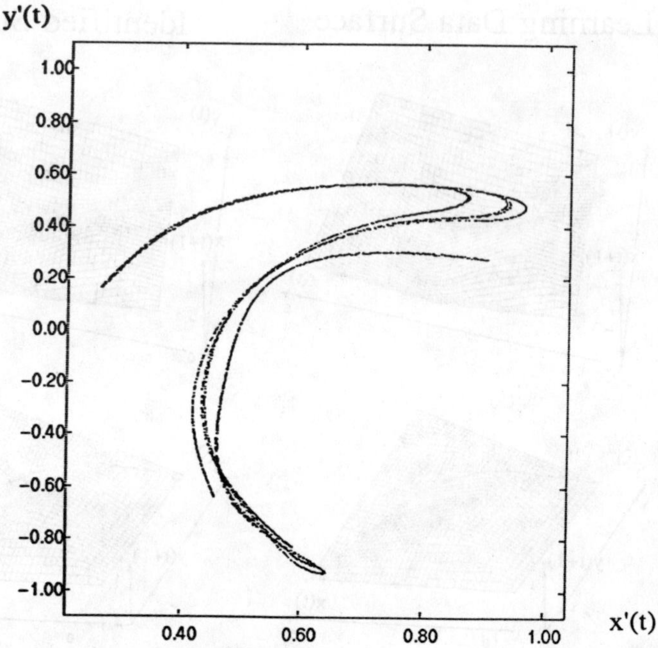


**Fig. 10.** Strange attractor of Ikeda map acquired by RBFN.

By comparing Fig. 10 with Fig. 4, identified RBFN model acquires the strucuture of original strange attractor generated by the dynamics (26) very well. Thus, iterated map of chaotic dynamics (26) is acquired by the RBFN or fuzzy rules distributed as in Fig. 7.

In order to evaluate the prediction result more quantitatively, we computed a correlation $\rho$ defined by (27) between the original time series $x(t)$ and predicted time series $x'(t)$, $t = 1, \ldots, M$,

$$\rho = \frac{\frac{1}{M}\sum_t (x(t) - x_m)(x'(t) - x'_m)}{\sqrt{\frac{1}{M}\sum_t (x(t) - x_m)^2}\sqrt{\frac{1}{M}\sum_t (x'(t) - x'_m)^2}} \tag{27}$$

where $x_m$, $x'_m$ are mean value of $x(t)$, $x'(t)$, $t = 1, \ldots, M$, respectively. Now we show the relation between $\rho$ and prediction step $s$ for $x(t)$ and $s$ step ahead predicted time series $x'(t)$ in Fig. 11, and the same relation for $y(t)$ and $y'(t)$

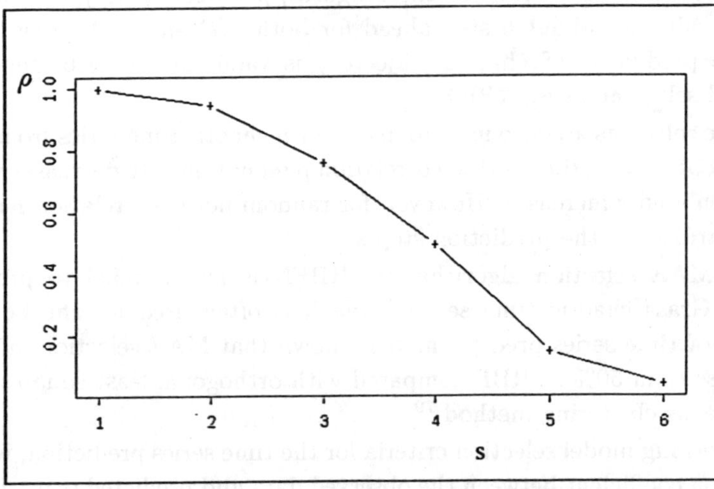


**Fig. 11.** **Correlation $\rho$ vs prediction step $s$ for $x(t)$ and $x'(t)$.**

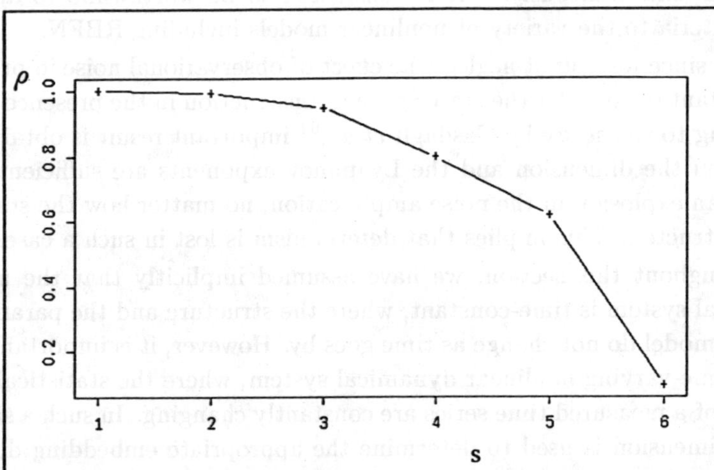


**Fig. 12.** **Correlation $\rho$ vs prediction step $s$ for $y(t)$ and $y'(t)$.**

in Fig. 12 when $M = 90$. By these figures, $\rho$ monotonically decreases with positive value for about 6 step ahead for both $x(t)$ and $y(t)$. This indicates that the prediction of chaotic trajectory is valid for about 6 steps by the proposed self-generating RBFN.

These relations are also used to distinguish chaotic time series from random noise.[43] For chaotic time series, correlation $\rho$ monotonically decreases when the prediction step $s$ increases. However, for random noise, correlation $\rho$ is almost zero regardless of the prediction step $s$.

The MAE selection algorithm for RBFN is also applied to predict the Mackey–Grass chaotic time series[61] which is often used for the benchmark problem of time series prediction. It is shown that MAE selection method requires less than 50% of RBF compared with orthogonal least squares method and $k$-means clustering method.[79]

Concerning model selection criteria for the time series prediction, we used a correlation coefficient between the observed data and predicted ones. However, another criteria such as Minimum Description Length (MDL) can be used to determine both of an appropriate number of units in the hidden layer of the three-layer network with sigmoidal function, and the suitable embedding dimension simultaneously.[65] It is considered to be worthwhile to apply this MDL criteria to the variety of nonlinear models including RBFN.

Also, since we cannot neglect the effect of observational noise in practice, it is important to consider the state space reconstruction in the presence of noise. According to the study by Casdagli *et al.*,[64] important result is obtained such that when the dimension and the Lyapunov exponents are sufficiently large, there is an explosion in the noise amplification, no matter how the state space is reconstructed. This implies that determinism is lost in such a case.

Throughout this section, we have assumed implicitly that the nonlinear dynamical system is time-constant, where the structure and the parameters of the true model do not change as time goes by. However, it is important to deal with a time-varying nonlinear dynamical system, where the statistical characteristics of a measured time series are constantly changing. In such a situation, fractal dimension is used to determine the appropriate embedding dimension in time-varying systems.[66] Therefore, one of the future research directions is to develop a method for identifying a time-varying nonlinear dynamical system while considering the fractal dimension of time series and simultaneously modifying the structure of the model accordingly. Moreover, concerning the structure identification of chaotic dynamics, Kajiwara *et al.* proposed a method to reconstruct a bifurcation diagram only from a chaotic time series by applying K–L transformation.[67]

One of the problems — not only for the time series analysis, but also for general nonlinear modeling — is that a large number of model parameters are required when the dimension of inputs increases. To overcome this difficulty, neuro-fuzzy GMDH[78] and a tree-structured adaptive radial basis function network called Adaptive Tree RBFN[68] are proposed. In particular, Adaptive Tree RBFN is composed of a tree-structured adaptive network and MAE selection method for basis function generation. By using Adaptive Tree RBFN, we can reduce about 30 or 40% of the model parameters compared to the conventional RBFN, while almost achieving the same model accuracy.[68] Therefore, it is effective to introduce this kind of macroscopic structure in the specific nonlinear identification models.

## 5. Applications of Chaos Theory

Since there are many published surveys on chaos engineering[4–9,11,75] and surveys on fuzzy and chaos,[10,26,27,69] we will list only the application areas of chaos engineering. The potential applicational fields of chaos theory are diverse as follows[5,9]:

(1) *Chaos computing*
Chaotic neural network,[5,9] computational universality and generalized shift map, optimization such as traveling salesman problem, artificial life, edge of chaos, antichaos and adaptation.

(2) *Deterministic prediction based on nonlinear dynamics*
Nonlinear prediction of chaotic time series.[39–49,51,53,54,59]

(3) *Identification and modeling of chaos*
Monitoring and diagnosis of complex plants such as electric power network or nuclear reactor,[6] nonlinear signal processing such as voice synthesis, noise reduction and adaptive equalizer,[17] and system identification.[41]

(4) *Bio chaos*
Analysis and diagnosis of EEG, MEG, cardiac chaos[61] and capillary chaos,[18] sensitivity and mental stress engineering.[18]

(5) *Chaotic memory*
Memory dynamics of brain and neural network.[12,13,32]

(6) *Chaotic coding and decoding*
Image data compression, fractal image processing,[8] chaotic cryptogram[75] and chaotic signal communication.[15]

(7) *Chaotic pattern recognition*
Taste sensor, fractal image recognition and feature extraction.[6]

(8) *Chaotic art*
Chaotic computer graphics, fractal computer graphics, chaotic music.

(9) *Utilization of chaotic fluctuation and application to consumer electronics*
Home appliances using chaos theory such as kerosene fan heater using chaotic fluctuation,[8–11] dish washer,[73] washing machine.[80]

(10) *Devices for generating chaotic signals*
Electrical and electronic circuits,[8,74] optical laser system[15,62] artificial membrane.[6]

(11) *Generation and control of chaotic signals in nonlinear engineering systems*
Control of chaotic signals, suppression of chaotic vibrations in mechanical systems and electrical systems,[14] path planning and wandering of autonomous robot using chaos,[6] control of chaotic plant with two degree of freedom controller.[16]

Here, we describe some of the applications which have been put into practice. Concerning the diagnosis and monitoring complex systems, Tsuda *et al.* developed a medical instrument to analyze chaotic pulse waves in capillary vessels of a finger.[18] They discussed the possibility that shapes of reconstructed strange attractor reflect mental or physical conditions.

With regard to the applications to the home appliances, Sanyo Electric Company announced a kerosene fan heater shown in Fig. 13 in June 1992, which is the first consumer electronic product in the world using chaos theory.[8,9,11] Up to that time, it was believed desirable to keep room temperature constant. On the contrary, studies by Wyon[71] and other researchers suggest that subjects actually prefer temperature swings about the optimal set point to the constant temperature. On the basis of this idea, appropriate temperature swing patterns are computed by a simple equation which produces a kind of $1/f$-like fluctuations generated by intermittent chaos.[72] Psychological experiments with the PMV (Predictive Mean Vote) values of about 30 subjects showed that sensation with appropriate chaotic fluctuations is more comfortable than with conventional control.[11]

Another application to home appliances is a dishwasher with a 2-link nozzle by Matsushita Electric Industry.[73] A nozzle of usual dishwashers is composed of one link. However, to increase washing efficiency, they adopted a 2-link mechanism which enables the nozzle to move with higher degrees of freedom than conventional ones. They analyzed the behavior of the 2-link nozzle, with

Fig. 13. Kerosene fan heater using chaotic fluctuations[11].

Poincare sections, circle map, the largest Lyapunov exponent and the correlation dimension, and washing experiments with real tableware suggested that the behavior of the 2-link nozzle is quasi-priodic or possibly chaotic with washing capability higher than conventional ones.

Goldstar Electric Company developed a chaos-controlled washing machine[80] in 1993. In this washing machine, the shape of the attractor, which is reconstructed from the time series of water flow, is used to judge whether clothes are in the state of tangling or untangling. If the state is judged as tangling, a pulsator pushed up the clothes to make them untangled.

## 6. Conclusion

We have reviewed the fusion researches with chaos and fuzzy, and neuro-fuzzy technology. These fusion researches are classified into two categories such as (1) fuzzy inference scheme is introduced to the nonlinear dynamical systems, and

(2) law of crisp chaotic dynamical systems is acquired by fuzzy inference rules.

However, it seems that the extent of fusion of these researches is still superficial.

Fuzzy theory is a top down paradigm where language and symbols are combined with numeric and pattern information. On the contrary, in biological systems, pattern information processing and symbol processing are combined by the dynamics of massively parallel neural networks. Chaotic information processing is now regarded as one of the promising computational model of this biological information processing.[12,13] Chaos can create diversity from very simple deterministic rules, and this mechanism is considered to be used to realize the structual change of autonomous systems, as a mechanisim to alter the sub-goals of creatures. More intensive studies on fusion of chaos and fuzzy theory are expected such that subjective performance index or preferences, namely the objectives to survive for each individual in the dynamically changing environments, are described by the combination of fuzzy theory and information processing principle based on nonlinear dynamical systems.

## References

1. L. A. Zadeh, "Fuzzy logic, neural networks and soft computing," in *Proc. of 5th IFSA Congress*, plenary talk, 1993.
2. R. L. Devaney, *An Introduction to Chaotic Dynamical Systems*, second edition (Addison-Wesley, 1989).
3. J. M. T. Thompson and H. B. Stewart, *Nonlinear Dynamics and Chaos — Geometrical Methods for Engineers and Scientists* (John Wiley & Sons, 1986).
4. K. Horiuchi and T. Koga (eds.), special issue on Engineering Chaos, *Trans. of IEICE*, vol. E73, no. 6 (1990).
5. K. Aihara, Expanding applications of deterministic chaos in engineering, *Scientific American* (the Japanese language version) **22**, no. 3 (1992), pp. 26–33.
6. K. Aihara (ed.), *Applied and Applicable Chaos* (Science-sha, Tokyo, 1994).
7. K. Aihara (ed.), *Neuro, Fuzzy and Chaotic Computing: Towards a New Generation of Analog Computing* (Ohm-sha, Tokyo, 1993).
8. K. Aihara and R. Tokunaga (eds.), *Application Strategy of Chaos* (Ohm-sya, Tokyo, 1993).
9. K. Aihara and R. Katayama, "Chaos engineering In Japan," *Communications of the ACM* **38** (1995) (to appear).
10. R. Katayama, "Chaos and fuzzy control," *Systems, Control and Information* **38**, no. 11 (1994), pp. 619–624.
11. R. Katayama, Y. Kajitani, K. Kuwata, and Y. Nishida, "Developing tools and methods for applications incorporating neuro, fuzzy and chaos technology," *Computers and Industrial Engineering* **24**, no. 4 (1993), pp. 579–592.

12. C. A. Skarda and W. J. Freeman, "How brains make chaos in order to make sense of the world," *Behavioral and Brain Sciences* **10**, no. 1 (1987), pp. 161–195.
13. I. Tsuda, "Dynamic link of memory — chaotic memory map in nonequilibrium neural networks," *Neural Networks* **5** (1992), pp. 313–326.
14. E. Ott, C. Grebogi, and J. A. Yorke, "Controlling chaos," *Phys. Rev. Lett.* **64** (1990), pp. 1196–1199.
15. L. M. Pecora, "Overview of chaos and communications research," in *Proc. of SPIE — Chaos in Communications* **2038** (1993), pp. 2–25.
16. T. Kameda, K. Aihara, and Y. Hori, "Application of a TDOF controller to chaotic dynamical systems," *T. IEE Japan* **113-C**, no. 1 (1993), pp. 43–49.
17. K. Kuwata, M. Watanabe, and R. Katayama, "Adaptive equalizer using self generating radial basis function," in *Proc. of 3rd International Conference on Fuzzy Logic, Neural Nets and Soft Computing* (1994), pp. 493–496.
18. I. Tsuda, T. Tahara, and H. Iwanaga, "Chaotic pulsation in human capillary vessels and its dependence on mental and physical conditions," *International Journal of Bifurcation and Chaos*, vol. 2, no. 2 (1992), pp. 313–324.
19. J. B. Rosser, *From Catastrophe to Chaos, A General Theory of Economic Discontinuities* (Kluwer, 1991).
20. T. Terano, K. Asai, and M. Sugeno (eds.), *Fuzzy Systems Theory and Its Applications* (Academic, San Diego, California, 1992).
21. R. Katayama, "Recent trends on fuzzy hardware in consumer electronics products," *Journal of Japan Society for Fuzzy Theory and Systems* **6**, no. 3 (1994), pp. 481–488.
22. H. Takagi, "Fusion technology of fuzzy theory and neural networks — Survey and future directions," in *Proc. of the International Conference on Fuzzy Logic and Neural Networks* (1990), pp. 13–26.
23. P. Diamond, Fuzzy Chaos, Preprint (Mathematics Department, University of Queensland, 1977).
24. P. E. Kloeden, "Chaotic mappings on fuzzy sets," in *Proc. of 2nd IFSA Congress* (1987), pp. 20–25.
25. P. E. Kloeden, "Chaotic iterations of fuzzy sets," *Fuzzy Sets and Systems* **42** (1991), pp. 37–42.
26. P. Diamond, "Chaos and fuzzy representations of dynamical systems," in *Proc. of 2nd International Conference on Fuzzy Logic and Neural Networks* (1992), pp. 51–58.
27. H. N. Teodorescu, "Chaos in fuzzy systems and signals," in *Proc. of 2nd International Conference on Fuzzy Logic and Neural Networks* (1992), pp. 21–50.
28. K. Horiuchi and M. Takeoka, "On repeated fuzzy operators," in *Proc. of 7th Fuzzy System Symposium* (1991), pp. 529–532.
29. T. Yamakawa, E. Uchino, T. Miki, and T. Nakamura, "Transitions of fuzzy number through nonlinear dynamical systems", in *Proc. of 2nd International Conference on Fuzzy Logic and Neural Networks* (1992), pp. 145–148.
30. P. Grim, "Self-reference and chaos in fuzzy logic," *IEEE Trans. on Fuzzy Systems* **1**, no. 4 (1993), pp. 237–253.

31. Y. Ezawa, K. Haji, and T. Uemura, "Graphical display of fuzzy fractals," in *Proc. of 10th Fuzzy System Symposium* (1994), pp. 399–400.
32. T. Sato, H. Ushida, T. Yamaguchi, A. Imura, and T.Takagi, "Chaotic memory search in fuzzy associative interface," in *Proc. of 3rd International Conference on Fuzzy Logic, Neural Nets and Soft Computing* (1994), pp. 203–206.
33. O. Katai, T. Horiuchi, T. Sawaragi, S. Iwai, and T. Hiraoka, "Chaos structures of fuzzy symbolic dynamics and their relation to constraint-oriented problem solving," in *Proc. of 10th Fuzzy System Symposium* (1994), pp. 395–398.
34. H. N. Teodorescu, "Chaotic fuzzy models in economy," in *Proc. of 2nd International Conference on Fuzzy Logic and Neural Networks* (1992), pp. 153–155.
35. K. S. Narendra and A. M. Annaswamy, *Stable Adaptive Systems* (Prentice Hall, New Jersey, 1989).
36. K. Tanaka and M. Sugeno, "Stability analysis and design of fuzzy control systems," *Fuzzy Sets and Systems* **45**, no. 2 (1992), pp. 135–156.
37. H. Yamaguchi, "A stability of fuzzy control system near unstable point," in *Proc. of 8th Fuzzy Systems Symposium* (1992), pp. 505–508.
38. C. L. Karr and E. L. Gentry, "Control of a chaotic system using fuzzy logic," in eds. A. Kandel and G. Langholz, *Fuzzy Control Systems* (CRC Press, 1994), pp. 475–497.
39. M. Casdagli, "Nonlinear prediction of chaotic time series," *Physica D* **35** (1989), pp. 335–356.
40. F. Takens, "Detecting strange attractors in turbulence," *Lecture Notes in Mathematics* (1981), pp. 366–381.
41. S. Chen, S. A. Billings, C. F. N. Cowan, and P. M. Grant, "Practical identification of NARMAX models using radial basis functions," *Int. J. Control* **52**, no. 6 (1990), pp. 1327–1350.
42. P. Grassberger and I. Procaccia, "Characterization of strange attractor," *Phys. Rev. Lett.* **50** (1983), pp. 346–349.
43. G. Sugihara and R. May, "Nonlinear forcasting as a way of distinguishing chaos from measurement error in time series," *Nature* **344**, no. 19 (1990), pp. 734–741.
44. J. D. Farmer and J. J. Sidorowich, "Predicting chaotic time series," *Phys. Rev. Lett.* **59**, no. 8 (1987), pp. 845–848.
45. J. Jimenz, J. A. Moreno, and G. J. Ruggeri, "Forecasting on chaotic time series: A local optimal linear-reconstruction method," *Phys. Rev. A* **45**, no. 6 (1992), pp. 3553–3558.
46. A. I. Mees, "Dynamical systems and tesselations: Detecting determinism in data," *International Journal of Bifurcation and Chaos* **1**, no. 4 (1991), pp. 777–794.
47. T. Ikoma and K. Hirota, "Nonparametric regressive model by using membership function," in *Proc. of 8th Fuzzy System Symposium* (1992), pp. 281–284.
48. T. Iokibe, M. Kanke, Y. Fujimoto, and S. Suzuki, "Short-term prediction on chaotic time series by local fuzzy reconstruction method," in *Proc. of 3rd International Conference on Fuzzy Logic, Neural Nets and Soft Computing* (1994), pp. 491–492.

49. A. Lapedes and R. Farber, "Nonlinear signal processing using neural networks," technical report, no. LA-UR-87-2662 (Los Alamos National Laboratory, 1987).
50. M. J. D. Powell, "Radial basis functions for multivariable interpolation: a review," in J. C. Mason and M. G. Cox, eds., *Algorithms for Approximation* (Clarendon Press, Oxford, 1987), pp. 143–167.
51. J. Moody and C. J. Darken, "Fast learning in networks by locally-tuned processing units," *Neural Computations* **1** (1989), pp. 281–294.
52. T.Poggio and F. Girosi, "A theory of networks for approximation and learning," *A. I. Memo*, no. 1140, C.B.I.P. Paper (1989).
53. S. Chen, C. F. N. Cowan, and P. M. Grant, "Orthogonal least squares learning algorithm for radial basis function networks," *IEEE Trans. on Neural Networks* **2**, no. 2 (1991), pp. 302–309.
54. T. Ikoma and K. Hirota, "Non-linear autoregressive model by using fuzzy relation based approximate function," in *Proc. of 8th Fuzzy System Symposium* (1992), pp. 277-280.
55. K. Hornik, M. Stinchcombe, and H. White, "Multilayer feedforward networks are universal approximators," *Neural Networks* **2** (1989), pp. 359–366.
56. E. Hartman, J. Keeler, and J. Kowalski, "Layered neural networks with gaussian hidden units as universal approximations," *Neural Computation* **2** (1990), pp. 210–215.
57. F. Girosi and T. Poggio, "Networks and the best approximation property," *Biological Cybernetics* **63** (1990), pp. 169–176.
58. L. X. Wang and J. M. Mendel, "Fuzzy basis functions, universal approximation, and orthogonal least-squares learning," *IEEE Trans. on Neural Networks* **3**, no. 5 (1992), pp. 807–814.
59. R. Katayama, Y. Kajitani, K. Kuwata, and Y. Nishida, "Self generating radial basis function as neuro-fuzzy model and its application to nonlinear prediction of chaotic time series," in *Proc. Second IEEE International Conference on Fuzzy Systems* (1993), pp. 407–414.
60. R. Katayama, M. Watanabe, K. Kuwata, Y. Kajitani, and Y. Nishida, "Performance evaluation of self generating radial basis function for function approximation," in *Proc. of IJCNN'93 Congress* (1993), pp. 471–474.
61. M. C. Mackey and L. Grass, "Oscillation and chaos in physiological control systems," *Science* **197** (1977), pp. 287–289.
62. K. Ikeda, H. Daido, and O. Akimoto, "Optical turbulence: Chaotic behavior of transmitted light from a ring cavity," *Phys. Rev. Lett.* **45**, no. 9 (1980), pp. 709–712.
63. R. Katayama, K. Kuwata, Y. Kajitani, and M. Watanabe, "Embedding dimension estimation of chaotic time series using self generating radial basis function network," *Fuzzy Sets and Systems* (to appear).
64. M. Casdagli, S. Eubank, J. D. Farmer, and J. Gibson, "State space reconstruction in the presence of noise," *Physica D* **51** (1991), pp. 52–98.

65. M. Lehtokangas, J. Saarinen, P. Huuhtanen, and K. Kaski, "Chaotic time series modeling with optimum neural network architecture," in *Proc. of IJCNN'93 Congress* (1993), pp. 2276–2279.
66. I. Matsuba, "Application of neural sequential associator to long-term stock price prediction," in *Proc. of IJCNN'91 Congress* (1991), pp. 1196–1201.
67. S. Kajiwara, I. Tokuda, R. Tokunaga, and T. Matsumoto, "Reconstructing bifurcation diagrams only from chaotic time-wave forms," in *Proc. of 3rd International Conference on Fuzzy Logic, Neural Nets and Soft Computing* (1994), pp. 515–517.
68. M. Watanabe, K. Kuwata, R. Katayama, T. Kudou, and Y. Nishida, "A tree-structured adaptive radial basis function," in *Proc. of 16th International Conference on Computers and Industrial Engineering* (1994), pp. 202–205.
69. T. Yamakawa, H. N. Teodorescu, E. Unicho, and T. Miki, "A perspective on application of chaotic fuzzy logic systems," in *Proc. First European Congress on Fuzzy and Intelligent Technologies* (1993), pp. 987–991.
70. K. Kuwata, Y. Kajitani, R. Katayama, and Y. Nishida, "Chaos simulator as a developing tool for applications of chaos engineering," in *Proc. of 5th IFSA Congress* (1993), pp. 853–856.
71. D. P. Whon, "The role of the environment in buildings today: Thermal aspects (factors affecting the choice of a suitable room temperature)," *Build International* **6** (1973), pp. 39–54.
72. I. Procaccia and H. Schuster, "Functional renormalization group theory of universal $1/f$ noise in dynamical systems," *Phys. Rev.* **A28** (1983), pp. 1210–1212.
73. H. Nomura, E. Naito, N. Wakami, S. Kondo, and K. Aihara, "Analysis of chaotic behavior of a 2-link nozzle in a dishwasher," in *Proc. of Japan–U.S.A. Symposium on Flexible Automation, ISCIE/ASME* (1994), pp. 231–234.
74. T. Yamakawa, T. Miki, and E. Uchino, "A chaotic chip for analyzing nonlinear discrete dynamical network systems," in *Proc. of 2nd International Conference on Fuzzy Logic and Neural Networks* (1992), pp. 563–566.
75. H. N. Teodorescu, "Engineering applications of chaos theory," in *Tutorials of 3rd International Conference on Fuzzy Logic, Neural Nets and Soft Computing* (1994), pp. 33–69.
76. S. Hayashi, "Auto-tuning fuzzy PI controller," in *Proc. of 4th IFSA Congress*, vol. Eng. (1991), pp. 41–44.
77. H. Ichihashi, "Hierarchical and recurrent networks of fuzzy models," in *Proc. of the 2nd International Conference on Fuzzy Logic and Neural Networks* (1992), pp. 141–144.
78. H. Ichihashi and I. B. Turksen, "A neuro-fuzzy approach to data analysis of pairwise comparisons," *Int. J. Approximate Reasoning* **9** (1993), pp. 227–248.
79. R. Katayama, "Identification methods and their applications by use of a fuzzy neural network model," Ph.D. dissertation (Keio University, 1994).
80. H. S. Kim and Y. H. Roh, "Implementation and development of chaos theory in washing machine," *Journal of Korean Society of Mechanical Engineering* **34**, no. 6(1994), pp. 475–481.

# FUSION OF CHAOS AND FUZZY LOGIC, AND ITS APPLICATIONS

## Short-Term Prediction On Chaotic Time Series

TADASHI IOKIBE, SHOJI MURATA, MASAYA KOYAMA

*System Technology Division, Meidensha Corporation, 36-2 Nihonbashi Hakozaki-cho, Chuo-ku, Tokyo, 103, Japan*

**More than 50 years have passed since computer first appeared. And it is said that at the beginning of the 21st century, a supercomputer having the performance of information processing similar to a human brain will be developed, and a personal computer will have this performance in the 2030s. Our application technique of computer has progressed from the calculation to the realization of intelligent systems. Some important points leading to the realization of intelligent systems are "composition," "emergence," "fuzzy system," "parallel processing," etc. As new information processing technologies, "fuzzy logic," "neuro," "genetic algorithm," "chaos" and "artificial life" have been drawing wide attention.**

**In this paper we consider chaos and fuzzy logic. In Sec. 1, we describe chaos engineering. In Sec. 2, nonlinear short-term prediction on chaotic time series is described, in particular the Local Fuzzy Reconstruction Method is described in detail. And in Sec. 3, some applications of nonlinear short-term prediction are shown.**

***Keywords*: Chaos; Fuzzy Logic; Determinism; Chaos Engineering; Nonlinear Short-Term Prediction; Time Series; Embedding; Attractor; Logistic Map; Lorenz System; Henon Map; Local Fuzzy Reconstruction Method; Gram Schmidt' Orthogonal System Method; Tessellation Method; Exchange Rate; Temperature.**

## 1. Chaos

### 1.1. *Chaos engineering*

The original meaning of chaos is the complex, confused and unstable state, but the *chaos* which has attracted attention recently is *deterministic chaos.* Previously all non-periodic, complex and irregular phenomena were considered to be indeterministic phenomena subject to randomness. But it is clear that irregular, unstable and complex behavior can often be generated from a differential or difference equation governed by determinism (in which, once the initial value is given, the subsequent states are all determined in principle). This is the deterministic chaos of dynamical system.

The study of chaos can go back to the Poincaré's study in the beginning of the 20th century. In the 1960s, Lorenz discovered the chaotic behavior in the simulation of Lorenz's equation. Smale found out about the chaos in the differentiable dynamical system. In the 1970s, Ruell and Takens discovered the strange attractor in the turbulence. Li and York found out the occurrence condition of chaos, after that "chaos" has come into use generally.

Chaos has been considered to be related not only to the various fields of basic science, e.g., mathematics, physics and chemistry, but also to the engineering fields like mechanics, electric, electronics, information processing, robotics, etc., and to the non-engineering fields like medical science, biology, meteorology, music, art and so on.

In particular, the studies of chaos in the fields of basic science and the application technology of chaos are called *Chaos engineering*.[4,5]

### 1.2. *Availability of chaos engineering*

Many applications of chaos have been studied. For example, chaos neural network, generalized shift mapping, optimization and self-learning organization are applications of a chaos computing, power demand prediction,[7,23] water demand prediction,[8,9,13] traffic prediction,[15] exchange rate prediction, temperature and humidity prediction are that of deterministic nonlinear short-term prediction. Others are electrical and electronic circuit,[19,20] pattern recognition,[22] chaotic art, chaotic music,[21] biological chaos,[6] associative memory,[16,17,18] robotics,[24] etc.

This paper deals with the field of deterministic nonlinear short-term prediction.

## 2. Deterministic Nonlinear Short-Term Prediction

### 2.1. *Approach to the deterministic nonlinear short-term prediction*

As mentioned in the Subsec. 1.2, one of the available application we have is a short-term prediction of the time series under chaotic behavior. This is to find some deterministic law in a phenomenon which was considered noise or to be irregular and predict its state in the near future. Short-term prediction employs the following approach as shown in Fig. 1. In this approach to short-

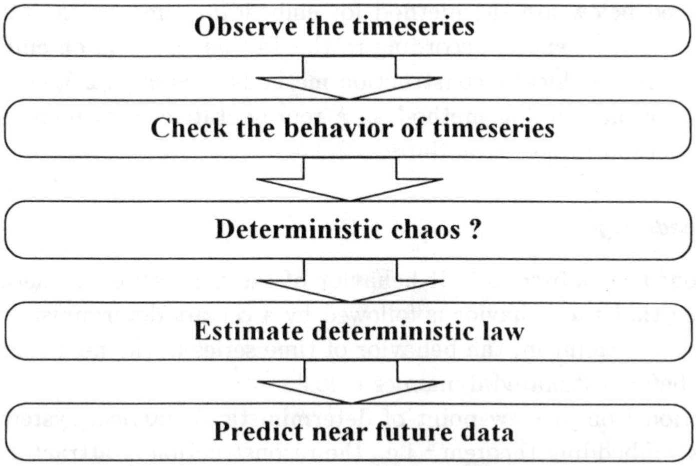


Fig. 1. Approach to deterministic nonlinear short-term prediction.

term prediction, time series data is observed with even sampling intervals at first. If the behavior of the observed time series data is chaotic, it can be assumed that the behavior follows a certain deterministic law. Then, if the nonlinear deterministic regularity (dynamics) can be estimated, data in the near future, until the deterministic causality is lost, can be predicted from the observed data at a certain time point because chaos has a "sharp dependency on initial condition".

Concretely, by employing Takens' embedding theorem,[1] the time series data observed at constant intervals is reconstructed into the $n$ dimensional state space to draw a trajectory. The key point of this approach lies in the presumption of dynamics in dynamical system. Next, the data vector including the latest time series data observed is plotted in the $n$ dimensional reconstructed state space. And by utilizing the data vector neighboring to the plotted data vector and the data vector at $s$ steps ahead, the target predictive data vector is presumed by a local reconstruction method. The Gram Schmidt' orthogonal system method,[3] the tessellation method,[2,8,9] and the local fuzzy reconstruction method[10–15] and others are proposed as local reconstruction methods to estimate the dynamics. Finally, the data vector presumed by any local reconstruction method is converted into the original time series to obtain the predicted value of the target time series data at $s$ steps ahead.

Explained below are the method for embedding time series data into the $n$ dimensional state space, according to the Takens' theorem on embedding in Subsec. 2.2 and the local reconstruction methods in Subsec. 2.3, especially the local fuzzy reconstruction method as a representative local nonlinear reconstruction method proposed in Subsec. 2.3.3.

### 2.2. *Embedding*

As mentioned in Subsec. 2.1, if behavior of the time series is chaotic, it can be assumed that the behavior is followed by a certain determinism. If we can assume the determinism, the behavior of time series in the near future can be predicted before estimated dynamics is lost.

Prediction from a view point of deterministic dynamical system is based on Takens' embedding theorem,[1] i.e., the reconstruction of attractor and state space of the original dynamical system from single observed time series data.

From the observed time series $y(t)$, vector $(y(t), y(t-\tau), y(t-2\tau), \cdots y(t-(n-1)\tau))$ is generated where $\tau$ is a time delay or lag. This vector indicates one point of an $n$ dimensional reconstructed state space $R^n$. Therefore, a trajectory can be drawn in the $n$ dimensional reconstructed state space by changing $t$. Let us assume that the target system is a deterministic dynamical system and that the observed time series is obtained through an observation system corresponding to $C^1$ continuous mapping from the state space of dynamical system to the one-dimensional Euclidean space $R$.

Then, the reconstructed trajectory is an embedding of the original time series when $n$ value is sufficiently large. Namely, if any attractor has appeared in the original dynamical system will appear in the reconstructed state space. $n$ is usually called an "embedding dimension."

In order that such reconstruction achieves "embedding," it has been proven that the dimension $n$ should satisfy $n \geq 2m + 1$, where $m$ represents the state space dimension of the original dynamical system. However, this is a sufficient condition. Depending on data, embedding can be established even when $n$ is less than $2m + 1$.

From the observed time series, the state space and attractor are reconstructed according to the Takens' theorem on embedding. Further, based on the attractor, such trajectory as the above is determined and nonlinear dynamics is presumed. This method of prediction is termed a deterministic nonlinear prediction.

Concretely, the time series $y(t)$ observed at constant sampling time intervals is embedded in a $n$ dimensional state space with the embedding dimension $n$ and time delay $\tau$ by the aforementioned Takens' embedding theorem as shown in Fig. 2.

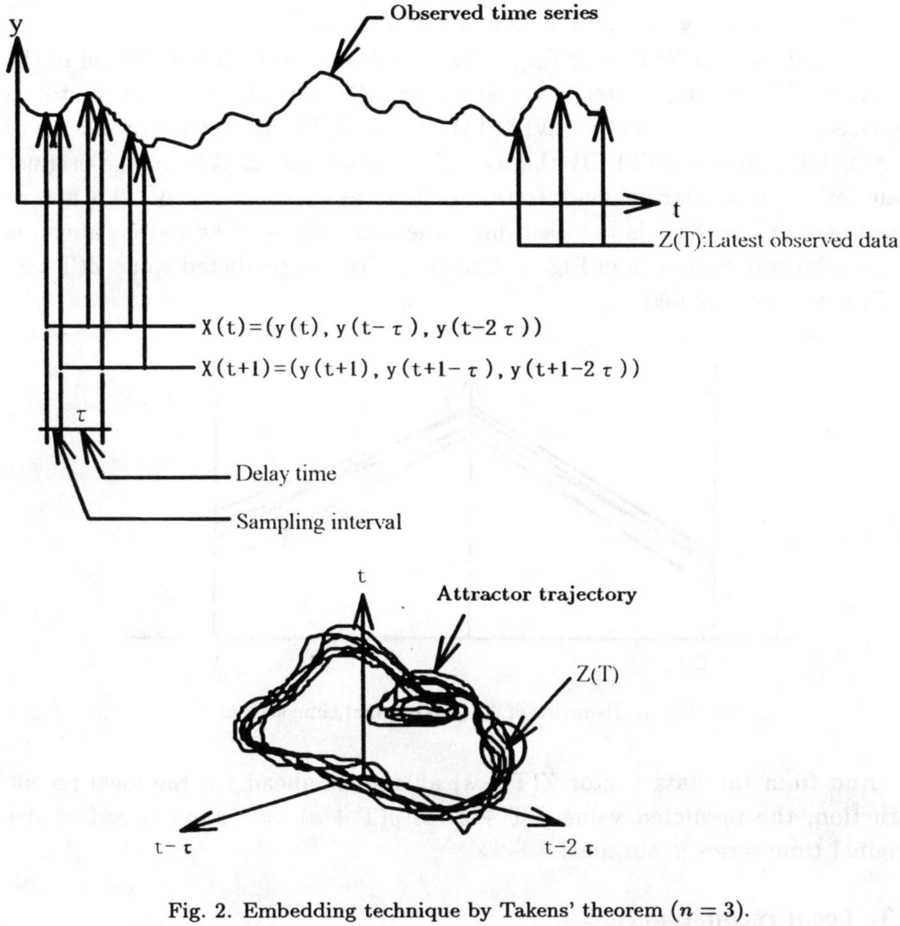


Fig. 2. Embedding technique by Takens' theorem ($n = 3$).

This process is called "reconstruction." As a result, the following vector is obtained.

$$\mathbf{X}(t) = (y(t),\, y(t-\tau),\, \cdots,\, y(t-(n-1)\tau)) \qquad (1)$$

When the reconstruction is repeated on a number of observed data, a smooth manifold consisting of a finite number of data vectors can be composed in the $n$ dimensional reconstructed state space. With regard to the state space and attractor trajectory reconstructed by the embedding procedure, the near-future trajectory of the data vector including the latest time series observed is assumed by using each trajectory of that data vector and the neighboring one, thereby determining the data vector at $s$ steps ahead.

The data vector $\mathbf{Z}(\mathrm{T})$ resulting from the latest observation is plotted in the $n$ dimensional reconstructed state space and the neighboring data vector is replaced with $\mathbf{X}(i)$ (where $i \in N(\mathbf{Z}(\mathrm{T}))$ and $N(\mathbf{Z}(\mathrm{T}))$ are the set of index $i$ of $\mathbf{X}(i)$ neighboring to $\mathbf{Z}(\mathrm{T})$. By the way, the characteristics $\mathbf{X}(i)$ in the original time series has similar behavior to that of $\mathbf{Z}(\mathrm{T})$ as shown in Fig. 3. This is very important. Since $\mathbf{X}(i)$ is the past data, the state $\mathbf{X}(i+s)$ at $s$ steps ahead is already known, as shown in Fig. 4. Utilizing this, the predicted value $\hat{\mathbf{Z}}(\mathrm{T}+s)$ of $\mathbf{Z}(\mathrm{T}+s)$ is obtained.

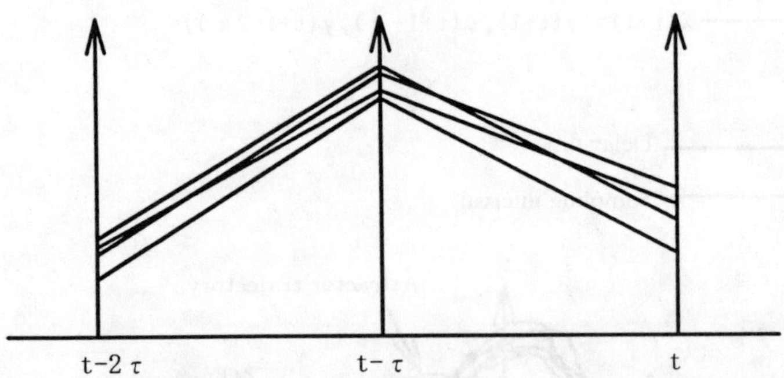


Fig. 3. Behavior of $\mathbf{X}(i)$ in original time series.

And from the data vector $\hat{\mathbf{Z}}(\mathrm{T}+s)$ at $s$ steps ahead via the local reconstruction, the predicted value $\hat{y}(\mathrm{T}+s)$ of $y(\mathrm{T}+s)$ at $s$ steps ahead of the original time series is obtained.

### 2.3. *Local reconstruction*

Several estimation methods of dynamics in the subspace in reconstructed state space shown in Fig. 4 are proposed, e.g., such as the Gram Schmidt' orthogonal system method,[2] and the tessellation method.[3,8,9] These are local linear reconstruction methods[10–15] and the fuzzy reconstruction method is a kind of nonlinear local reconstruction method.[10–15]

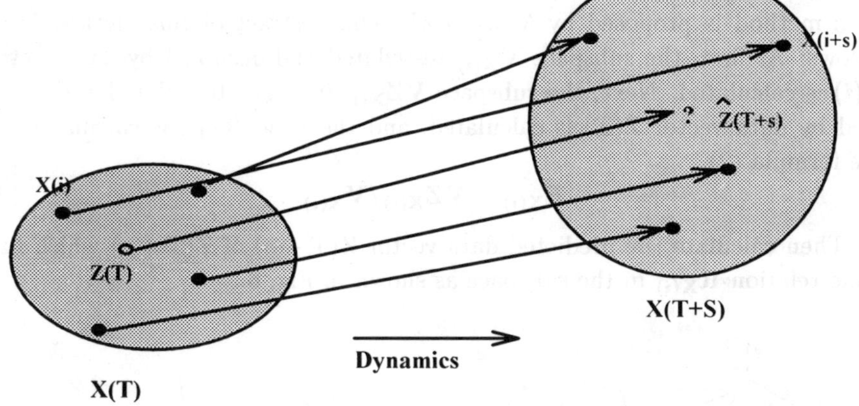


Fig. 4. Trajectory from $\mathbf{X(T)}$ to $\mathbf{X(T+s)}$.

### 2.3.1. *Gram Schmidt' orthogonal system method*

This method is proposed by Jimenez.[2] The abstract of this method is as follows: After the construction of the orthogonal coordinates by neighboring vectors $\mathbf{X}(i)$ of $\mathbf{Z}(\mathrm{T})$, the components of projection $\mathbf{r}_{\mathbf{Z}(\mathrm{T})}$ of $\mathbf{Z}(\mathrm{T})$ to each axis is calculated. Next, after the construction of the orthogonal coordinates by data vectors $\mathbf{X}(i+s)$ which are $s$ step after of $\mathbf{X}(i)$, multiply components of each axis by $\mathbf{r}_{\mathbf{Z}(\mathrm{T})}$ and construct a predicted data vector $\hat{\mathbf{Z}}(\mathrm{T}+s)$ of $\mathbf{Z}(\mathrm{T}+s)$ as shown in Fig. 5.

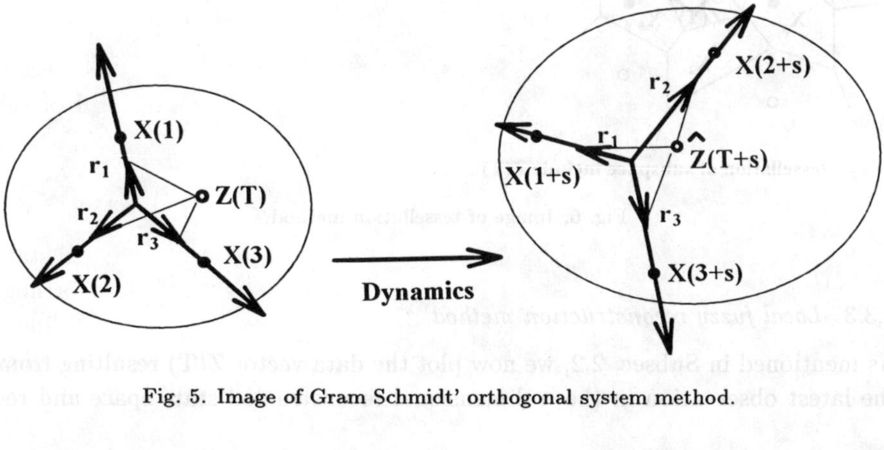


Fig. 5. Image of Gram Schmidt' orthogonal system method.

### 2.3.2. *Tesselation method*

This method is proposed by Mees *et al.* The abstract of this method is as follows: At first, the subspace $\mathbf{V}_{\mathbf{X}(i)}$ tessellated and occupied by data vector $\mathbf{X}(i)$ is calculated. Next, the subspace $\mathbf{VZ}_{\mathbf{X}(i)}$ in $\mathbf{V}_{\mathbf{X}(i)}$ tessellated and occupied by data vector $\mathbf{Z}(\mathrm{T})$ is calculated, and the ratio $\mathbf{R}_{\mathbf{X}(i)}$ is calculated by the formula (2).

$$R_{\mathbf{X}(i)} = \mathbf{VZ}_{\mathbf{X}(i)}/\mathbf{V}_{\mathbf{X}(i)} \tag{2}$$

Then calculate the predicted data vector $\hat{\mathbf{Z}}(\mathrm{T}+s)$ of $\mathbf{Z}(\mathrm{T}+s)$ which has same relation $\mathbf{R}_{\mathbf{X}(i)}$ in the subspace as shown in Fig. 6.

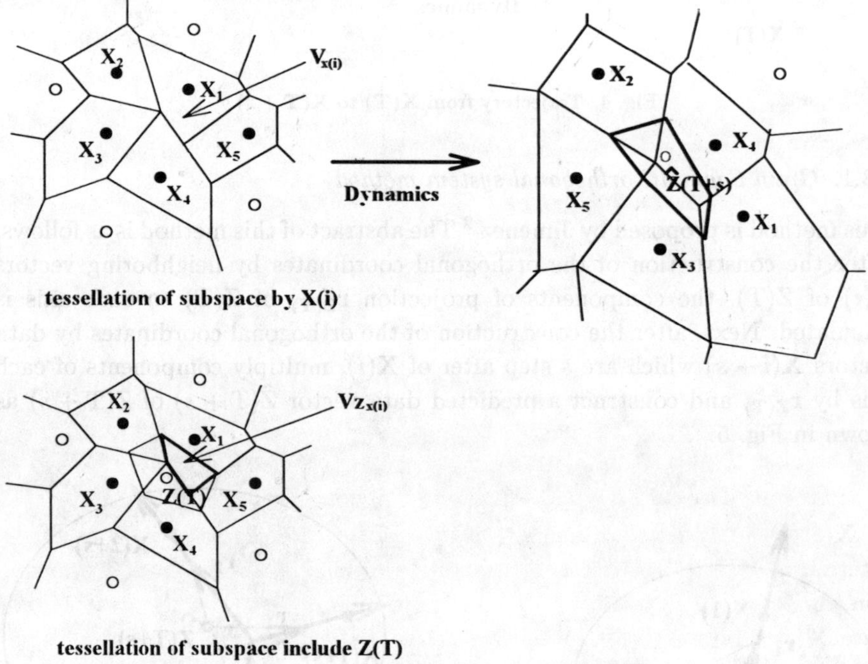


Fig. 6. Image of tessellation method.

### 2.3.3. *Local fuzzy reconstruction method*

As mentioned in Subsec. 2.2, we now plot the data vector $\mathbf{Z}(\mathrm{T})$ resulting from the latest observation in the $n$ dimensional reconstructed state space and replace the neighboring data vector with $\mathbf{X}(i)$. And, we replace the state of data $\mathbf{X}(i)$ at $s$ steps ahead with $\mathbf{X}(i+s)$ and similarly the predicted value of data vector $\mathbf{Z}(\mathrm{T}+s)$ at $s$ steps ahead with $\hat{\mathbf{Z}}(\mathrm{T}+s)$.

If the behavior of the observed time series corresponds to deterministic chaos, the transition from state $\mathbf{X}(i)$ to state $\mathbf{X}(i+s)$ after $s$ steps can be assumed to be dependent on the dynamics subjected to determinism as shown in Fig. 7. This dynamics can be described in a linguistic form as shown below.

$$\text{IF } \mathbf{X}(\mathrm{T}) \text{ is } \mathbf{X}(i) \text{ THEN } \mathbf{X}(\mathrm{T}+s) \text{ is } \mathbf{X}(i+s) \tag{3}$$

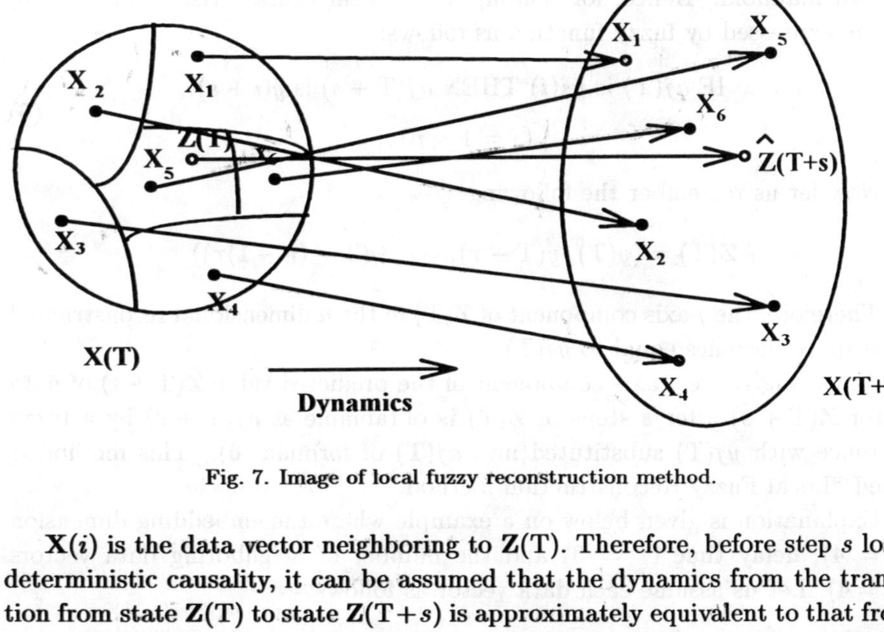


Fig. 7. Image of local fuzzy reconstruction method.

$\mathbf{X}(i)$ is the data vector neighboring to $\mathbf{Z}(\mathrm{T})$. Therefore, before step $s$ loses deterministic causality, it can be assumed that the dynamics from the transition from state $\mathbf{Z}(\mathrm{T})$ to state $\mathbf{Z}(\mathrm{T}+s)$ is approximately equivalent to that from state $\mathbf{X}(i)$ to state $\mathbf{X}(i+s)$ because chaos has a "sharp dependency on initial value." When the attractor embedded in the $n$ dimensional reconstructed state space is smooth manifold as mentioned in Subsec. 2.1, the trajectory from $\mathbf{Z}(\mathrm{T})$ to $\mathbf{Z}(\mathrm{T}+s)$ is influenced by the Euclidean distance from $\mathbf{Z}(\mathrm{T})$ to $\mathbf{X}(i)$.

Now let us remember the following relations.

$$\begin{aligned} \mathbf{X}(i) &= (y(i),\, y(i-\tau),\, \cdots,\, y(i-(n-1)\tau)) \\ \mathbf{X}(i+s) &= (y(i+s),\, y(i+s-\tau),\, \cdots,\, y(i+s-(n-1)\tau)) \end{aligned} \tag{4}$$

This formula can be rewritten as follows when focusing attention on the $j$ axis in the $n$ dimensional reconstructed state space.

$$\text{IF } aj(\mathrm{T}) \text{ is } yj(i) \text{ THEN } aj(\mathrm{T}+s) \text{ is } yj(i+s) \qquad (j = 1 \sim n) \tag{5}$$

where $aj(\mathrm{T})$: $J$ axis component of $\mathbf{X}(i)$ value neighboring to $\mathbf{Z}(\mathrm{T})$ in $n$ dimensional reconstructed state space; $aj(\mathrm{T}+s)$: $J$ axis component of $\mathbf{X}(i+s)$ in $n$ dimensional reconstructed state space; $n$: Dimension of embedding.

Also, the trajectory from $\mathbf{Z}(\mathrm{T})$ to $\mathbf{Z}(\mathrm{T}+s)$ is influenced by Euclidean distance from $\mathbf{Z}(\mathrm{T})$ to $\mathbf{X}(i)$. This influence becomes nonlinear because of a smooth manifold. Hence, for rending a nonlinear characteristic, formula (5) can be expressed by fuzzy function as follows:

$$\text{IF } aj(\mathrm{T}) \text{ is } \tilde{y}j(i) \text{ THEN } aj(\mathrm{T}+s) \text{ is } \tilde{y}(i+s) \qquad (j = 1 \sim n)\,. \tag{6}$$

Now let us remember the following:

$$\mathbf{Z}(\mathrm{T}) = (y(\mathrm{T}), y(\mathrm{T}-\tau), \cdots, y(\mathrm{T}-(n-1)\tau))$$

Therefore, the $j$ axis component of $\mathbf{Z}(\mathrm{T})$ in the $n$ dimensional reconstructed state space becomes equal to $yj(\mathrm{T})$.

Accordingly, the $j$ axis component of the predicted value $\hat{\mathbf{Z}}(\mathrm{T}+s)$ of data vector $\mathbf{Z}(\mathrm{T}+s)$ after $s$ steps of $\mathbf{Z}(\mathrm{T})$ is obtainable as $aj(\mathrm{T}+s)$ by a fuzzy inference with $yj(\mathrm{T})$ substituted into $aj(\mathrm{T})$ of formula (6). This method is called "Local Fuzzy Reconstruction method."

Explanation is given below on a example where the embedding dimension ($n = 4$), delay time ($\tau = 3$) and the number of neighboring data vectors ($N = 4$). Let us assume each data vector as follows:

$$\mathbf{Z}(\mathrm{T}) = (y1(\mathrm{T}), y2(\mathrm{T}-3), y3(\mathrm{T}-6), y4(\mathrm{T}-9))$$

$$\mathbf{X}(a) = (y1(a), y2(a-3), y3(a-6), y4(a-9))$$

$$\mathbf{X}(b) = (y1(b), y2(b-3), y3(b-6), y4(b-9))$$

$$\mathbf{X}(c) = (y1(c), y2(c-3), y3(c-6), y4(c-9))$$

$$\mathbf{X}(d) = (y1(d), y2(d-3), y3(d-6), y4(d-9))$$

$$\hat{\mathbf{Z}}(\mathrm{T}+s) = (y1(\mathrm{T}+s),\ y2(\mathrm{T}+s-3),\ y3(\mathrm{T}+s-6),\ y4(\mathrm{T}+s-9))$$

$$\mathbf{X}(a+s) = (y1(a+s),\ y2(a+s-3),\ y3(a+s-6),\ y4(a+s-9))$$

$$\mathbf{X}(b+s) = (y1(b+s),\ y2(b+s-3),\ y3(b+s-6),\ y4(b+s-9))$$

$$\mathbf{X}(c+s) = (y1(c+s),\ y2(c+s-3),\ y3(c+s-6),\ y4(c+s-9))$$

$$\mathbf{X}(d+s) = (y1(d+s),\ y2(d+s-3),\ y3(d+s-6),\ y4(d+s-9))$$

On this assumption, the fuzzy rule given in formula (6) can be represented by formulae (7), (8), (9) and (10).

Regarding the first axis of reconstructed state space:

$$\begin{array}{l} \text{IF } a1(\mathrm{T}) \text{ is } \tilde{y}1(a) \text{ THEN } a1(\mathrm{T}+s) \text{ is } \tilde{y}1(a+s) \\ \text{IF } a1(\mathrm{T}) \text{ is } \tilde{y}1(b) \text{ THEN } a1(\mathrm{T}+s) \text{ is } \tilde{y}1(b+s) \\ \text{IF } a1(\mathrm{T}) \text{ is } \tilde{y}1(c) \text{ THEN } a1(\mathrm{T}+s) \text{ is } \tilde{y}1(c+s) \\ \text{IF } a1(\mathrm{T}) \text{ is } \tilde{y}1(d) \text{ THEN } a1(\mathrm{T}+s) \text{ is } \tilde{y}1(d+s) \end{array} \tag{7}$$

Regarding the second axis of reconstructed state space:

$$\begin{array}{l} \text{IF } a2(\mathrm{T}) \text{ is } \tilde{y}2(a-3) \text{ THEN } a2(\mathrm{T}+s) \text{ is } \tilde{y}2(a+s-3) \\ \text{IF } a2(\mathrm{T}) \text{ is } \tilde{y}2(b-3) \text{ THEN } a2(\mathrm{T}+s) \text{ is } \tilde{y}2(b+s-3) \\ \text{IF } a2(\mathrm{T}) \text{ is } \tilde{y}2(c-3) \text{ THEN } a2(\mathrm{T}+s) \text{ is } \tilde{y}2(c+s-3) \\ \text{IF } a2(\mathrm{T}) \text{ is } \tilde{y}2(d-3) \text{ THEN } a2(\mathrm{T}+s) \text{ is } \tilde{y}2(d+s-3) \end{array} \tag{8}$$

Regarding the third axis of reconstructed state space:

$$\begin{array}{l} \text{IF } a3(\mathrm{T}) \text{ is } \tilde{y}3(a-6) \text{ THEN } a3(\mathrm{T}+s) \text{ is } \tilde{y}3(a+s-6) \\ \text{IF } a3(\mathrm{T}) \text{ is } \tilde{y}3(b-6) \text{ THEN } a3(\mathrm{T}+s) \text{ is } \tilde{y}3(b+s-6) \\ \text{IF } a3(\mathrm{T}) \text{ is } \tilde{y}3(c-6) \text{ THEN } a3(\mathrm{T}+s) \text{ is } \tilde{y}3(c+s-6) \\ \text{IF } a3(\mathrm{T}) \text{ is } \tilde{y}3(d-6) \text{ THEN } a3(\mathrm{T}+s) \text{ is } \tilde{y}3(d+s-6) \end{array} \tag{9}$$

Regarding the fourth axis of reconstructed state space:

$$\begin{array}{l} \text{IF } a4(\mathrm{T}) \text{ is } \tilde{y}4(a-9) \text{ THEN } a4(\mathrm{T}+s) \text{ is } \tilde{y}4(a+s-9) \\ \text{IF } a4(\mathrm{T}) \text{ is } \tilde{y}4(b-9) \text{ THEN } a4(\mathrm{T}+s) \text{ is } \tilde{y}4(b+s-9) \\ \text{IF } a4(\mathrm{T}) \text{ is } \tilde{y}4(c-9) \text{ THEN } a4(\mathrm{T}+s) \text{ is } \tilde{y}4(c+s-9) \\ \text{IF } a4(\mathrm{T}) \text{ is } \tilde{y}4(d-9) \text{ THEN } a4(\mathrm{T}+s) \text{ is } \tilde{y}4(d+s-9) \end{array} \tag{10}$$

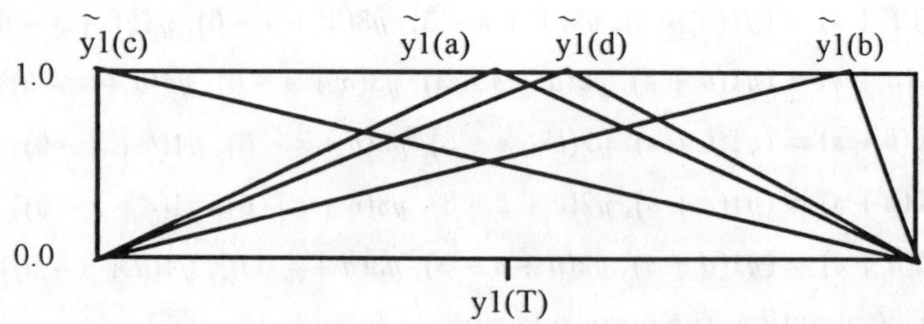


Fig. 8. Membership functions in antecedent statement.

Also, since $\mathbf{X}(a)$, $\mathbf{X}(b)$, $\mathbf{X}(c)$ and $\mathbf{X}(d)$ are neighboring data vectors to $\mathbf{Z}(\mathrm{T})$, each axis of the reconstructed state space in the antecedent part of fuzzy rules (7), (8), (9) and (10). The examples of membership functions corresponding to rules (7) are shown in Fig. 8.

Of course the membership functions in the subsequent part are all functions of a singleton for speeding up computation.

For the dynamics expressed by the linguistic fuzzy rules and membership functions, we conduct a fuzzy inference with the following taken as input data, $a1(\mathrm{T}) = y1(\mathrm{T})$, $a2(\mathrm{T}) = y2(\mathrm{T})$, $a3(\mathrm{T}) = y3(\mathrm{T})$ and $a4(\mathrm{T}) = y4(\mathrm{T})$.

As a result, we obtain:

$$\begin{aligned} \hat{y}1(\mathrm{T}+s) &= a1(\mathrm{T}+s) \\ \hat{y}2(\mathrm{T}+s-3) &= a2(\mathrm{T}+s) \\ \hat{y}3(\mathrm{T}+s-6) &= a3(\mathrm{T}+s) \\ \hat{y}4(\mathrm{T}+s-9) &= a4(\mathrm{T}+s) \end{aligned} \tag{11}$$

Thus, the predicted value $y1(\mathrm{T}+s)$ after the original time series data $y1(\mathrm{T})$ has advanced $s$ steps is available as $a1(\mathrm{T}+s)$.

## 3. Application to the Practical Time Series Data

Logistic map, Lorenz system, Hénon map are well known as the time series data generated from arithmetic expressions and characterized by a chaotic behavior, and we have already applied the deterministic nonlinear short-term prediction methods on these time series.[7–15] However, the target of our interest in prediction is not the chaotic time series data generated from the arithmetic equations, but the social phenomena such as demands for power[23] and tap water,[8,9,13] traffic density,[15] local weather prediction, stock price, and exchange rate whose determinism cannot be known easily.

### 3.1. *Application to the time series of exchange rate*

This subsection shows the short-term prediction of several exchange rates.

To show concrete results, 699(1992.12.31 ~ 1994.11.2) time series data of the exchange rates (JPY/USD, JPY/DM, JPY/FFR, JPY/STG) are tested. The first half of the observed time series data is embedded in an $n$ dimensional reconstructed state space according to the Takens' theorem on embedding as stated in Subsec. 2.2. The latest embedded data vector is replaced with $\mathbf{Z}(\mathrm{T})$ and the time series data after $s$ steps is predicted by the local fuzzy reconstruction method explained in Subsec. 2.3.3. Next, the time series data after each step is fetched, the data vector including this data is replaced with $\mathbf{Z}(\mathrm{T}+1)$ and the value after $s$ steps is predicted by the local fuzzy reconstruction method. The sequence is iterated up to the last data.

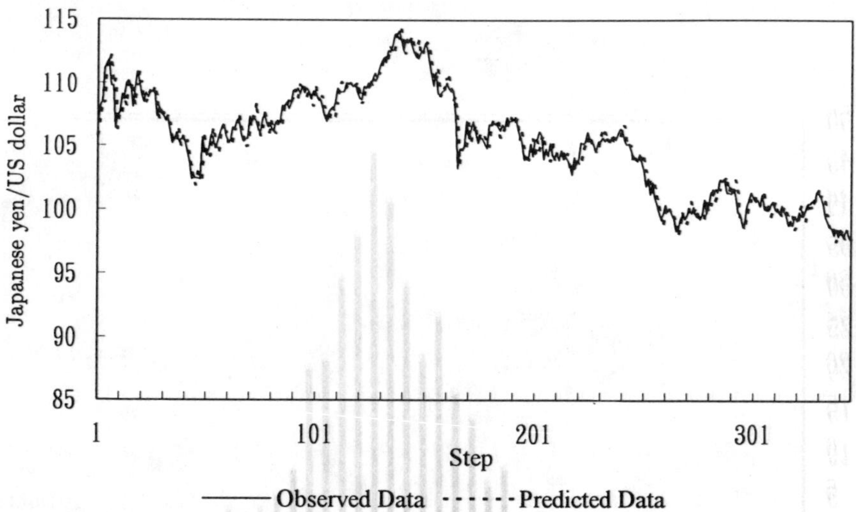


Fig. 9. Result of prediction one step before of Japanese yen/US dollar (time series).

Figures 9–12 show the result of predicting the time series of the exchange rate of JPY/USD in the case where the dimension of embedding ($n$), delay time ($\tau$) and the number of neighboring data ($N$) are 7, 3 and 4, respectively.

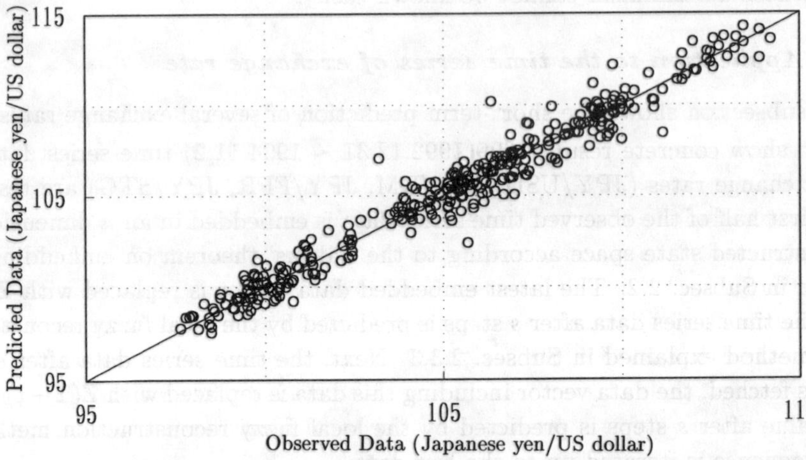


**Fig. 10.** Result of prediction one step before of Japanese yen/US dollar (correlation).

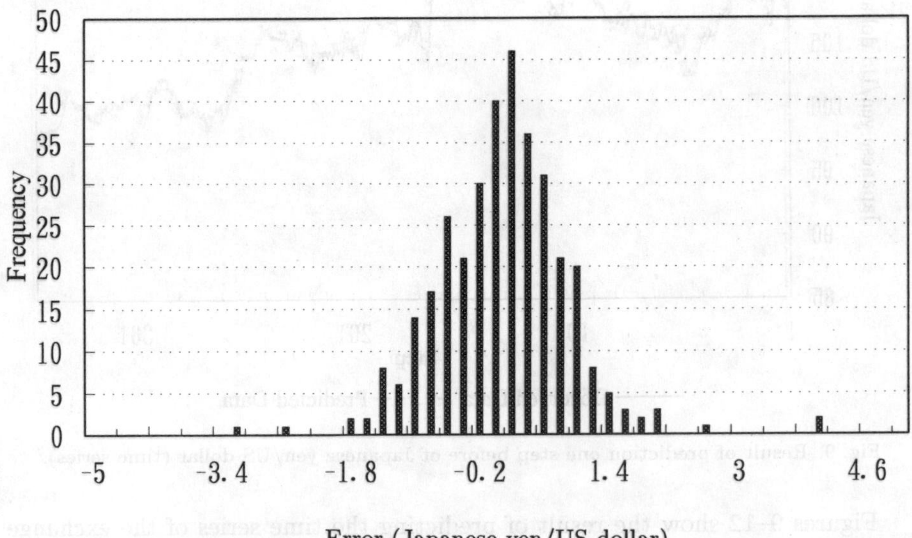


**Fig. 11.** Result of prediction one step before of Japanese yen/US dollar (error distribution).

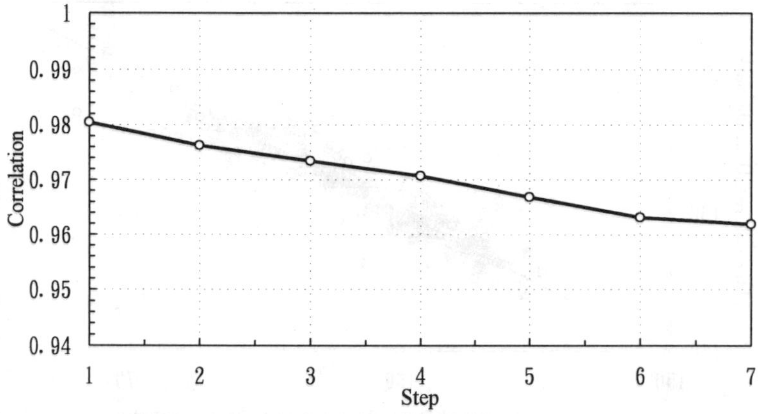


Fig. 12. Result of prediction of Japanese yen/US dollar. (Change of correlation coefficient by prediction step).

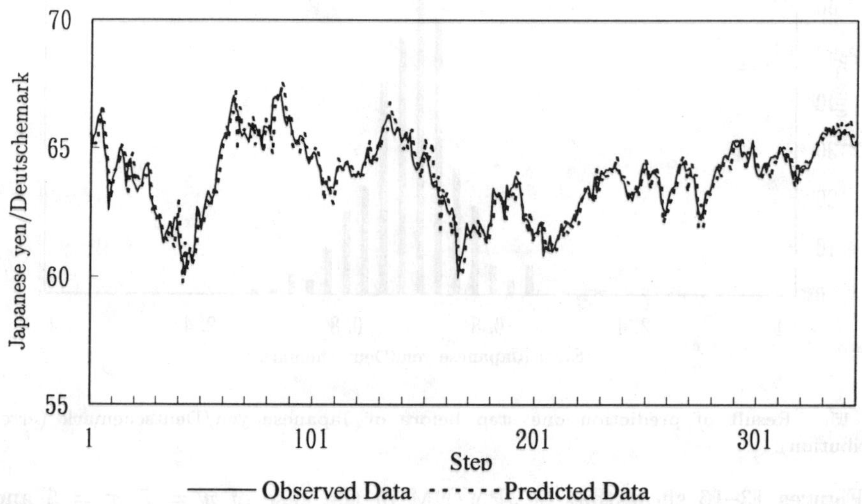


Fig. 13. Result of prediction one step before of Japanese yen/Deutschemark (time series).

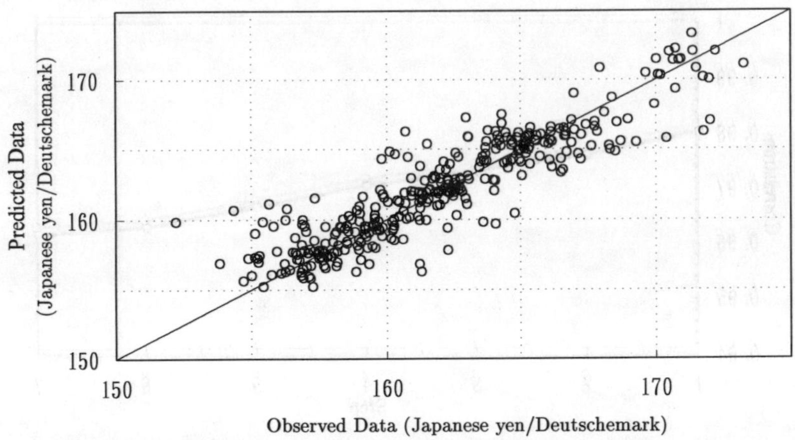


**Fig. 14. Result of prediction one step before of Japanese yen/Deutschemark (correlation).**

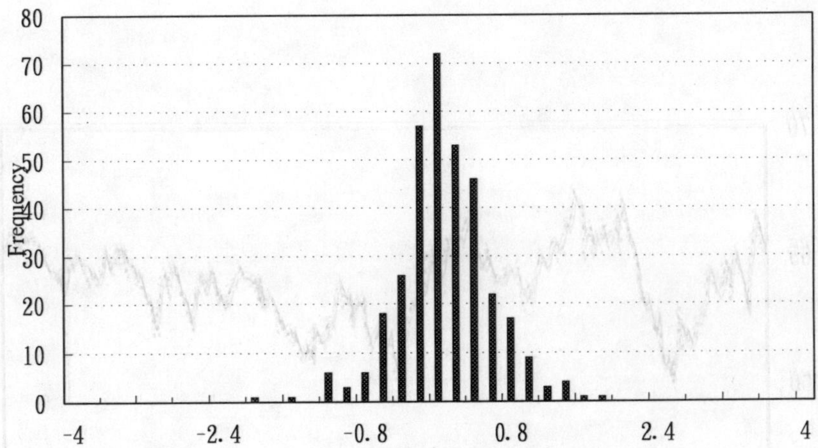


Fig. 15. **Result of prediction one step before of Japanese yen/Deutschemark (error distribution).**

Figures 13–16 show that of JPY/DM in the case of $n = 7$, $\tau = 5$ and $N = 2$; Figs. 17–20 show that of JPY/FFR in case of $n = 3$, $\tau = 4$ and $N = 3$; and Figs. 21–24 show that of JPY/STG in case of $n = 7$, $\tau = 6$ and $N = 3$ respectively.

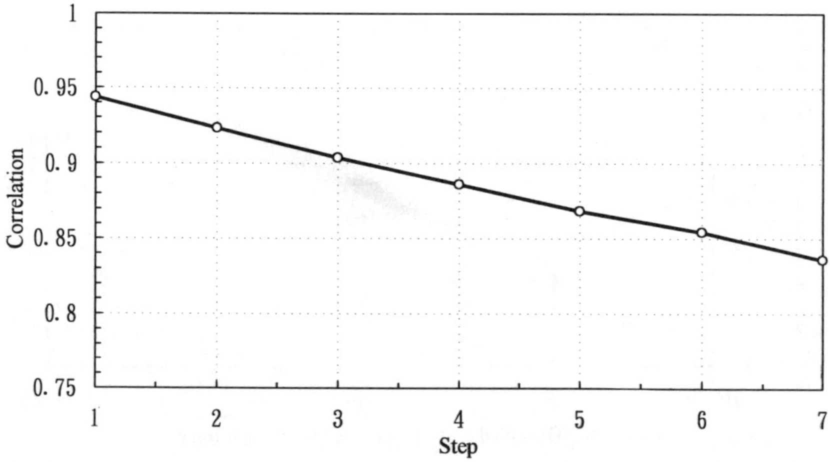


Fig. 16. Result of prediction of Japanese yen/Deutschemark (change of correlation coefficient by prediction step).

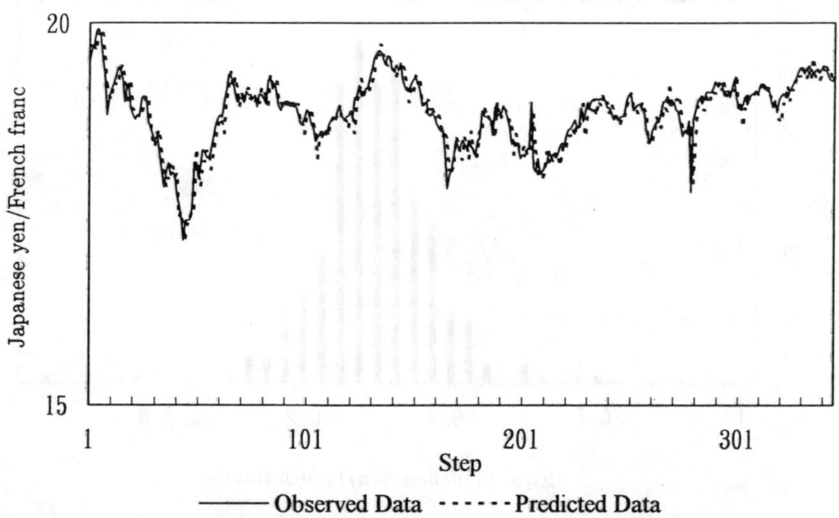


Fig. 17. Result of prediction one step before of Japanese yen/French franc (time series).

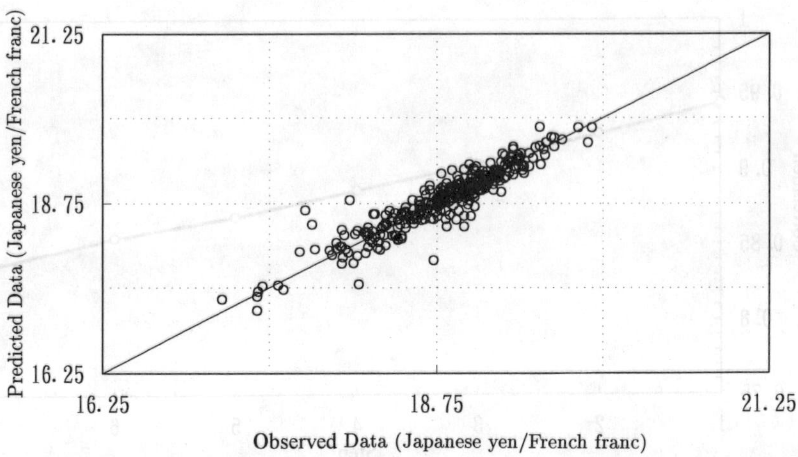


Fig. 18. Result of prediction one step before of Japanese yen/French franc (correlation).

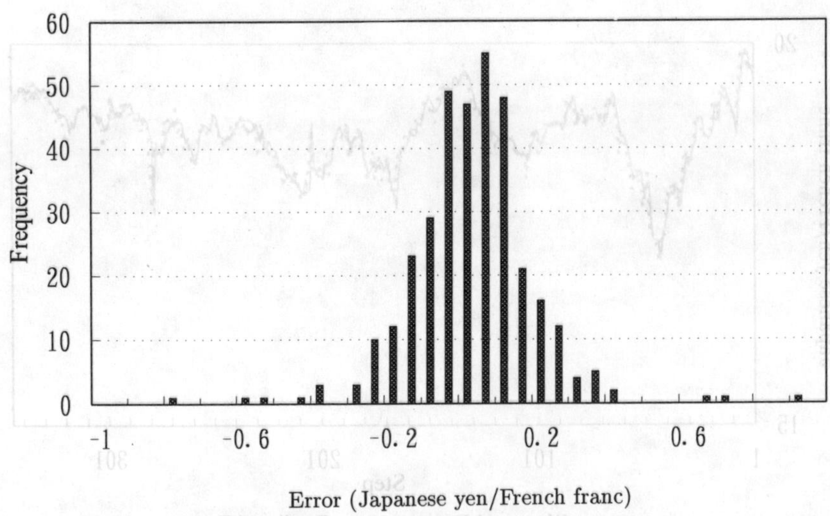


Fig. 19. Result of prediction one step before of Japanese yen/French franc (error distribution).

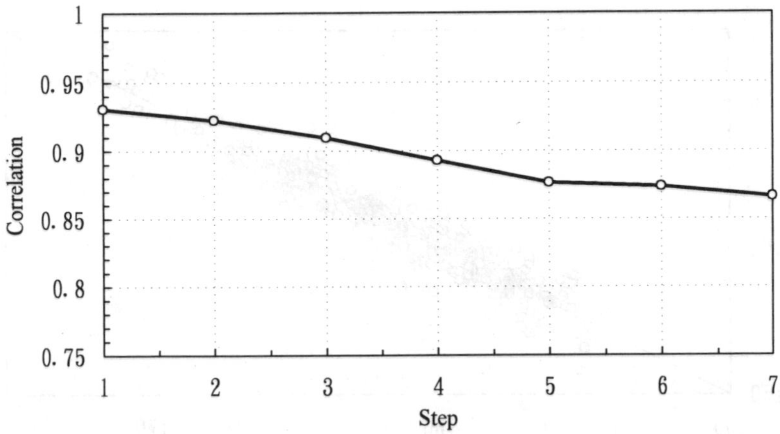


Fig. 20. Result of prediction of Japanese yen/French franc (change of correlation coefficient by prediction step).

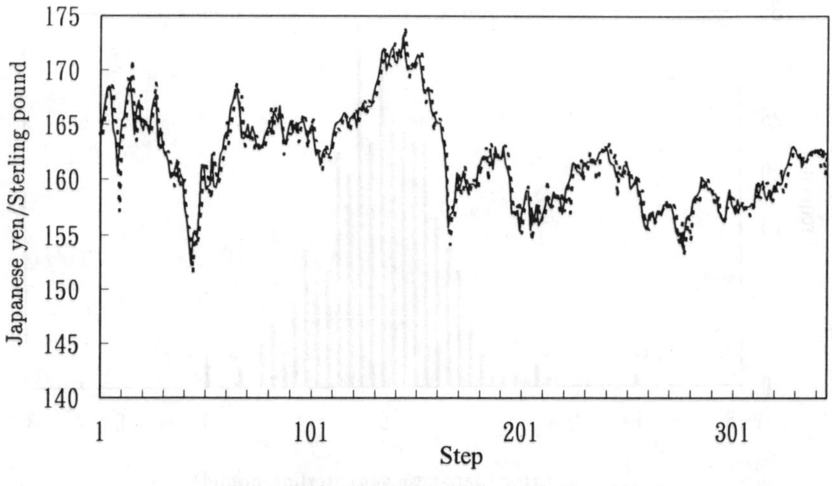


Fig. 21. Result of prediction one step before of Japanese yen/Sterling pound (time series).

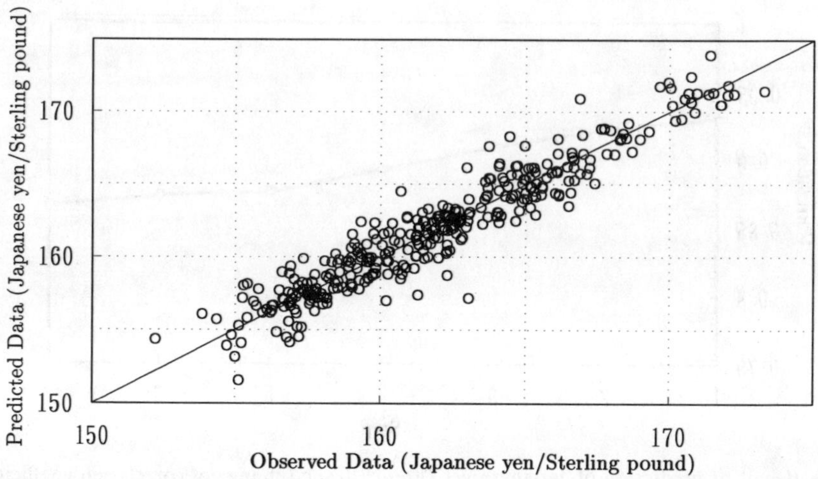


Fig. 22. Result of prediction one step before of Japanese yen/Sterling pound (correlation).

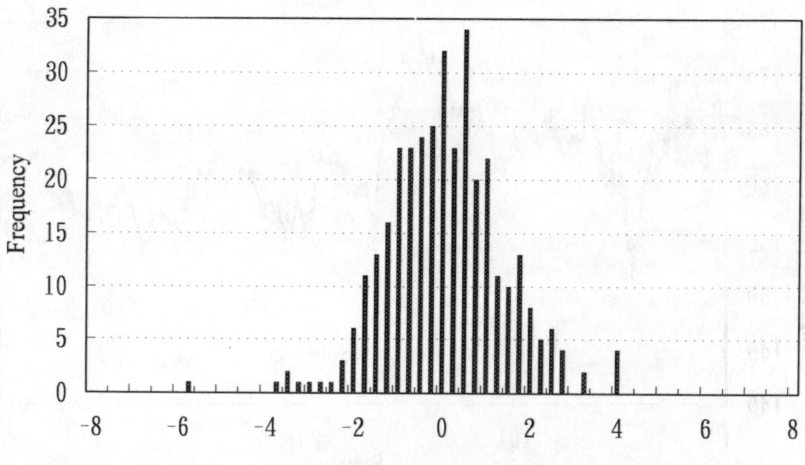


Fig. 23. Result of prediction one step before of Japanese yen/Sterling pound (error distribution).

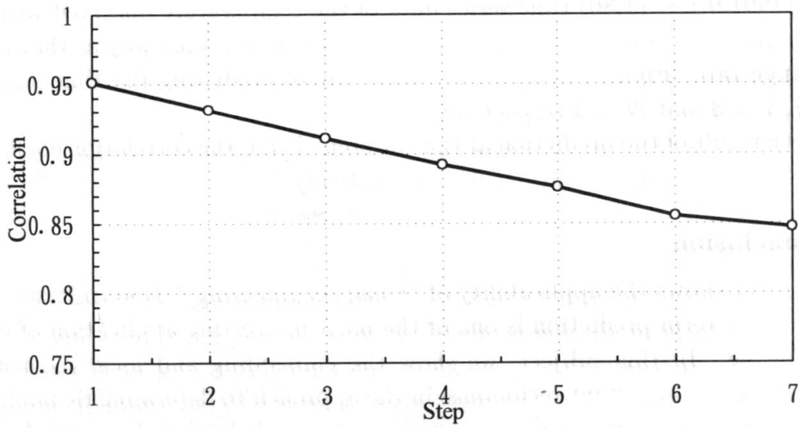


Fig. 24. Result of prediction of Japanese yen/Sterling pound (change of correlation coefficient by prediction step).

As a result of the prediction of each step ahead, the correlation coefficient and RMSE (Root Mean Square Error) of the JPY/USD were 0.980 and ¥0.803, that of the JPY/DM were 0.943 and ¥0.499, that of the JPY/FFR were 0.930 and ¥0.166, that of the JPY/STG were 0.951 and ¥1.321 respectively.

RMSE was obtained by the formula (12), and correlation was obtained by the formula (13).

$$RMSE = \sqrt{\frac{\sum_{i=0}^{N}(b_i - a_i)^2}{N}} \tag{12}$$

$$Correlation = \frac{\frac{1}{N}\sum_{i=0}^{N}(b_i - b_{ave})(a_i - a_{ave})}{\sqrt{\frac{1}{N}\sum_{i=0}^{N}(b_i - b_{ave})^2}\sqrt{\frac{1}{N}\sum_{i=0}^{N}(a_i - a_{ave})^2}} \tag{13}$$

where $a_i$ = observed data, $b_i$ = predicted data, $a_{ave}$ = average of observed data, $b_{ave}$ = average of predicted data, $N$ = number of data.

### 3.2. *Applicaton to the time series of temperature*

2182 (1991.9.1 ~ 11.30) time series data of the temperature observed with one hour interval at certain point in Tokyo is tested in the same way as that of the exchange rate. Figures 25–28 show the result of predicting the time series at $n = 7$, $\tau = 3$ and $N = 3$ respectively.

As a result of the prediction at the one step ahead, the correlation coefficient and RMSE were 0.983 and 0.930°C respectively.

## 4. Conclusion

This paper shows the applicability of "Chaos engineering." Deterministic nonlinear short-term prediction is one of the most interesting application of chaos engineering. In this subject, we show the embedding and local reconstruction which are important techniques in the approach to deterministic nonlinear short-term prediction on time series having chaotic behavior. In particular, the local fuzzy reconstruction method which linguistically expresses the dynamics of target dynamical system according to fuzzy rules is explained in detail.

We also applied the deterministic nonlinear short-term prediction method stated above the time series of exchange rates (JPY/USD, JPY/DM, JPY/FFR

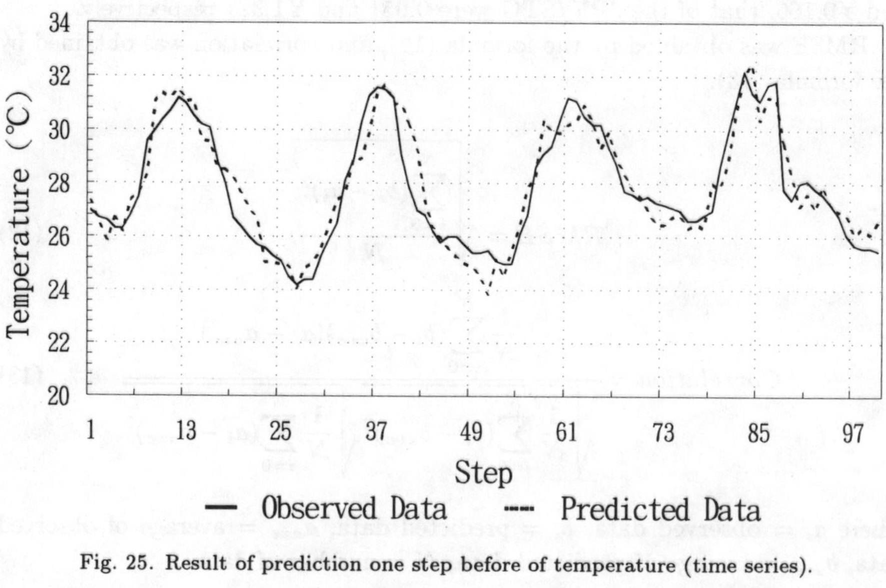


Fig. 25. Result of prediction one step before of temperature (time series).

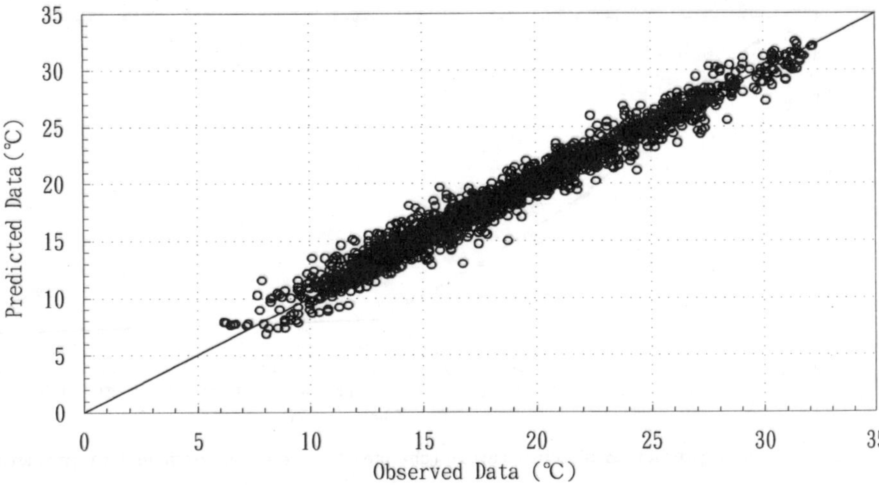


**Fig. 26. Result of prediction one step before of temperature (correlation).**

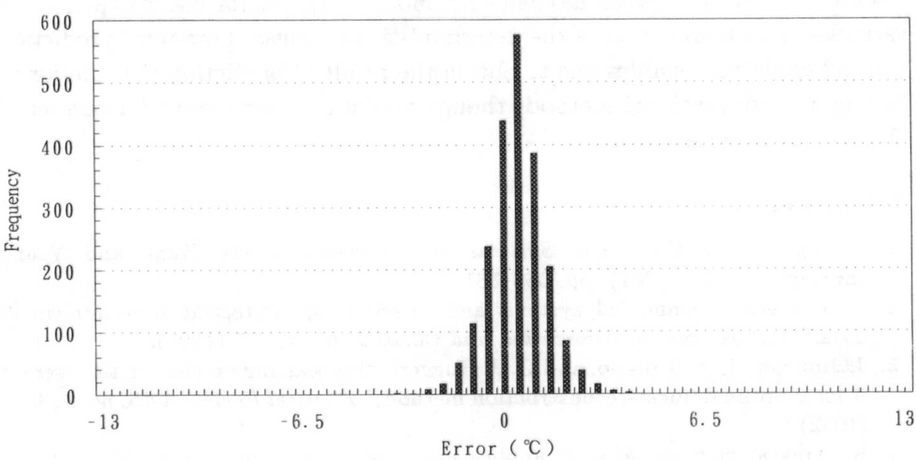


**Fig. 27. Result of prediction one step before of temperature (error distribution).**

and JPY/STG) and temperature whose dynamics are not clear. It was confirmed that the results of these predictions are satisfactory as shown in Sec. 3 separately.

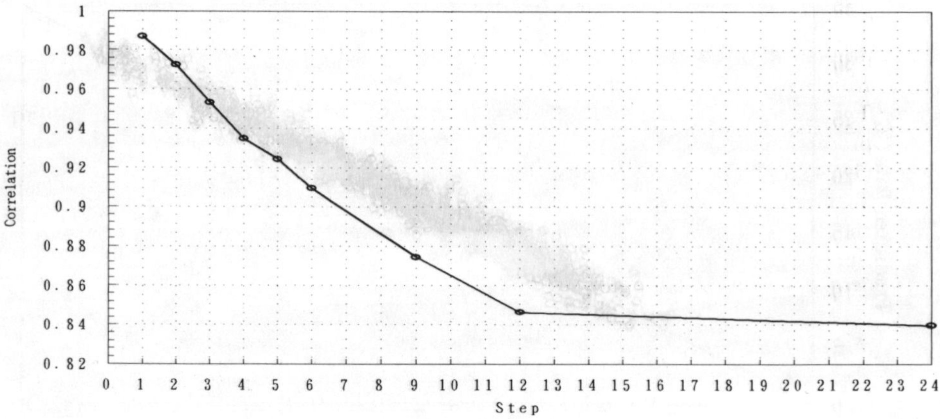


Fig. 28. Result of prediction of temperature (change of correlation coefficient by prediction step).

The prediction dependent on the conventional linear stochastic process represented by auto-regression model (ARX model) requires the use of explanatory variables. But application of the deterministic nonlinear short-term prediction method explained enables one to obtain the result of prediction not inferior to that by the conventional methods though no data is used except for one set of observed time series data.

## References


1. F. Takens, *In Dynamical Systems and Turbulence*, eds. Rand and Young (Springer, Berlin, 1981), pp. 366–381.
2. A. I. Mees, "Dynamical systems and tessellations: Detecting determinism in data," *Int. Journal of Bifurcation and Chaos* **1**, no. 4, 777 (1991).
3. J. Jiménez, J. A. Moreno, and G. J. Ruggeri, "Forecasting on chaotic time series: A local optimal linear-reconstruction method," *Physical Review A* **45**, no. 6, 3553 (1992).
4. K. Aihara, "Chaos: Aim at applications," *Mathematical Science* **30**, no. 6, 5 (1992).
5. K. Aihara, "Chaos: Expanding industrial applications," *Nikkei Science* **22**, no. 3, 26 (1992).
6. I. Tsuda, T. Tahara, and H. Iwanaga, "Chaotic pulsation in human capillary vessels and its dependence on metal and physical conditions," *Int. Journal of Bifurcation and Chaos* **2**, no. 2, 313 (1992).
7. T. Iokibe, "Fuzzy and chaos," in *Proceedings of the 30th Electronics Industry Technical Convention, EIAJ* (1993), pp. 37–42.

8. T. Iokibe, T. Kimura, and K. Aihara, "An application of deterministic nonlinear prediction to water demand data," in *Proc. of 1993 International Symposium on Nonlinear Theory and its Applications* (*NOLTA'93*) (1993), pp. 787–790.
9. T. Iokibe, T. Kimura, and K. Aihara, "An application of deterministic nonlinear prediction to water demand data," *Transactions of IEEJ* **114-D**, no. 4, 409 (1994).
10. T. Iokibe, M. Kanke, Y. Fujimoto, and S. Suzuki, "Short-term prediction on chaotic time series by local fuzzy reconstruction method," in *Proc. of the 10th Fuzzy System Symposium, SOFT* (1994), pp. 391–394.
11. T. Iokibe, M. Kanke, Y. Fujimoto, and S. Suzuki, "Short-term prediction on choatic time series by local fuzzy reconstruction method", in *Proc. of Brazil–Japan Joint Symposium on Fuzzy Systems* (1994), pp. 136–139.
12. T. Iokibe, M. Kanke, Y. Fujimoto, and S. Suzuki, "Short-term prediction on chaotic time series by local fuzzy reconstruction method", in *Proc. of the 3rd International Conference on Fuzzy Logic, Neural Nets and Soft Computing* (1994) pp. 491–492.
13. T. Iokibe and N. Ohshima, "Industrial application based on chaos and fuzzy", in *1994 National Convention Record IEEJ* (Industrial Applications Society, 1994), S.6-3, pp. S121–S126.
14. T. Iokibe, "Application based on chaos and fuzzy," in *Proc. of the 3rd Fuzzy Workshop, SOFT* (1994) pp. 52–62.
15. T. Iokibe and N. Mochizuki, "Short-term prediction based on fuzzy and chaos," in *Proc. of Intelligent System Symposium, SICE* (1994) pp. 77–82.
16. T. Sato, T. Yamaguchi, H. Ushida, A. Imura, and T. Takagi, "Chaotic associative memories of fuzzy rules sets," in *Proc. of the 10th Fuzzy System Symposium, SOFT* (1994), pp. 387–390.
17. T. Sato, H. Ushida, T. Yamaguchi, A. Imura, and T. Takagi, "Chaotic memory search in fuzzy associative memory system," *Journal of Japan Society for Fuzzy Theory and Systems* **6**, no. 4 (1994), pp. 765–774.
18. T. Sato, H. Ushida, and T. Yamaguchi, "Facial expressions model construction using chaotic retrieval," in *Proc. of Intelligent System Symposium, SICE* (1994), pp. 83–88.
19. H. Murakami, "Experimental control of chaos in Yamakawa's chaotic chip," in *Proc. of the 3rd International Conference on Fuzzy Logic, Neural Nets and Soft Computing* (1994), pp. 603–604.
20. H. Takakubo, K. Takakubo, and K. Shono, "Observation of dynamics in CMOS chaos generator," in *Proc. of the 3rd International Conference on Fuzzy Logic, Neural Nets and Soft Computing* (1994), pp. 605–608.
21. H. J. Kang and M. Park, "Music autocomposition using chaotic dynamics," in *Proc. of the 3rd International Conference on Fuzzy Logic, Neural Nets and Soft Computing* (1994), pp. 431–434.
22. K. Toko, T. Matsuno, and K. Yamafuji, "Chaos in recognition of taste," in *Proc. of the 3rd International Conference on Fuzzy Logic, Neural Nets and Soft Computing* (1994), pp. 519–520.

23. H. Mori and S. Urano, "Chaotic time series analysis for short-term power system load forecasting," in *Proc. of the 3rd International Conference on Fuzzy Logic, Neural Nets and Soft Computing* (1994), pp. 441–442.
24. J. Tani, "Experiment of learning and chaotic planning of a mobile robot," in *Proc. of the 3rd International Conference on Fuzzy logic, Neural Nets and Soft Computing* (1994), pp. 521–522.

# APPLICATIONS OF FUZZY LOGIC AND NEURAL NETWORKS IN SPACE OPERATIONS

YASHVANT JANI
*Hitachi America Ltd.*
*Brisbane, CA 94005*
*jani_y@halsp.hitachi.com*

ROBERT N. LEA* and ROBERT H. BROWN*
*NASA Johnson Space Center*
*Houston, TX 77058*
*blea@gothamcity.jsc.nasa.gov*

Fuzzy logic deals with the imprecision of natural world data which cannot be described by precise equations and traditional logic. It is constructed around the idea of a set in which the membership is a matter of degree rather than yes or no. Though known for more than 30 years, the value of fuzzy logic has been appreciated only recently through applications, mainly in consumer appliances, in Japan. Fuzzy logic principles have been investigated at the NASA Lyndon B. Johnson Space Center for several space applications such as translational and rotational control of spacecraft, camera tracking systems, auto focusing of microscopes, temperature control in a payload module, tether length control and robotics manipulator control.

Neural networks, known for almost a half-century, deal with the complexities of the natural world data by generalizing from prototypes or samples in much the same way as humans extrapolate from one experience to similar situations. Neural networks have been studied for fault diagnosis and health monitoring of space operations and have been extended to include time correlated data and create the Space–Time Neural Network. Neural networks have been combined with fuzzy logic to investigate the adaptive decision making systems for translational and rotational control. In this chapter, some of the space applications of fuzzy logic and neural networks are described including tether skip-rope identification using Space–Time Neural Networks. Also described are tools developed at the Lyndon B. Johnson Space Center and other NASA centers to support such investigations.

*Retired from NASA Lyndon B. Johnson Space Center.

## 1. Introduction

Neural networks and Fuzzy logic present the next step in computerizing the human thought processes. Rule-based expert systems allow one to develop computer programs in a manner that relates rules to numbers. Fuzzy logic takes the next step by relating rules to fuzzy sets. This fuzzy logic attribute facilitates the capture of human thought processes in an optimal manner for automation. Fuzzy logic provides the ability to handle control problems where there is uncertainty due to complex dynamics of an environment. Fuzzy logic can be used to intuitively control the system as long as insight about the system behavior exists in the operators' mind. Neural networks, the first step in reproducing the behavior of the human brain, provides a method to develop control parameters by observation or training and bring with them the ability to pattern match in the imprecise and noisy world. An example is the human's ability to comprehend the original text on a storm-damaged sign. Neural networks and fuzzy logic jointly provide the ability to address the mechanization of very difficult control and pattern matching problems in computer software and hardware in a more natural and optimal manner.

Lotfi Zadeh first developed the concept of fuzzy sets[1] in 1965 and established these concepts firmly into fuzzy logic during the 70's with other pioneers.[2–7] Contrary to its name, fuzzy logic is a precise sub-discipline in mathematics that enables mathematicians and engineers to utilize human-like thinking in decision making processes. Handling imprecise information is easier in a fuzzy logic architecture as compared to conventional logic, however, it was not accepted in the United States for a long time simply because the word 'fuzzy' implied unclear. It was the impression that this logic explains decision making processes by using magic rather than sound mathematical principles. Control engineers argued that the desired control could be achieved using existing control theory principles until the triple inverted pendulum was balanced using fuzzy logic principles. This task could not be performed adequately using conventional logic. The real utility of this logic was shown by Japanese engineers[8–10] when they applied these principles to subway control, automatic transmission control, camera focusing, and many other applications. These applications illustrated the most interesting aspect to U.S. business people — using this logic saved developmental time and costs. It is only recently that U.S. business has taken any interest in this field.[11–13]

Investigations in the areas of fuzzy logic and neural networks have been underway since 1984 (as reported at Tech2000 conference[14]) at several NASA centers. The utility of this logic in several space applications especially in autonomous operations has been shown utilizing high fidelity simulations.[15] The NASA objectives are to investigate new technologies for control and decision making processes. Several NASA centers are evaluating fuzzy logic techniques, neural network methods, genetic algorithms, and learning techniques based on fuzzy sets and neural networks for building expert systems and robust controllers. Particularly, NASA is investigating the feasibility of applying these technologies to space operations to achieve desired operational efficiency and reduce overall life cycle costs.

In the following sections, several NASA applications of fuzzy logic, neural networks and fuzzy-neuro techniques are described. Some of the software tools developed for research and investigation are also described. Advantages and disadvantages of fuzzy logic, neural network and fuzzy-neuro approaches are discussed in the summary.

## 2. Fuzzy Logic Applications

In this section several NASA applications are described. These include a six-degree-of-freedom robotic arm controller developed and implemented at the Goddard Space Flight Center, a six-degree-of-freedom proximity operations controller developed to demonstrate feasibility for shuttle operations, and a camera tracking system developed for application to crew activity monitoring around the space station and to assist in the performance of proximity operations efficiently. Other applications described are collision avoidance algorithms that were developed for application to vehicles such as Mars Rover where automatic navigation and guidance for the vehicle trajectory is envisioned. All of these algorithms were developed to ensure easy transfer to on-orbit shuttle and space station operations. Additional applications are a remote microscope focusing, developed at McDonnell Douglas Space Systems Company for a crew health care system, and a temperature control application, developed at Space Industries for middeck payload carrier experiments. Tether length control has been developed at the Johnson Space Center to support new techniques for fuzzy and fuzzy-neuro decision making processes. Information regarding other applications can be found in references provided at the end of this chapter.

### 2.1. *Remote manipulator system*

In this section a system under development at the Goddard Space Flight Center for the control of a remote manipulator system is described. This system generates the joint rate commands required to move the point of resolution of a robotic arm based on commands generated by remote manual operation of translational and rotational hand controllers. A second function of the system is to control the joint rates at the required levels based on feedback from the joint angle encoders. This section concentrates on the closed loop control function.

The Goddard Space Flight Center (GSFC) Remote Manipulator System (RMS) consists of a six-degree-of-freedom arm, a set of software[16] modeled on the Shuttle RMS Flight Software Systems Requirements (FSSR), and a fuzzy logic controller for controlling the six joint rates to the required levels as generated by the FSSR software.

The controller developed is designed to control the rate of each of six joints to match the commanded rate as computed by the FSSR. This software accepts translational and rotational rates of the point of resolution (POR) of the arm (for this discussion we assume the POR is the end of the wrist roll joint) generated by a remote manipulator system operator who utilizes translational and rotational hand controllers. Related work on the concept of fuzzy logic for control of the point of resolution of robotic arms was developed and reported in Ref. 17. However, that work concentrated on the problem of specifying an end position and orientation of the point of resolution and the fuzzy logic control of the transition. In this work the goal is to take commands from an operator that specifies a linear motion of the point of resolution, or rotations to achieve a new orientation of the end effector, convert them into joint rates required to achieve these end effector rates through the FSSR software, and to achieve and maintain the commanded rates through fuzzy logic control based on feedback from angle position encoders. The angle position encoder measurements and the time tag associated with the measurements are the only feedback to the control system.

This method controls the joint rates based on the feedback of the joint angle measurements. From these measurements estimated joint rates are computed. From the joint rates and the Rate_Cmd that is generated by the FSSR, joint rate errors, E, and changes in joint rate errors, DE, are computed. Next E and DE are treated as fuzzy input variables to a fuzzy rulebase that outputs a fuzzy variable D_Rate_Cmd which is defuzzified and added, initially to the Rate_Cmd, and then to the previous Rate command, to create a Rate_Cmd_Out to be sent to the joint. The equations are,

$$\text{Rate\_Cmd\_Out} = \text{Rate\_Cmd} + \text{D\_Rate\_Cmd\_Prev} + \text{D\_Rate\_Cmd} \quad (1)$$

$$\text{D\_Rate\_Cmd\_Prev} = \text{Rate\_Cmd\_Out} - \text{Rate\_Cmd}\,. \quad (2)$$

The rulebase (Table I) and the fuzzy membership functions were built using the Togai Infralogic TIL_SHELL.[18] Seven levels of membership functions, NL, NM, NS, Z, PS, PM, and PL, were used for E, DE, and D_Rate_Cmd.

Table I. Rulebase for Delta_rate_cmd.

| D_ Rate Cmd | | E | | | | | | |
|---|---|---|---|---|---|---|---|---|
| | | NL | NM | NS | Z | PS | PM | PL |
| DE | NM | PL | PL | PM | PM | PS | Z | NL |
| | N | PL | PM | PS | PS | NS | NM | NL |
| | Z | PL | PM | PS | Z | NS | NM | NL |
| | P | PM | PS | PS | NS | NM | NL | NL |
| | PM | PS | Z | Z | NM | NL | NL | NL |

An additional computation of a Delta, based on an integral like algorithm, was introduced to smoothen the commanded rate. Experiments were used with varying lengths of history of actual rate errors and methods of performing the integration like operations. It was decided that actual integral operators, such as the trapezoidal or Simpson's rule, place too much emphasis on past history. This system is constantly changing and past information, although quite useful for trending, will cause too much of a delay in taking required action. It was decided that the best results were obtained by simply adding rate errors over the previous $k$ cycles to determine if the actual rate is consistently high or low and then adding a delta to the rate command to compensate for the bias. It was also necessary to filter out abnormally large rate errors that occur when a new rate command has been issued by the FSSR.

The parameter which was computed as input to the integral rule base was,

$$\text{Rate\_Error\_Integral} = \sum_{i=0}^{k} \text{Rate\_Error}_{n-i}\,. \tag{3}$$

For the tuned Goddard RMS system $k$ equals to six was chosen. Output from the system is a second delta to the Rate_Cmd, Delta, that is generated by a second fuzzy rulebase and fuzzy membership functions for the variables Rate_Error_Integral and Delta.

The primary objective of the system is to control the motion of the point of resolution of the arm consistent with the operator's hand controller commands. For example, if he commands a translation in the $x$ direction, the POR should move essentially along a linear path in the $x$ direction. This will only occur if the commanded rates are achieved over approximately the same time span and with rate buildup in each joint proportional to joint rate change required for the joint. If some joints have slower response to rate commands than others, then the motion will not be in the desired direction and will require additional commands to achieve the desired arm position. Therefore it is desirable to increase the acceleration of the lagging joints in order that they reach the commanded level at the same time as the other joints.

Since a scalar multiple of translational velocity is equivalent to the same scalar multiple of joint rates, if joint rates are increased or decreased proportional to the joint rate errors we will achieve motion in the correct direction although possibly slower or faster than the actual commanded rates. In order to add this feature, inputs of all six angle positions are used to compute estimates of joint rates and then joint rates are used to compute rate errors. Next the maximum joint rate error, $K$, is computed, and all six ratios $K_i = E_i/K$ are computed. These $K_i$ are then input to another rulebase that outputs a gain, $G$, that weights the Rate_Cmd_Out proportional to the size of the errors in the six joint rates. Thus, those joints that are slow to respond are given priority. The final equation for computation of Rate_Cmd_Out is given by

$$\begin{aligned}&\text{Rate\_Cmd\_Out}\\ &\quad = G^{*}\ (\text{Rate\_Cmd} + \text{D\_Rate\_Cmd\_Prev} + \text{D\_Rate\_Cmd} + \text{Delta})\end{aligned} \tag{4}$$

where,

$$\text{D\_Rate\_Cmd\_Prev} = \text{Rate\_Cmd\_Out} - \text{Rate\_Cmd\_Prev} \tag{5}$$

and Rate_Cmd_Prev = Rate_Cmd on the previous cycle.

This software has been integrated into the hardware/software system at the Goddard Space Flight Center. This system has been tuned and demonstrated at the WNN/FNN '94 conference held in Washington D.C. in December 1994.[19] Although it performs well (as well as or better than a PID controller), the initial tests have indicated the need for other sensors, such as strain guages or accelerometers, to indicate oscillations of the flexible links caused by accelerations and braking. Currently a software model is being built to use in the design and tuning of a near optimal system that will significantly reduce the ringing effects of the arm during operations.

As soon as the new robotic arm model has been completed, tests will be designed to evaluate performance of the system with the addition of simulated sensors such as the strain guages or accelerometers and to determine the optimal placements of such sensors and to evaluate their effectiveness in controlling the flexing and undesired oscillations of the links.

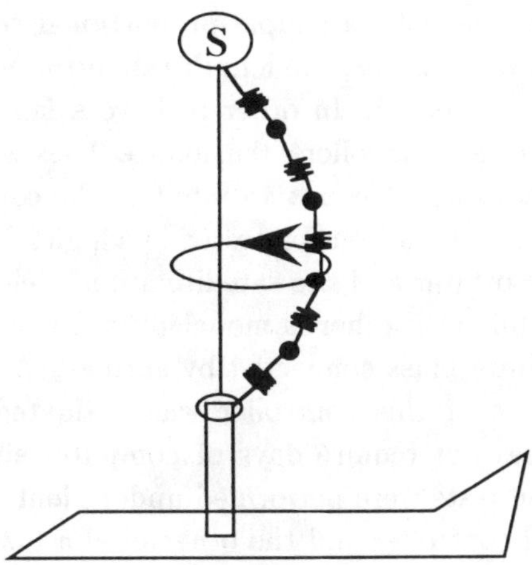


Fig. 1. Tether circular transverse motion known as "skip rope" effect.

## 2.2. *Tether length control*

The Tethered Satellite System (TSS),[20] consisting of the Shuttle, tether, and an Italian satellite, has complex and non-linear dynamics because of the spring-like behavior of the tether and its interaction with two six-degrees-of-freedom end-bodies. The TSS has natural librational and longitudinal oscillations, whose characteristics change as the tether is deployed and retrieved. When the finite mass of the tether is properly modeled, the TSS also exhibits 'skip rope' motion, particularly if the motion is excited by the interaction between the geomagnetic field and electric current in the tether (see Fig. 1 and Fig. 25 later in this chapter). A conventional control system based on the linearization of the system dynamics is employed in the onboard software. Its inputs are sensed length, length rate, and tension parameters and it is expected to maintain not only the length but also the in-plane librational angle.

Due to the complexity of this problem and the recent success of fuzzy logic techniques in controlling systems with non-linear dynamics using imprecise measurements as input from sensors, a fuzzy logic based tether length controller has been developed and implemented into the TSS simulation to investigate the usefulness and robustness that can be achieved with fuzzy control. A concept which utilizes fuzzy membership functions for length, length rate, and voltage parameters, and the rulebase used in generating the control voltage signal was developed. Initially, a simple proportional controller was created based on length error only. Later, the length rate error was included as input, and the rulebase was expanded. In order to have a fair comparison between the conventional and fuzzy controllers, tension feedback was not included as an input to the controllers since it is not an input to the conventional controller.

The fuzzy controllers[21,22] have been tested in a high fidelity tether dynamics simulation, where the orbiter and the satellite are modeled with six-degree-of-freedom dynamics, while the tether is modeled first as a massless spring, and then as beads with finite mass connected by springs. As an initial feasibility study, the performance of this controller was evaluated using the massless tether model that does not require days of computer simulation time. Two full mission simulation tests were performed under identical set-up except one used the conventional controller and the other used a fuzzy controller.

The performance of the linear proportional controller for the massless tether simulation was very encouraging, resulting in a length error of around 16 m during deployment compared to 10.5 m for the conventional controller. During the on-station phase, the fuzzy controller maintained length error at about 2.5 m compared to 2.2 m for the conventional controller. During the retrieval phase, the fuzzy controller reached a maximum length error of 8 m while the conventional controller reached 6.5 m length error. Tension error was about the same throughout the mission for both controllers.

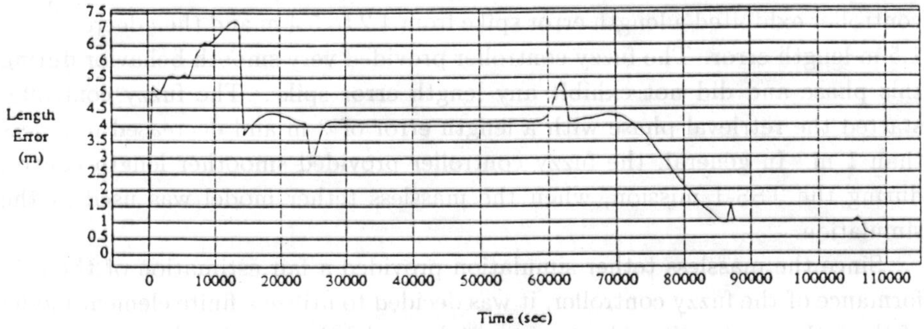


SIMULATION APPLICATION: TSS SIMULATION WITH MASSLESS TETHER
RUN IDENTIFICATION: SYS-17: Deployer CDMS Reel Control Software Model (proportional only)
MODEL: TETHER_ML
DATE: Tue Feb 05 1991 10:27:07 AM
NUMBER OF DATA POINTS: 184
DATA SAMPLING FREQUENCY: 0.002 Hz

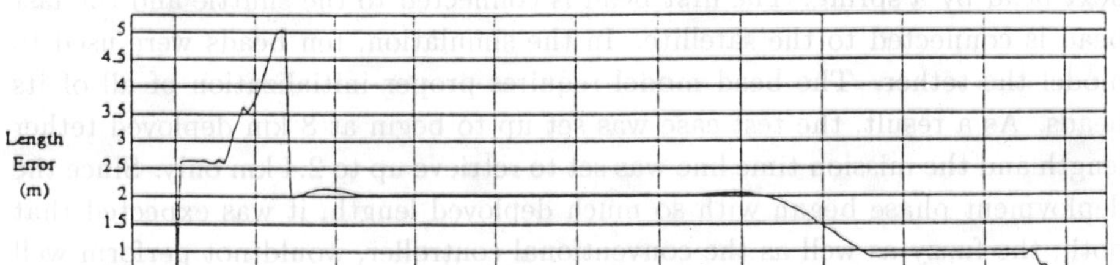


SIMULATION APPLICATION: Tether With Fuzzy Reel Control
RUN IDENTIFICATION: SYS-17: Deployer CDMS Reel Control Software Model
MODEL: TETHER_ML
DATE: Tue Feb 05 1991 02:12:25 PM
NUMBER OF DATA POINTS: 184
DATA SAMPLING FREQUENCY: 0.002 Hz

Fig. 2. Length error for conventional and fuzzy controllers for massless tether model.

When length rate error was included in the fuzzy controller for the massless tether, the overall length error was reduced to within 5 m (Fig. 2), while the conventional controller resulted in a length error of about 7.5 m during the deployment phase. During the on-station phase, the fuzzy controller (PD type) maintained a maximum length error of 2.1 m as compared to 4.2 m length error for the conventional controller. During the retrieval phase, the conventional controller exhibited a length error spike from 4.2 to 5.4 m and then decreased to 1.5 m length error. The fuzzy controller provided very smooth behavior during this phase and did not exhibit any length error spike. The fuzzy controller started the retrieval phase with a length error of 2 m and decreased it to less than 1 m. In general, the fuzzy controller provided smoother length control during the TSS-1 mission, when the massless tether model was used in the simulation.

Since the massless tether simulation provided a fair estimation of the performance of the fuzzy controller, it was decided to utilize a finite element model of the tether system in order to accurately model the complex dynamics of the tethered system. The finite element simulation allowed us to study the dynamic effects that can lead to a transverse oscillatory motion known as the skiprope phenomenon. The finite element simulation models the tether as a string of beads, each bead being a 3-degree-of-freedom object connected to the next bead by a spring. The first bead is connected to the shuttle and the last bead is connected to the satellite. In the simulation, ten beads were used to model the tether. The bead model requires proper initialization of all of its beads. As a result, the test case was set up to begin at 8 km deployed tether length and the mission time line was set to retrieve up to 2.4 km only. Since the deployment phase began with so much deployed length, it was expected that both, the fuzzy as well as the conventional controller, would not perform well and both would saturate. Therefore, in evaluating the controller performance, emphasis was placed only on the retrieval phase.

The performance of the fuzzy controller with the bead model simulation was evaluated in terms of in-plane and out-of-plane libration amplitude as well as length error, length rate error, and tension error. As expected, the length error saturated beyond 15 m and in fact it went over 100 m during deployment. Then, during on-station phase the length error reduced to a reasonable value. Both controllers maintained nearly the same value of the length error. However, during the retrieval phase of the mission, the fuzzy controller resulted in 2 to 4 m less length error than the conventional controller as shown in Fig. 3. The tension error plots shown in Fig. 4 indicated that the fuzzy controller provided better overall tension control than the conventional controller. From 12,000 to 45,000 seconds of on-station mission time, the tension error in the fuzzy controller varied from 3 to slightly over 9 N, with most of the activity between 6 and 8 N as indicated by the dark band. The fuzzy controller compared to the conventional controller showed a smoother tension error control from about 45,000 to 65,000 seconds during the retrieval of the tether from 20 km to 2.4 km length.

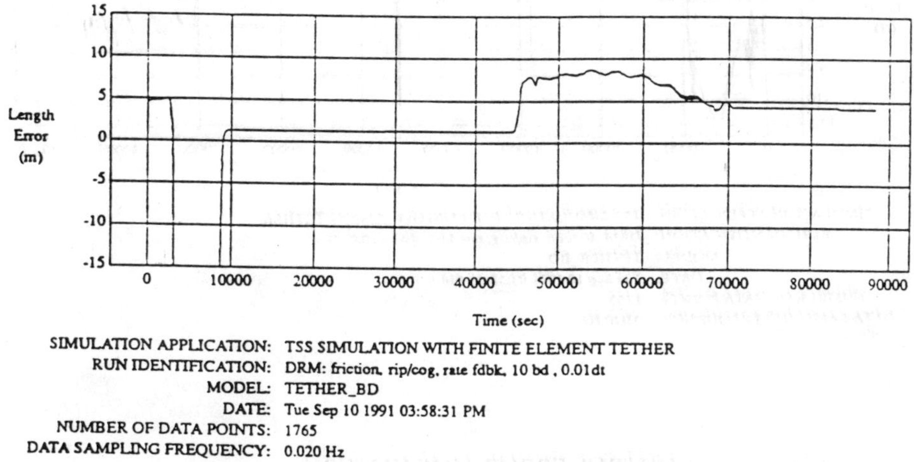


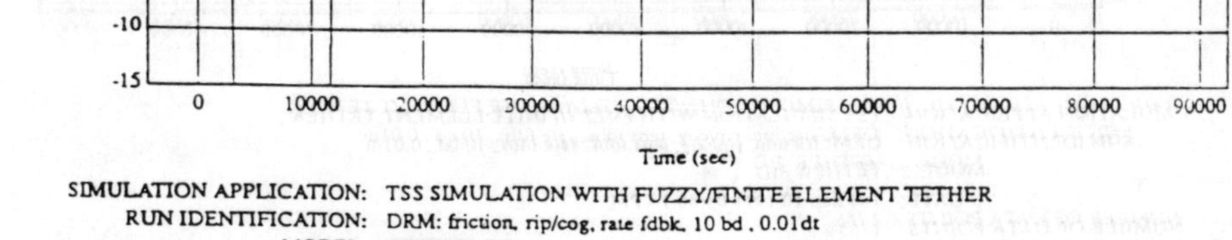


Fig. 3. Length error for conventional and fuzzy controllers for bead model.

CONTROL ERROR TIME HISTORY

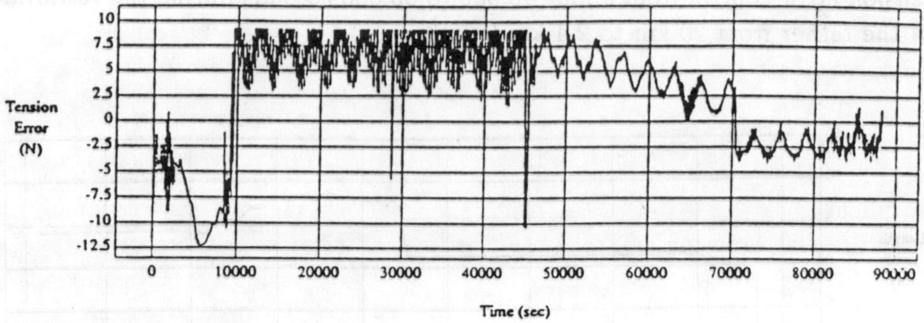


SIMULATION APPLICATION: TSS SIMULATION WITH FINITE ELEMENT TETHER
RUN IDENTIFICATION: DRM: friction, rip/cog, rate fdbk, 10 bd , 0.01dt
MODEL: TETHER_BD
DATE: Tue Sep 10 1991 03:58:31 PM
NUMBER OF DATA POINTS: 1765
DATA SAMPLING FREQUENCY: 0.020 Hz

CONTROL ERROR TIME HISTORY

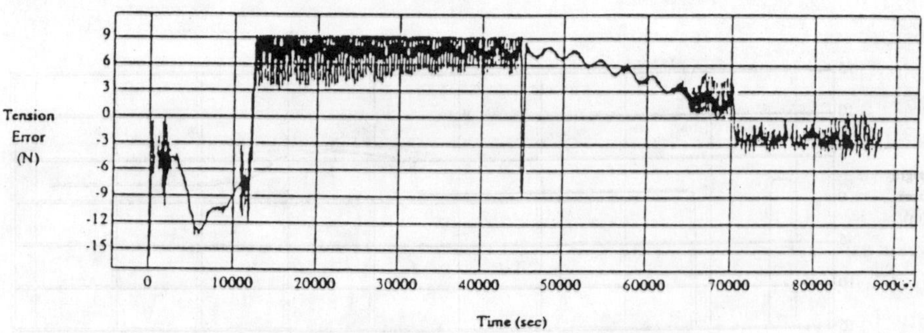


SIMULATION APPLICATION: TSS SIMULATION WITH FUZZY/FINITE ELEMENT TETHER
RUN IDENTIFICATION: DRM: friction, rip/cog, tens fdbk, rate fdbk, 10 bd , 0.01dt
MODEL: TETHER_BD
DATE: Tue Sep 10 1991 03:45:14 PM
NUMBER OF DATA POINTS: 1765
DATA SAMPLING FREQUENCY: 0.020 Hz

Fig. 4. Tension error for conventional and fuzzy controllers for bead model.

While it is important to control length error and tension error as tightly as possible, it may be more important to control the in-plane and out-of-plane libration amplitudes, especially during the retrieval. If the in-plane librations increase beyond 45 deg, the satellite motion can result in a situation known as wrap-around due to the orbital forces. It is very important to avoid such situations for the safety of the crew and the shuttle. The fuzzy control shows real advantage over conventional control in this case as shown in Fig. 5 of in-plane libration amplitude. During the on-station phase, the fuzzy control results in about half the in-plane libration amplitude in comparison to the conventional control. During retrieval, between 70,000 and 90,000 seconds, the conventional control results in a librational amplitude of $-12$ to $+12$ deg. The fuzzy control during the same time shows a libration amplitude of only $-4.5$ to $+4.5$ deg.

Results based on the massless as well as finite element bead models of the tethers show promise regarding the applicability of fuzzy-logic-based length control to space-based tethers. Deployed length of tethers can be easily controlled using length error and length rate error as input and imposing the output delta voltage in addition to the guidance command on the reel motor drive. Better length control by the fuzzy controller has the additional advantage of small tension oscillation during on-station and retrieval phases. Simulation results also indicate that the fuzzy controller outperforms the conventional controller in control of in-plane and out-of-plane libration amplitudes.

The examination of whether librational oscillations, particularly in-plane oscillations, can also be controlled with the fuzzy logic approach will continue. It is planned to study the fuzzy controller's ability to handle different mission profiles such as a 100 km downward deployment. Studies thus far have dealt with upward deployments. If a fuzzy controller can be easily modified to handle multiple mission profiles, overall life cycle costs for space tethered operations can be drastically reduced, allowing space tethers to be a useful technology for future space activities. One problem that neither conventional nor fuzzy controllers has addressed with respect to TSS-1 is the problem of controlling the skiprope motion.

One surprising benefit, in addition to better length control, is smaller in-plane librational oscillations during the retrieval phase. Overall the results of the study have indicated that fuzzy-logic-based control can provide the simplicity and robustness desired for complex tethered operations in space.

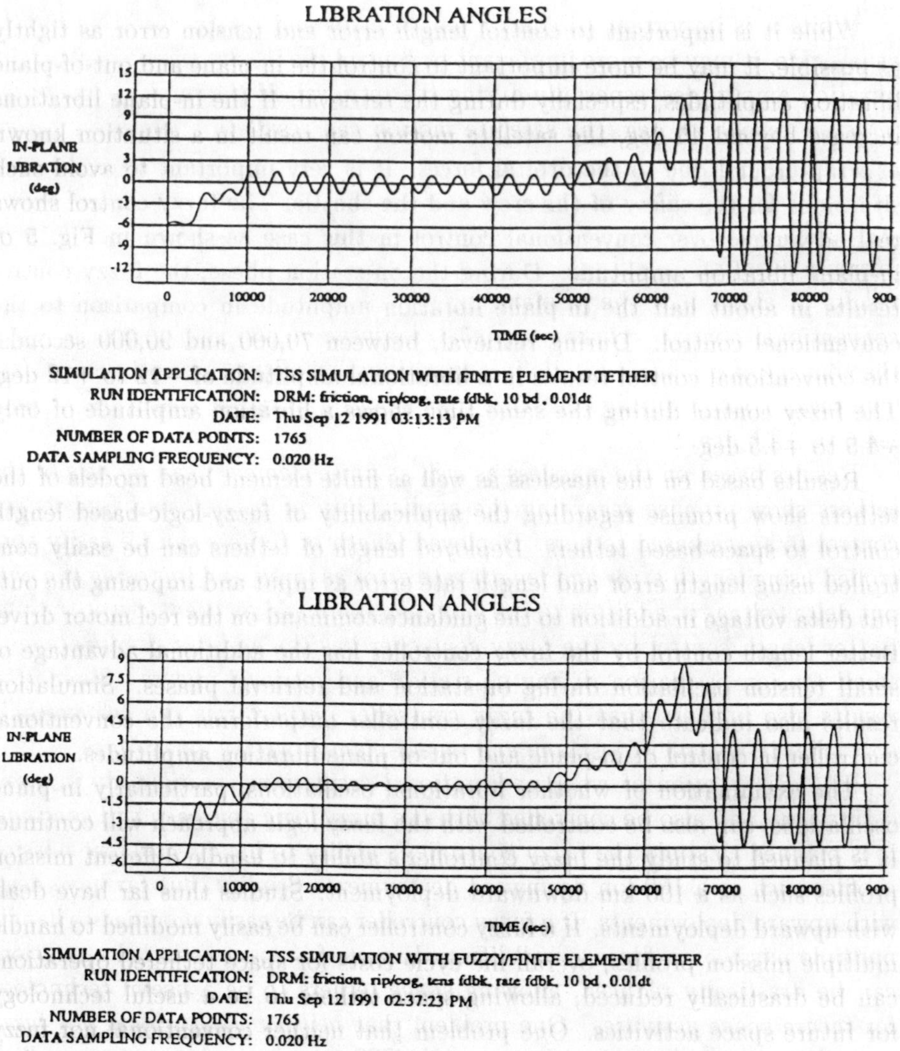


Fig. 5. Comparison of conventional and fuzzy tether length controllers for libration angles.

## 2.3. *Collision avoidance*

Future unmanned missions to Mars[23] will investigate the terrain and collect soil samples in advance of manned missions. Path planning is a crucial element in the activities to be undertaken by an autonomous rover. Obstacles such as boulders or troughs may block the shortest path from the current position to the target position for the next sample acquisition. Collision avoidance algorithms have been developed to receive fuzzified sensor data and generate short-term path decisions. These algorithms utilize range to obstacles, current velocity and orientation to generate steering and velocity commands as shown in Fig. 6. For rover trajectory control, the range parameter has been limited to 20 m, and the speed limited to 5 m per second. The azimuth was completely covered by taking a universe discourse from −200 to 200 deg. These membership functions are shown in Fig. 7(a). The steering angle range for the rover was

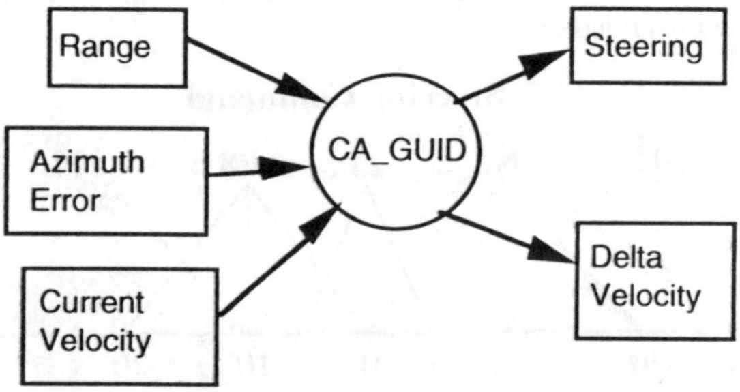


Fig. 6. Fuzzy-logic-based collision avoidance guidance.

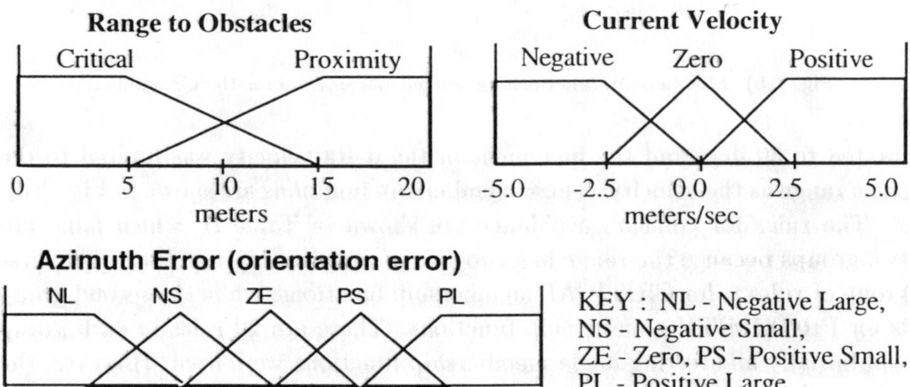


Fig. 7(a). Membership functions for input parameters for CA guidance.

**Delta Velocity Command**

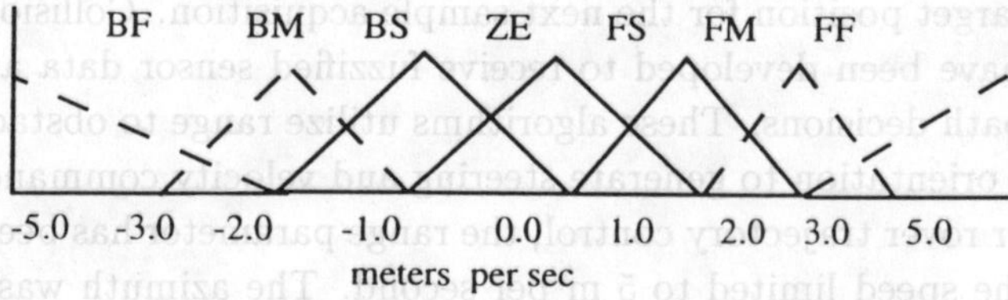


KEY : BF - Backward Fast, BM - Backward Medium,
BS - Backward Small, ZE - Zero, FS - Forward Small,
FM - Forward Medium, FF - Forward Fast
Functions shown by dashed line are defined but not used in the final rulebase

**Steering Command**

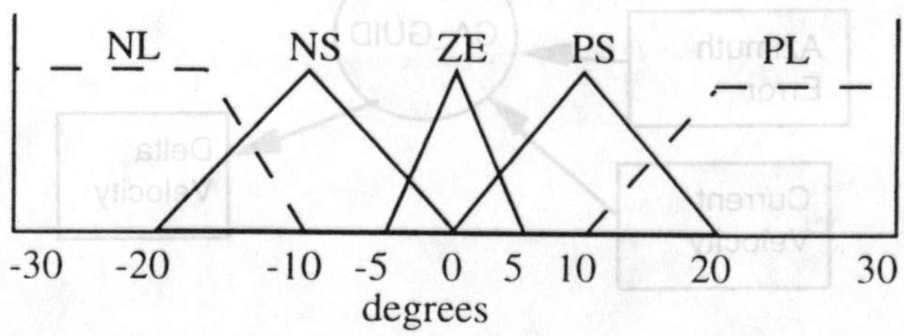


KEY : NL- Negative Large, NS - Negative Small, ZE - Zero,
PS - Positive Small, PL - Positive Large,
Functions shown by dashed line are defined but not used in the final rulebase

Fig. 7(b). Membership functions for output parameters for the CA guidance.

limited to 30 deg, and the increment in the delta velocity was limited to the same range as the velocity. These membership functions are shown in Fig. 7(b).

The rules for collision avoidance are shown in Table II, which falls into two groups because the range has two sets of membership functions. The first group of rules is for CRITICAL membership functions, while the second group is for PROXIMITY membership functions. There are 20 rules in each group and initially all steering angle membership functions were used. However, the rules and membership functions were tuned during performance testing. As a result, the steering and delta velocity command have only three membership functions. The membership functions that were eliminated in the tuning process are shown with a dashed line.

Table II. Rulebase for collision avoidance.

| When RANGE equals CRITICAL | | | | | | |
|---|---|---|---|---|---|---|
| | Azimuth Error (orientation error) | | | | | |
| | NL | NS | ZE | PS | PL | |
| Current Velocity | | | | | | |
| POS | NS | ZE | ZE | ZE | PS | STEERING |
| NEG | PS | ZE | ZE | ZE | NS | |
| POS | FS | ZE | ZE | ZE | FS | DELTA |
| NEG | BS | ZE | ZE | ZE | BS | VELOCITY |

| When RANGE equals PROXIMITY | | | | | | |
|---|---|---|---|---|---|---|
| | Azimuth Error (orientation error) | | | | | |
| | NL | NS | ZE | PS | PL | |
| Current Velocity | | | | | | |
| POS | NS | PS | ZE | NS | PS | STEERING |
| NEG | PS | NS | ZE | PS | NS | |
| POS | FS | FM | ZE | FM | FS | DELTA |
| NEG | BS | BM | ZE | BM | BS | VELOCITY |

KEY: NL–Negative Large, NS–Negative Small, ZE–Zero, PS–Positive Small, PL–Positive Large, POS–Positive, NEG–Negative, FS–Forward Small, BS–Backward Small, FM–Forward Medium, BM–Backward Medium

Simulation testing was performed for a set of representative test cases, and performance of the guidance algorithms were evaluated in a variety of obstacle scenarios. Five test cases were designed to test the capabilities of the collision avoidance system. The first test case shows that an obstacle directly in the center can be avoided very easily by basic steering commands. The second test case is designed to avoid a large central obstacle. The third and fourth test cases show that the obstacles can be successfully avoided one by one and the rover can still reach the desired destination. The fifth test case shows that the rover can avoid many obstacles even if they are lined up, leaving only a small opening. Trajectories for all five cases are shown in Fig. 8.

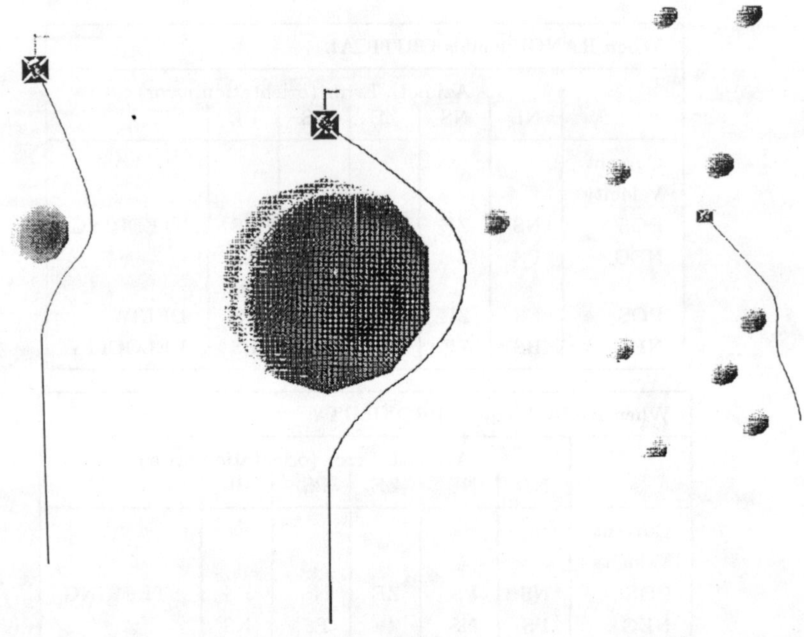


Fig. 8. (a) Collision avoidance for an object. (b) Collision avoidance for a large object. (c) Collision avoidance for a set of objects (easy). (d) Collision avoidance for a set of objects (hard). (e) Collision avoidance for a simple maze.

Some important points of this study were as follows: a) Trajectory guidance[24,25] includes not only position control but also orientation control. When the collision avoidance algorithms[26] were integrated with trajectory control, the orientation control could not be managed properly without major modification to the system. Proper orientation could be achieved only if there was sufficient distance between the last obstacle avoided and the target point so that the rover can turn itself. Since this distance cannot be guaranteed, the desired orientation of the rover at arrival was not addressed; b) the 'back-off' situation requires some knowledge about the obstacles just avoided, or some information about the obstacles in the path. Since our cameras are looking forward, the task to study the 'back-off' situation has been postponed and, c) when the rover is going forward, if it encounters an obstacle that can be avoided only by going back, it must remember that the distance it must go

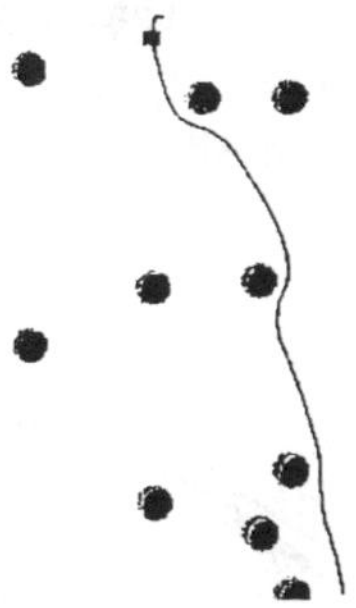

(d)

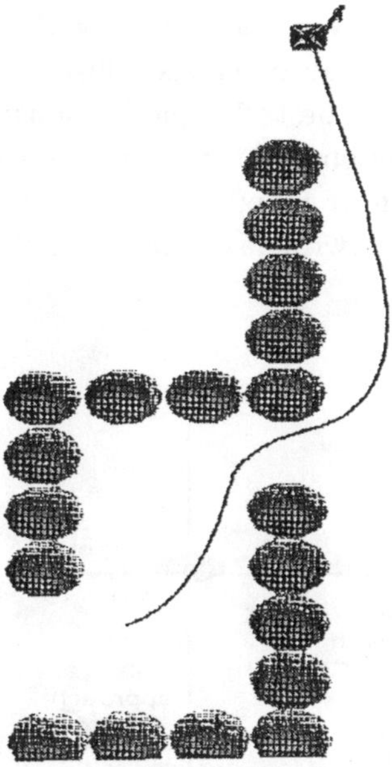

(e)

Fig. 8. (*Continued*)

back is a crucial factor. Otherwise, it can get into a situation where it continues to go back and forth when the critical distance for collision avoidance is not altered. Thus, the transition from forward to backward motion requires special care.

A higher-level path planner is needed when the vehicle is caught in a back-off setting, i.e., when it is not possible for the vehicle to pursue a "forward" path. It is significant to note that the method employed does not depend on object identification, but rather, detection of the degree to which an object (where present) or (more generally) an angular sector represents an obstacle. This is a significant relaxation over most collision avoidance schemes. Simulation results and planned enhancements point towards refinements in the algorithm, the possibility of adaptive tuning of the system and, as expected, the need for a higher-level path planner to handle cases that involve backoff, sensor fusion, positioning of the vehicle at the destination, moving obstacles, and other situations that involve radically changing environments of operation.

Fuzzy logic control together with straightforward algorithms yield an effective system for autonomous collision avoidance in an environment of uncertain information. The technique is robust and avoids complexity in initial stages where simple obstacle avoidance is a key element rather than involved object identification or mapping of a complete world model. This technique could be integrated with fast and sophisticated object identification algorithms if desired.

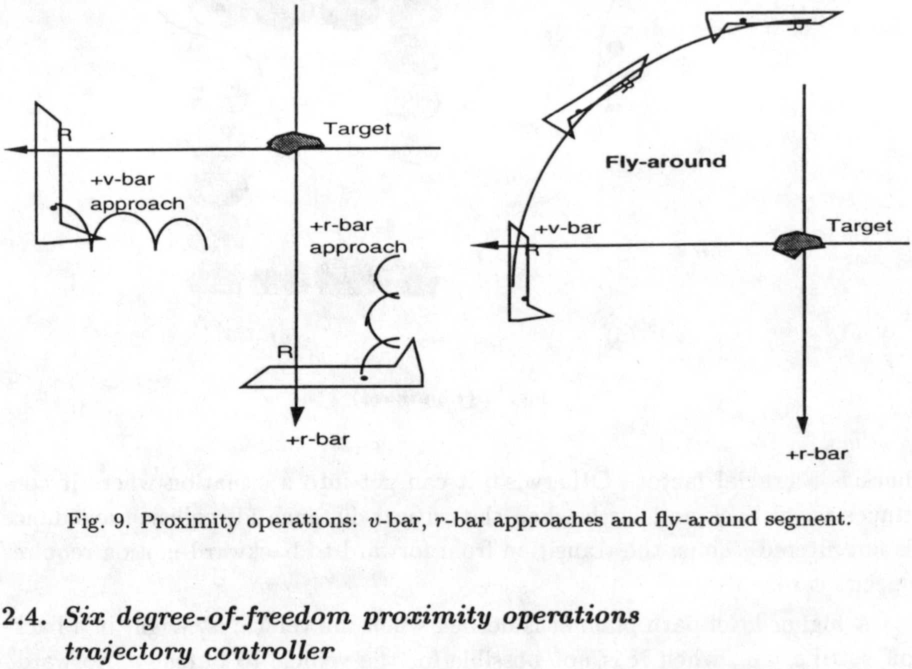


Fig. 9. Proximity operations: $v$-bar, $r$-bar approaches and fly-around segment.

## 2.4. *Six degree-of-freedom proximity operations trajectory controller*

A six-degree-of-freedom (6 DOF) controller for a spacecraft has been designed and tested[27,28] in a shuttle simulation for proximity operations (Fig. 9). When the shuttle approaches the satellite in an orbit, a local vertical local horizontal frame[15] is used as a reference frame. The $x$-axis of this frame generally along the velocity vector of the satellite is called $v$-bar, and the $z$-axis pointing towards the center of the orbit is called $r$-bar. When the shuttle approaches along the $x$-axis, the approach is known as a $v$-bar approach. This approach being in the same orbit saves fuel and provides operational flexibility. An approach along the $z$-axis is known as an $r$-bar approach as shown in Fig. 9.

During the station-keeping phase, a constant distance is maintained from the shuttle to the satellite to correct for any relative position and to establish the ground communication for information update. The preferred axis for the station-keeping is $v$-bar due to fuel savings. A fly-around phase consists of the shuttle moving from the $v$-bar to negative $r$-bar to inspect the satellite or to find out the position of the grapple mechanism. These operations performed manually by a crew member requires an extensive training in the real-time mission simulators to achieve proficiency.

The 6 DOF controller uses sensor measurements such as range, elevation and azimuth angles directly as input and generates the commands for the jet select logic to null out the errors. For each degree of freedom, there is a rulebase that requires the parameter error and its rate as input. For example, the pitch attitude is controlled by a rulebase of 31 rules given in Table IIIa. This rulebase requires pitch error and error rate whose membership functions are given in Fig. 10(a). The same rulebase and membership functions are used for the two other attitude (roll and yaw) controls.

The elevation, azimuth and range control are performed by the rules given in Table IIIb, IIIc, and IIId. Each parameter has different rules tuned to achieve desired performance. Corresponding membership functions are given in Figs. 10(b), 10(c), and 10(d). It should be noted that these membership functions are tuned for the shuttle operations specifically. Therefore, application of this software to other spacecraft operations requires minor modifications.

Table III(a). Fuzzy Rulebase for attitude control.

| | | Angle Error | | | | | | |
|---|---|---|---|---|---|---|---|---|
| | | NB | NM | NS | ZO | PS | PM | PB |
| | NB | PM | PM | PS | PM | | | |
| | NM | PM | PM | PS | PM | | | |
| Rate | NS | PS | PS | PS | PS | | | |
| Error | ZO | PS | PS | PS | ZO | NS | NS | NS |
| | PS | | | | NS | NS | NS | NS |
| | PM | | | | NM | NS | NM | NM |
| | PB | | | | NM | NS | NM | NM |

KEY: NB–Negative Big, NM–Negative Medium, NS–Negative Small, ZO–Zero, PS–Positive Small, PM–Positive Medium, PB–Positive Big

Table III(b). Rulebase for elevation control.

| | | ELEVATION RATE ERROR | | | | |
|---|---|---|---|---|---|---|
| | | NM | NS | ZO | PS | PM |
| | PM | | | NM | NM | NM |
| ELEV. | PS | | ZO | NS | NS | NM |
| ANGLE | ZO | PM | PS | ZO | NS | NS |
| ERROR | NS | PM | PS | PS | ZO | |
| | NM | PM | PM | PM | | |

Table III(c). Rulebase for azimuth control.

| | | AZIMUTH RATE ERROR | | | | |
|---|---|---|---|---|---|---|
| | | NM | NS | ZO | PS | PM |
| | PM | | | NM | NM | NM |
| AZIMUTH | PS | | | NS | NS | NM |
| ANGLE | ZO | PM | PS | ZO | NS | NM |
| ERROR | NS | PM | PS | PS | | |
| | NM | PM | PM | PM | | |

Table III(d). Rulebase for range control.

| | | RANGE RATE ERROR | | | | | | |
|---|---|---|---|---|---|---|---|---|
| | | NB | NM | NS | ZO | PS | PM | PB |
| | PB | | | | | NB | NB | NB |
| | PM | | | | | NM | NB | NB |
| | PS | | | | | NS | NM | NB |
| RANGE | ZO | PM | PS | PS | ZO | NS | NS | NM |
| ERROR | NS | PB | PM | PS | | | | |
| | NM | PB | PB | PM | | | | |
| | NB | PB | PB | PB | | | | |

KEY: NB–Negative Big, NM–Negative Medium, NS–Negative Small, ZO–Zero, PS–Positive Small, PM–Positive Medium, PB–Positive Big

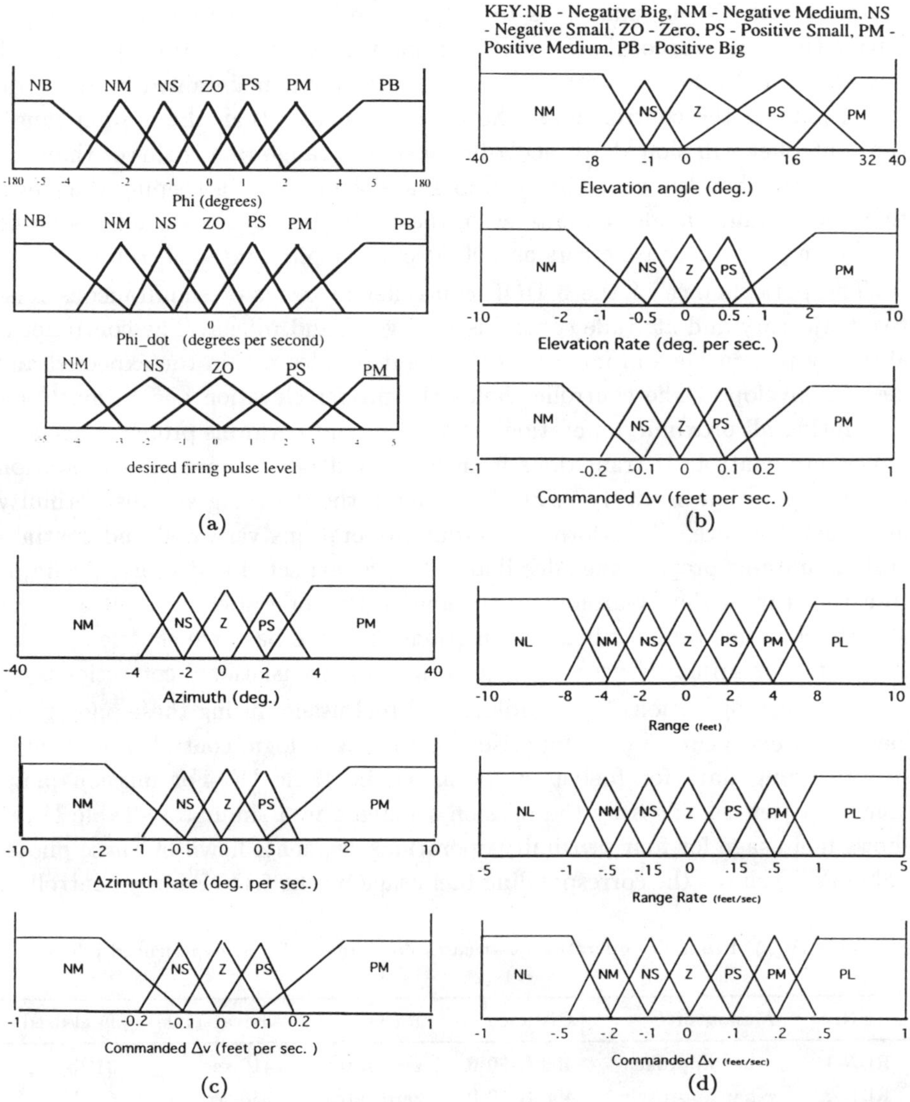


Fig. 10. (a) Membership functions for Phi, Phi_dot and desired pulse level. (b) Membership functions for elevation angle control. (c) Membership functions for azimuth angle control. (d) Membership functions for range control.

For a given mission profile, the controller maintains a proper range and range rate. The elevation and azimuth angle measurements are used in conjunction with the angular rates to follow a desired trajectory. For example, during the *v*-bar approach, the controller maintains zero elevation and azimuth angles, desired range and range rate. The attitude is maintained by the rotational part of the controller. If the range is smaller than the desired range, the controller will slow down accordingly. If the range rate is slower than the desired rate then the controller will increase the speed. In keeping the elevation and azimuth angles close to zero, the controller adjusts its actions based on the angle errors, rate errors as well as current pitch rate and roll rate.

The performance of the 6 DOF controller in terms of simultaneous relative trajectory and attitude control is very good and robust. The controller is also very responsive and maintains the flight profiles within the expected and desired envelope. The controller holds the proper elevation and azimuth angles during all proximity operations test cases, and performs proper range and range rate control. It transitions from *v*-bar and *r*-bar approaches to station keeping operation in a very natural manner without having any discontinuity in its performance. It performs fly-around operations very well and continuously maintains proper range deadbands for the expected trajectory. Again, it transitions from the fly-around to station-keeping operation without any loss of control. Since the membership functions and rule base are shuttle specific, the controller primarily uses single jet firings for translational corrections.

Proximity operations flight profiles and fuel usage during these operations have been excellent. The comparison of the fuzzy logic controller with mission planning data for fuel usage is shown in Table IV. For mission planning, experienced pilots fly the mission segments in a simulator. Table IV(a) shows fuel usage for four proximity operations segments flown by these pilots. Table IV(b) shows the corresponding fuel usage by the 6 DOF fuzzy controller.

Table IV(a). Proximity operations test cases flown manually by experienced pilots.

| RUN # | Maneuver | Distance | Rates | Arrival Time | Fuel used |
|---|---|---|---|---|---|
| RUN 1 | *v*-bar approach | 400 ft-50 ft | zero rates | 1410 sec | 31 lb |
| RUN 2 | *r*-bar approach | 400 ft-50 ft | zero rates | 1560 sec | 65 lb |
| RUN 3 | 1/4 Fly Around | @200 ft, | 0.2 rate | 1380 sec | 50 lb |
| RUN 4 | Station Keep | @200 ft | zero rates | 1800 sec | 36 lb |

Table IV(b). Test results with 6 DOF fuzzy controller for the same test cases.

| RUN # | Maneuver | Distance | Rates | Arrival Time | Fuel used |
|---|---|---|---|---|---|
| RUN 1 | $v$-bar approach | 400 ft-50 ft | zero rates | 1436 sec | 33 lb |
| RUN 2 | $r$-bar approach | 400 ft-50 ft | zero rates | 1686 sec | 58 lb |
| RUN 3 | 1/4 Fly Around | @200 ft | 0.2 rate | 1800 sec | 47 lb |
| RUN 4 | Station Keep | @200 ft | zero rates | 1800 sec | 12 lb |

Fuel usage for $v$-bar, $r$-bar and fly around test cases is comparable, but the fuzzy controller uses significantly less fuel for the station keeping task. Further tests of this 6 DOF controller should compare its performance with other manual crew procedure test cases flown in the shuttle mission simulator, and possibly flight data.

## 2.5. *Camera tracking system*

During proximity operations, satellites and space vehicles typically approach the space station with a well defined trajectory or they remain in a passing orbit. The relative trajectory of the approaching vehicle must be monitored for traffic management. For vehicles in a passing orbit, the trajectory must be monitored for collision avoidance. Extra Vehicular Activities (EVA) by crew members may also be in progress at the same time. Thus, the space station crew will be required to perform traffic management and collision avoidance functions, as well as tracking the crew members that are performing EVA.

The function of a camera tracking system is to track an object that is assigned to it by the central traffic management system. Tracking of an object means keeping the object in the center of the field of view (FOV) by rotating the camera axis along the line of sight (LOS) using the pan and tilt gimble drives appropriately. The line of sight is approximately defined by the $x$ and $y$ measurements in the FOV as shown in Fig. 11. These positions are usually computed using an image processing algorithm that utilizes the pixel map provided by the digitizer attached to the camera. There are also range measurements available to the controller from sensors such as radar or laser range finders. Using these measurements, the controller generally provides the pan and tilt rate commands to the gimble drives, and when the gimble drives rotate, the camera axis is aligned with the LOS. As a result, the object remains in the center of FOV. If the FOV is divided into 512 pixels, then the object should remain in the vicinity of 255 with a spread of about 15 to 20 pixels.

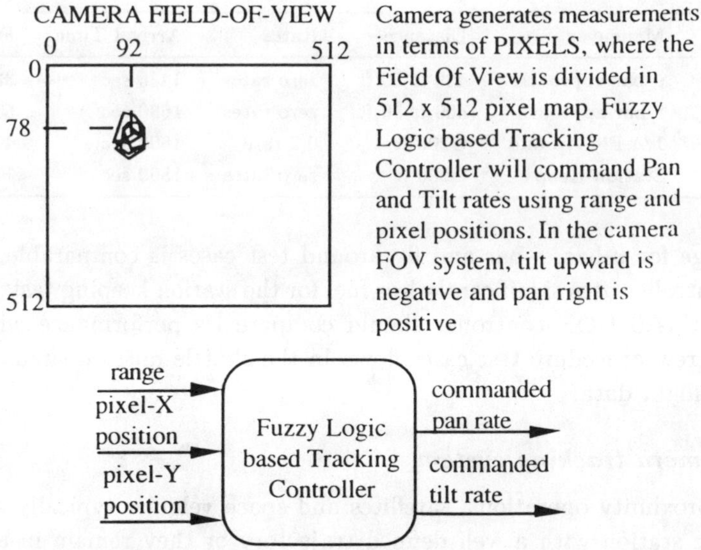


Fig. 11. Concept of the camera tracking controller.

The tracking controller thus has range, and $x$ and $y$ coordinates, measured in pixels, as input and pan and tilt rate commands as output. Several other factors needed to be considered in the design of the rulebase. The range measurement comes from a different source, and thus it may not be available simultaneously with the $x$ and $y$ measurements. The processing time required for the image processing may be large or small. Proximity trajectories are sensitive to angular motion, when the object is at close range. Thus it was better to design the rulebase to process range and position separately. With this in mind, the first controller was designed in such a way that the range measurement rules were in one group and $x$ and $y$ coordinates were in second group as shown in Fig. 12. The rules were divided into three groups as shown in Table V(a). The first group handles the range measurement and computes the scale factor. The next two groups use pixel measurements along with scale_factor to compute pan and tilt rate commands. This design is called Design 1 in this chapter. The membership functions for the input and output parameters for this design are shown in Fig. 13.

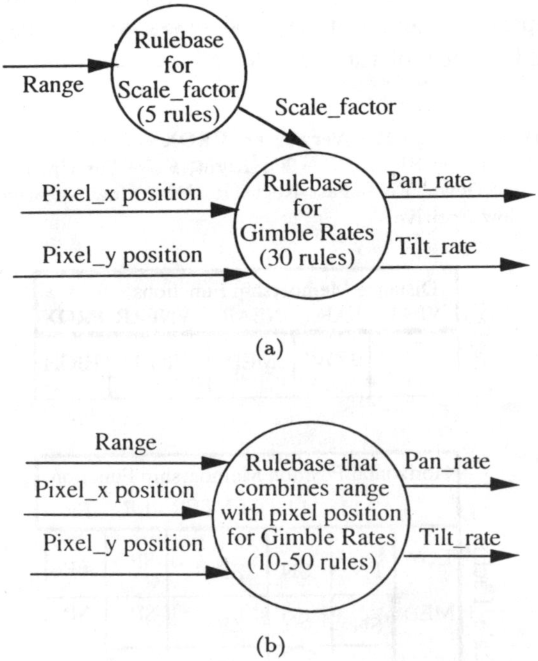


Fig. 12. (a) Design with three rulebases, (b) Design with one rulebase.

Originally there were plans to do hardware testing of this controller in the communication and tracking laboratory where components like cameras, digitizers, gimble drives and range sensors were already available. One choice of hardware considered for the fuzzy controller would accommodate a maximum of 12 rules. Thus, Design 1 had too many rules to implement in this hardware and it was necessary to reduce the rulebase. This process resulted in the second design which is referred to as Design 2 in this chapter. The controller flow is very simple as shown in Fig. 12, and the membership functions are given in Fig. 14 for this design. In Table V(b), the rules used for this design are provided.

Thus, two different camera tracking controllers[29,30] based on fuzzy logic principles have been designed and tested in software simulation at the Johnson Space Center. Seven test cases, as shown in Table VI, have been designed to test the performance of the controllers for proximity operations where approaches like $v$-bar, fly-around and station keeping are performed. The controllers are compared in terms of responsiveness, and ability to maintain the object in the field-of-view of the camera.

KEY: VFAR – Very Far, VNEAR – Very Near, PROX – Proxomity zone, FL – Far Left, LL – Little Left, LR – Little Right, FR – Far Right, FU – Far Up, LU – Little Up, LD – Little Down, FD – Far Down, FN – Fast Negative, SN – Slow Negative, ZR – Zero, FP – Fast Positive, SP – Slow Positive

| Distance Membership Functions | VFAR | FAR | NEAR | VNEAR | PROX |
|---|---|---|---|---|---|
| Scale_Factor | LOW | LOW | MED | HIGH | HIGH |

| Scale_Factor \ Horizontal Position Membership Functions | FL | LL | CENTER | LR | FR |
|---|---|---|---|---|---|
| LOW | FN | SN | ZR | SP | FP |
| MED | SN | SN | ZR | SP | SP |
| HIGH | SN | SN | ZR | SP | SP |

Pan_Rate Membership Functions

| Scale_Factor \ Vertical Position Membership Functions | FD | LD | CENTER | LU | FU |
|---|---|---|---|---|---|
| LOW | FP | SP | ZR | SN | FN |
| MED | SP | SP | ZR | SN | SN |
| HIGH | SP | SP | ZR | SN | SN |

Tilt_Rate Membership Functions

TABLE V(a)

Table V(b). Compact rulebase for the design 2 for camera tracking controller.

| | Pixel Position | | | | |
|---|---|---|---|---|---|
| | NL | NS | ZE | PS | PL |
| SMALL | NB | NM | ZE | PM | PB |
| LARGE | NM | NS | ZE | PS | PM |

KEY: NL–Negative Large, NM–Negative Medium, NS–Negative Small, ZE–Zero, PS–Positive Small, PM–Positive Medium, PL–Positive Large, PB–Positive Big, NB–Negative Big

Table VI. Software simulation testcases for camera tracking system.

| No. | Test Case Description | Initial Position | | | Initial Velocity | | |
|---|---|---|---|---|---|---|---|
| 1 | $V$-bar Approach | 150 | 0 | 0 | −0.4 | 0 | 0 |
| 2 | Station-keep, Out-of-plane Motion | 50 | 0 | 0 | 0 | −0.4 | 0 |
| 3 | Separation-Fly around | 150 | 0 | 0 | 0.4 | 0 | 0 |
| 4 | Fly around segment | 200 | 0 | 50 | 0 | 0 | −1.0 |
| 5 | Station keeping | 250 | 50 | 0 | 0 | 0 | 0 |
| 6 | Passing Orbit | 300 | −100 | 100 | 0 | 0 | −1.0 |
| 7 | Slow $V$-bar Approach | 150 | 0 | 0 | −0.3 | 0 | 0 |

Results showed that the Design 1 controller tracks the object well. Results for test case number 6 are shown in Fig. 15, where the objects relative positions as a function of time are given. Also, the pan and tilt rates and $x$ and $y$ coordinates are given to correlate the camera tracking activity. There was initially a zero zone in the $x$ and $y$ membership functions, measured in pixels, from 240 to 270. Therefore, the object remained on one side of the FOV when it was moving and the camera was tracking. When the object reversed direction, the controller stopped moving the gimble for a while until the object reached the other boundary. For all test cases, the controller had no problem in responding to changing trajectories or varied measurements. It could track the object at large range as well as short range and for very high angular speeds. The final version of the membership functions, shown in Fig. 13,

KEY: VFAR–Very Far, FAR–Far, NEAR–Near, VNEAR–Very Near, PROX–Proximity, FL–Far Left, LL–Little Left, CENTER–Center, LR–Little Right, FR–Far Right, FU–Far UP, LU–Little Up, CENTER–Cnter, LD–Little Down, FD–Far Down

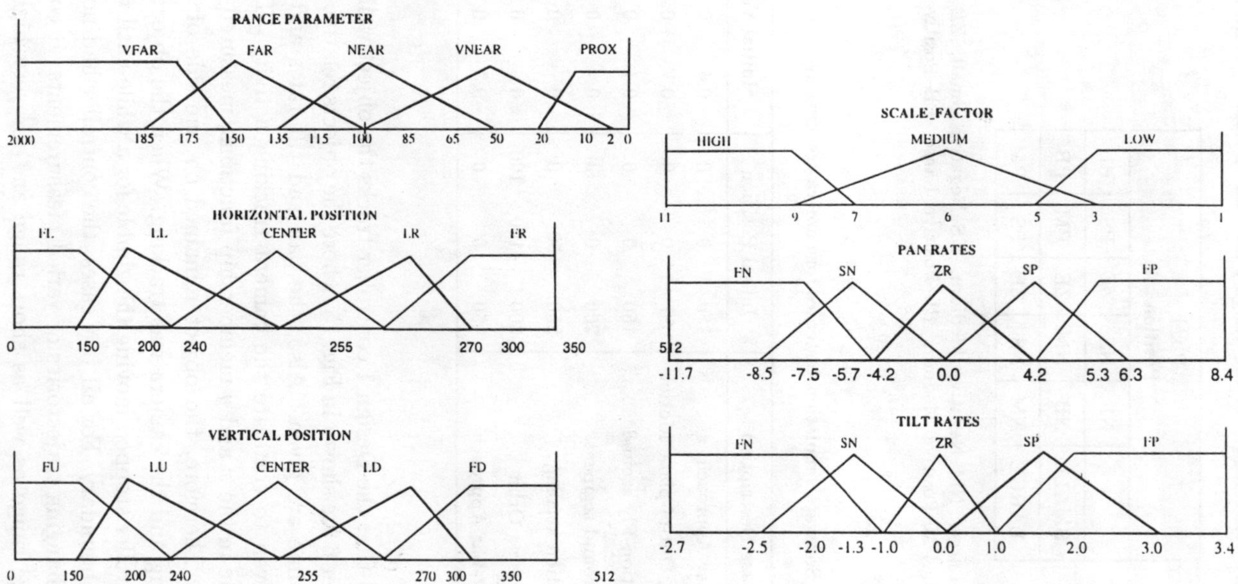


Fig. 13. Final values of the membership functions for the Design 1 Input/Output Parameters.

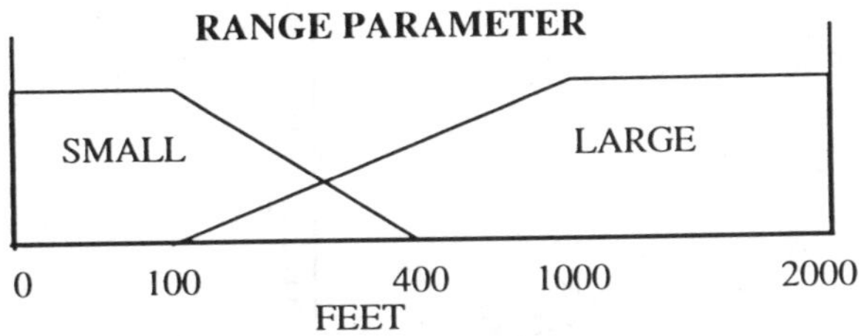


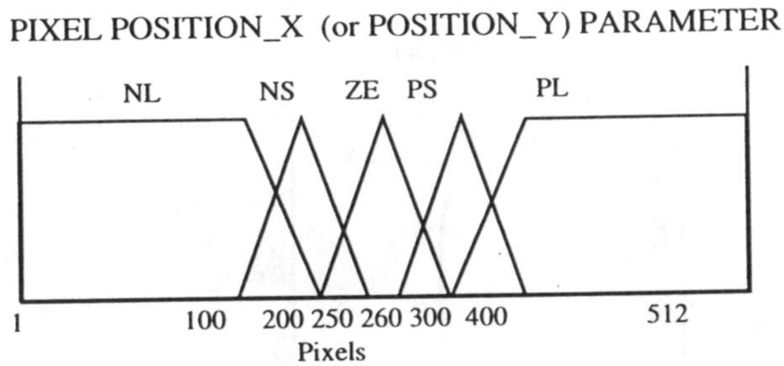


KEY: NL–Negative Large, NS–Negative Small, ZE–Zero, PS–Positive Small, PL–Positive Large

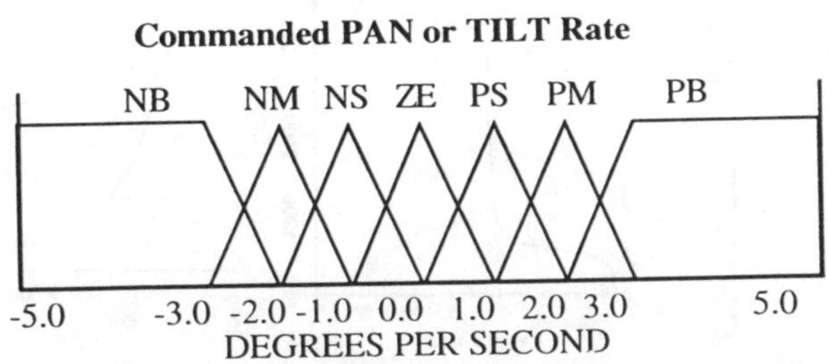


KEY: NB–Negative Big, NM–Negative Medium, NS–Negative Small, ZO–Zero, PS–Positive Small, PM–Positive Medium, PB–Positive Big

Fig. 14. Membership Functions for the Design 2 Input/Output Parameters.

indicates the gap between the LL and LR functions was eliminated for the $x$ pixel measurement. Similarly, the gap between LU and LD functions was eliminated for the $y$ measurement. The tilt rate ZR (zero) function is made narrow and the SP and SN functions are expanded to have a large overlap with the FN and FP functions. All other functions are not changed at all. Test case 1

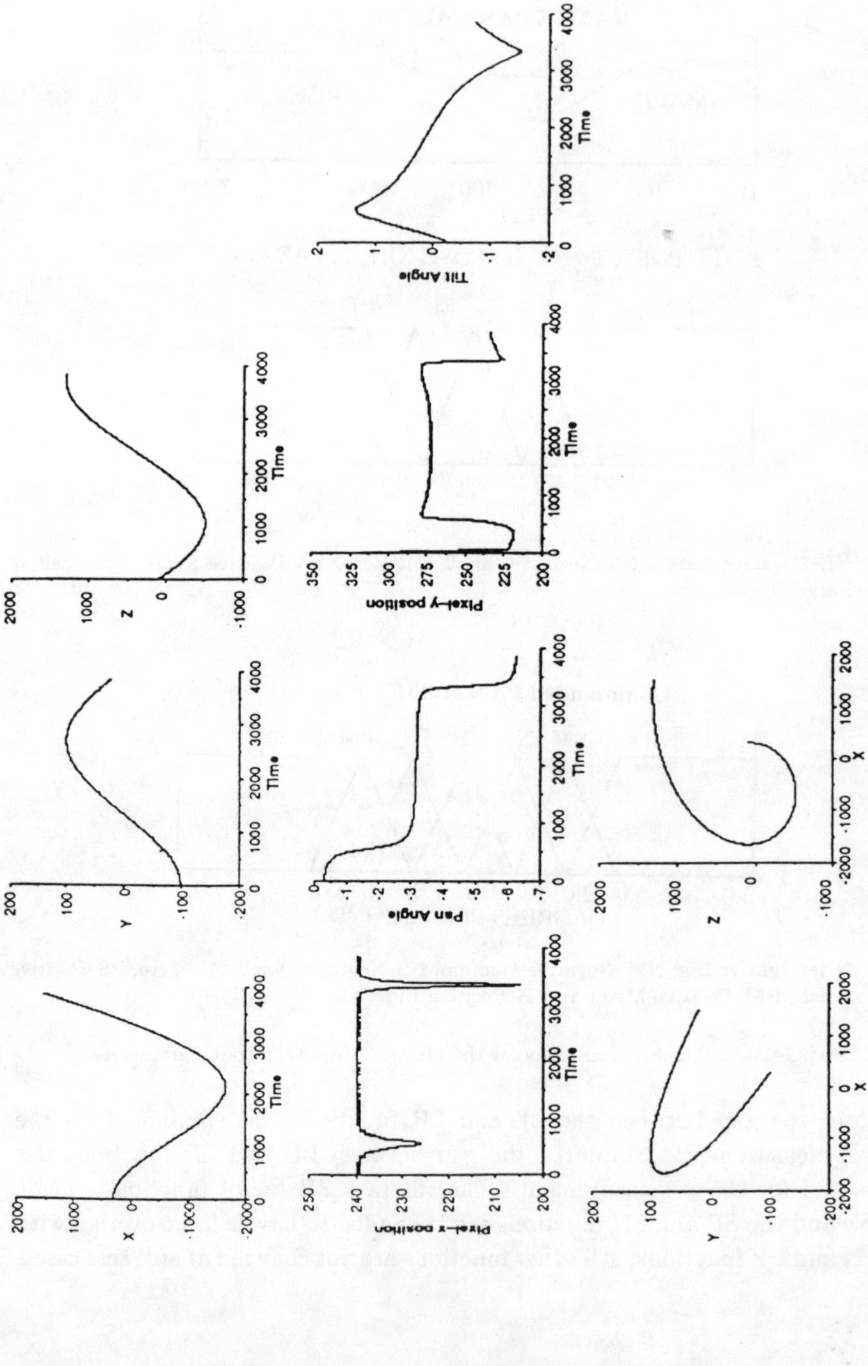


Fig. 15. Tracking controller performance for test case #6.

is performed with this final version of the controller and results are shown in Fig. 16. The final version of the Design 1 controller tracks the object very well during proximity operations. The $x$ and $y$ pixel input have no zero zone in their membership functions. The LL and LR overlap the CENTER membership for pixel-$x$ input. Similarly, the LU and LD overlap the CENTER membership function of pixel-$y$ input. Thus the object is maintained very closely in the center.

Design 2 does not have a scale_factor variable. Its processing flow is as shown in Fig. 12, where range, $x$, and $y$ are input and pan and tilt rates are output. It has a small rulebase as shown in Table V(b). The rules combine the range and $x$ position to compute the pan rate. The same rulebase is used to compute the tilt rate by using the $y$ position instead of $x$. The $x$ and $y$ coordinates have the same membership functions, and pan and tilt rate membership functions are also identical. Range, $x(y)$ and pan (tilt) rate membership functions are shown in Fig. 13. Because of hardware limitations, the pan and tilt rate membership functions had to be the same, and therefore the tilt rate range was expanded from $-5.0$ to $5.0$. Results of test case 1, with this controller in the software simulation, are shown in Fig. 16 for comparison with the design 1 controller.

Performance of the design 2 controller is also very good for the first test case. The $y$ position is maintained between 248 and 268 pixel values. There is a no-action zone (250 to 260) in the $y$ membership functions as shown in Fig. 16. Thus, the controller does not rotate when $y$ is between 250 and 260. When the object position is below 260 then the controller tilts downward to keep the object in the center. It continues to increase the tilt rate until a position of 268 is reached. Maximum tilt rate is less than 0.5 deg per sec. If the PS and NS membership functions are as wide as the Design 1 membership functions, then the controller response would be the same as the design 1.

In Fig. 16, the tilt rate and $y$ position of the object for test case 1 for both designs are shown. For the design 1, $y$ varies from 261 to 248, while for the design 2, it varies from 268 to 248. Thus both controllers track the object very well within a narrow zone and do not show any major difference. The tilt rates are also very comparable and do not show any significant deviations. In fact, the tilt rate for design 2 seems to be better than design 1. Because of the no zero zone, the controller in design 1 keeps a much tighter control on the tilt rate resulting in an over correction. It should be noted that the rulebase as well as the membership functions are tuned for design 1, while no such tuning has been performed for design 2.

Performance comparison for the first test case indicates that the rulebase for the tracking controller can be decreased to a small number without losing any performance capability. This is a very significant result, because the onboard computers for the space station have many functions to perform, and savings in terms of core/memory and computing time are always important.

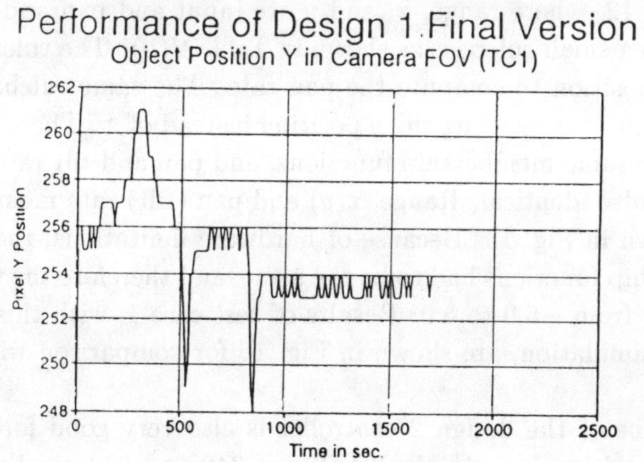


**Fig. 16. Test case 1 results with final version of the controller.**

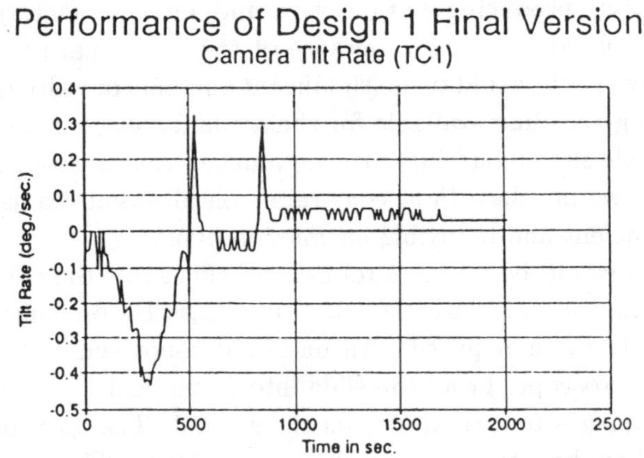


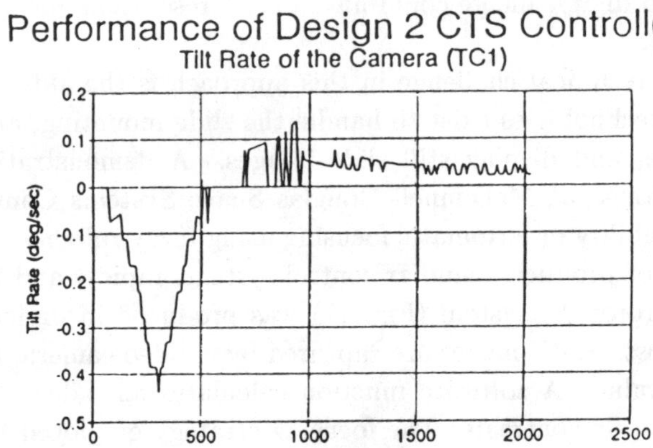


Fig. 16. (*Continued*)

## 2.6. *Remote microscope focusing*

The Space Station Crew Health Care System periodically collects microscope slides that must be analyzed by microbiologists on the ground within hours of collection. It would be desirable to minimize the time required for the crew to mount and focus microscope slides. However, simple telerobotic control by the microbiologist under visual feedback would entail several problems: first, a transmission delay of about two seconds in each direction will disrupt the focusing process, producing both slower and less satisfactory results than hands-on control would; second, because of the additional time required, the focusing process itself would tie up significant amounts of video downlink bandwidth, reducing the time available for the actual analysis; finally, the requirement to schedule slide mounting, focusing, image transmission, and analysis at the same time would place tight constraints on all resources used, and would be highly inefficient and inflexible for mission planning.

The system would be far more robust and efficient if the crew could collect the slide specimens and place the slides in a cassette or "jukebox," with no further crew attention required. An on-board robot could then mount each slide on the microscope, bring the slide into focus, and transmit one or more digitized still images to a computer on the ground. The microbiologists could then view and analyze the slides at their convenience. The resulting digitized images would require far less bandwidth to transmit than a continuous video image, yet each digital image could have better resolution and less noise than standard video.

The major technical challenge in this approach is the automatic focusing itself; mature technologies exist to handle the slide mounting, and to digitize, transmit, store, and display still slide images. A demonstration prototype system[31] developed at McDonnell Douglas Space Systems Company demonstrates the feasibility of automatic focusing using fuzzy rule-based control, and the possibility of producing similar control systems rapidly and inexpensively. The original prototype system (Fig. 17) was produced in under two months and at a low cost. Slide images are captured by a video camera, then digitized by gray-scale value. A software function calculates an index of "sharpness" based on gray-scale contrasts. The focusing strategy employed by the system attempts to maximize sharpness using fuzzy rules. At each step in the focusing process (Fig. 18), the current value of the normalized sharpness index and the difference between the current and the previous values are inputs to a fuzzy inference process. The output of that process is a suggested magnitude and direction for adjusting the microscope. For certain values (corresponding to well-focused images that do not show significant improvement with further adjustment), the rule-based system will suggest a change of zero; at that point, the focusing process stops.

In the current system, the range of possible *Sharpness* values is mapped onto six overlapping fuzzy sets. These sets roughly correspond to linguistic labels such as "very poor," "poor," "medium," "moderately sharp," "sharp," and "very sharp." Each of these sets is simply a function that assigns an integer value from 0 to 127 to any *Sharpness* value. Similarly, *Change_in_Sharpness* has six membership functions. Thus, the system has a total of 36 rules.

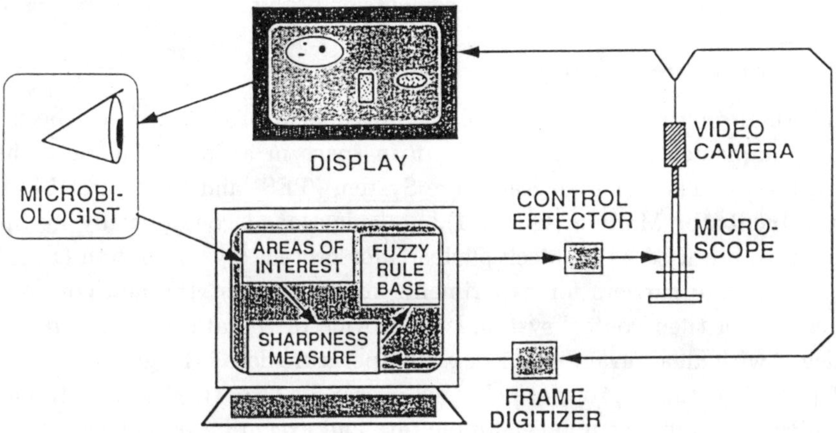


Fig. 17. Remote microscope focusing hardware setup.

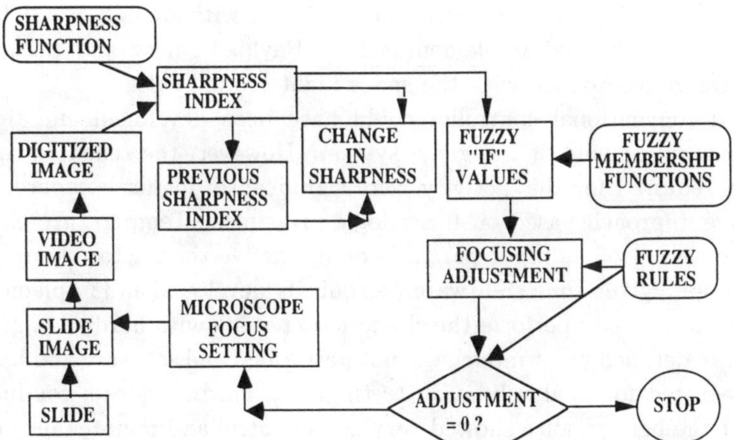


Fig. 18. Process flow in microscope focusing system.

To summarize, the current demonstration prototype has succeeded in its primary goal, which was to demonstrate the feasibility, low cost, and relative simplicity of a fuzzy logic approach to a complex control problem with space system applications. The system can focus most slides quickly with adequate or better sharpness, although improvements on both factors should be possible.

### 2.7. *Temperature control*

A Thermal Control System based on fuzzy logic principles[32] has been designed, implemented, tested and flown in space in a Space Shuttle flight in August 1992. The Thermal Enclosure System (TES) and Commercial Refrigerator/Incubator Module (C-R/IM), both developed commercially by Space Industries, are used as middeck payloads for protein crystal growth (Fig. 19). Since these are carriers for experiments requiring tight thermal control, requirement for their control system is to provide temperature control from 1 to 40 deg C with an accuracy of 0.1 deg. Temperature data is logged in EEPROM and passed to the appropriate channel over RS-422 port. A user interface is provided to change set points, sample rate and execution frequency.

Commercially available off-the-shelf conventional control systems could not provide this accuracy due to long delay times and system behavior. Because of vacuum conditions in space, heating is achieved purely by radiation resulting in a considerable delay in the system and thus the plant behavior is unique. The performance envelope is provided by the customer with enough design freedom to achieve cost effective implementations. Payload safety and performance reviews are in accordance with the space flight ground rules.

A new conventional controller could have been developed and tested to meet the requirements of the given system. However, the schedule time and resources required for this activity were significantly higher compared to the alternative approach based on fuzzy logic principles. Temperature error and error rate were used as input parameters and active thermistor channels were turned on for a given time. Software was quickly developed and implemented in the flight hardware to perform the closed loop testing with hardware. Development of the detailed plant model was not performed and software and hardware were integrated for final validation testing. All hardware tests conducted to check out the performance showed very good control and maintenance of temperature within the enclosures in the presence of perturbations from outside. The approach allowed the development team to meet the scheduled milestone within the resources and time allocated.

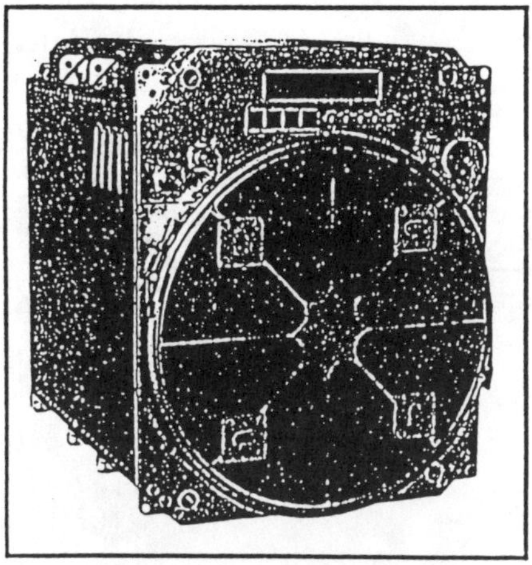

Thermal Enclosure System
(TES)

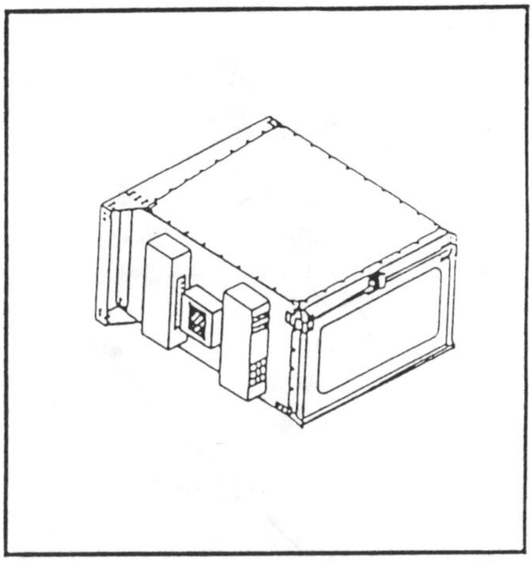

Commercial
Refrigerator/Incubator Module
(C-R/IM)

Fig. 19. Thermal Enclosure System (TES) and Commercial Refrigerator/Incubator Module (C-R/IM).

## 3. Applications of Neural Networks

Neural networks[7,33,34] is an alternative to fuzzy logic in computerizing the human thought process where dynamic mapping from the inputs to the outputs of the system is learned using parallel distributed processing models consisting of very simple neurons. Neural networks have been proven useful in mapping non-linear processes for which mathematical models do not exist or are very cumbersome to create. Neural networks are trained using suitable numerical algorithms and training examples for adjusting the weights of neurons. This black-box type approach for learning is much better compared to the approach of deriving fuzzy rules using performance data. Several advantages offered by neural networks include adaptivity (ability to adjust to new information), speed (via parallel processing), fault tolerance (ability to handle missing, confusing and noisy data). Very high speed neural networks that can be achieved through a specially designed silicon hardware make it possible to create real time monitoring and diagnostics systems.

An artificial neural system (ANS) or also referred to as a computational neural network (CNN), is basically a computer that uses simple processing

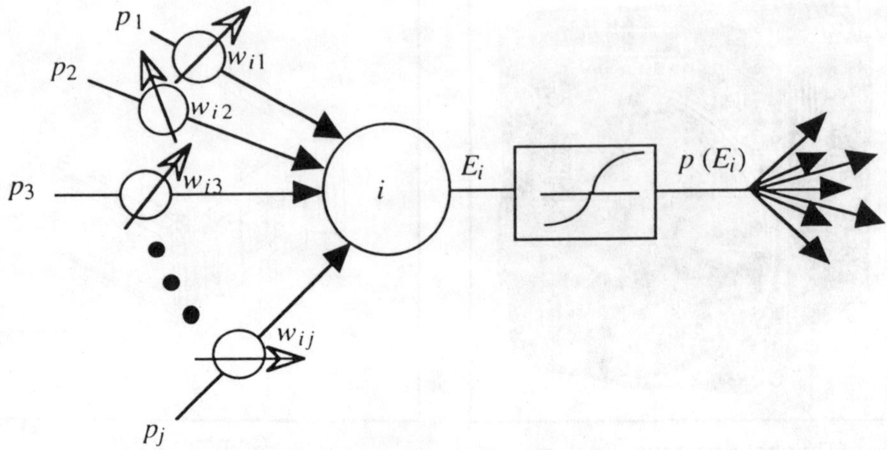


Fig. 20. A Neuron Processing Element.

elements connected in a highly parallel manner. The processing elements called neurons perform very simple Boolean or arithmetic functions on their inputs. The key to the functioning of an ANS are the weights associated with each element, which vary to determine the level of association between neurons. The collection of weights are what represent the information stored in the system. A typical artificial neuron is shown in Fig. 20. Each neuron is characterized by an internal threshold and the type of non-linearity. The neuron has multiple inputs but only one output. The input signals to the neuron are multiplied by the weights ($w_{ij}$) and summed to yield the total neuron input. The weights can be represented as a matrix and identified by subscripts. The neuron output is commonly taken as a sigmoid function of the input. In the early development of neural networks, hard limiters and threshold logic elements[35] were also used. However, the sigmoidal non-linearity are considered representative of real neurons, which approach limits for very small and very large inputs. Exponential functions as shown in the equation below as well as hyperbolic tangent function are used to represent sigmoidal non-linearity. Neuronal damping of the outputs is controlled by this exponential activation function,

$$\mathrm{P}(E_i) = (1 + \mathrm{e}^{-E_i})^{-1} \tag{6}$$

where

$$E_i = \Sigma w_{ij} p_j - T \tag{7}$$

and $p_j$ are the inputs typically normalized to one and $T$ is the internal threshold or offset that characterizes the neuron.

An ANS architecture is very different from a conventional von Neumann computer architecture. In a conventional computer, it is possible to correlate discrete information with memory cells. An ANS is modeled after certain theories of the neuronal characteristics in which information is represented by weights, and there is no direct correlation between a specific weight and stored data. This distributed representation of data is similar to that of a hologram in which the lines of the hologram act as a diffraction grating to reconstruct the stored image when laser light is passed through.

A neural network consists of many neurons that process inputs in parallel and form hidden layers and an output layer as shown in Fig. 21. These neurons, also called nodes, accept multiple inputs and generate an output as shown in Fig. 20. These outputs then become input for the next layer and so on until the output layer is reached. At each level, there are weights $w_{ij}$ for each connection and the same sigmoid function is used at each neuron. In Fig. 21, the most popular neural network based on the back propagation scheme is represented and therefore the output of each neuron does not feed other neurons in the same layer.[35] (paper 5.1)

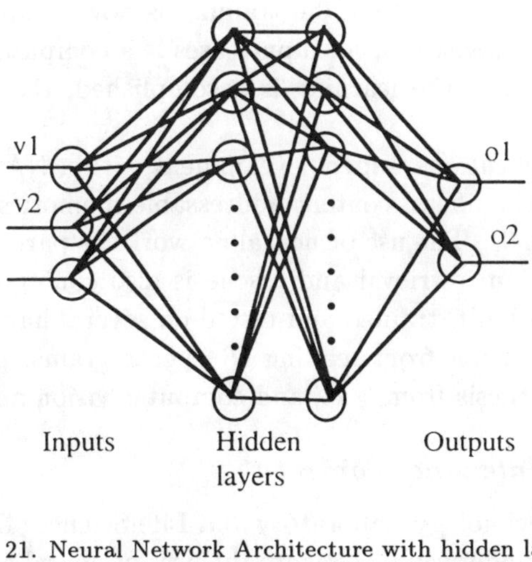


Fig. 21. Neural Network Architecture with hidden layers.

Neural networks are not programmed in the conventional sense. Instead they are trained using the training examples consisting of input and corresponding output data. There are several known neural net learning algorithms, such as back-propagation and counter-propagation, which are used to train networks. The net learns by automatically adjusting weights in the network that connect the neurons. The weights and the threshold values of neurons determine the propagation of data through the net and its response to the input. A typical back propagation algorithm for updating the weights $w_{ij}$ is

$$w_{ij}(t+1) = w_{ij}(t) + \eta\delta_j x'_j\,, \tag{8}$$

where, $\eta$ is the gain term, and $\delta_j$ is an error term for node $j$. Experiments have shown that the weights converge to a value faster if a momentum term is added and weight changes are smoothened by

$$w_{ij}(t+1) = w_{ij}(t) + \eta\delta_j x'_j + \alpha\big(w_{ij}(t) - w_{ij}(t-1)\big) \tag{9}$$

with $0 < \alpha < 1$. The network is considered as trained when the weights do not change or the difference between the network output and the example output is very small for all training examples. In other words, the root mean square (RMS) error must be very small. For complex dynamical problems, this is only one criteria to estimate that the network is trained. Many other criteria must be used to really determine that the training is now complete. Training the net to the correct responses for all input cases is a computationally expensive process. However, once the learning is accomplished, the net responds very quickly.

Anticipated applications under development within NASA for neural networks are pattern classifiers, content addressable memories and general nonlinear adaptive models. The use of neural networks as part of computer vision systems for inspection, retrieval and rescue is also contemplated. Prototype networks have been built, trained and tested on several hard problems including finding concentration from reading IR spectrograms, prediction of solar activity, speech synthesis from text, and computer vision applications.

### 3.1. *Posture maintenance control*

A joint Software Technology Laboratory and Life Sciences Directorate project at JSC is currently underway investigate the uses of neural network technology to discover specific sites in the central nervous system which are vital for the maintenance of posture. The network will be used to analyze posture platform data collected from patients with lesions in the central nervous system and post- and pre-flight data from astronauts. Neural networks are expected to formulate the appropriate correlation in the data.

Control of postural equilibrium is a basic physiological function that is mediated by multiple neurosensory feedback systems. Comparisons of pre- and post-flight data have shown alterations thought to be caused by sensory adaptation to microgravity. These alterations will affect the ability of astronauts to perform certain tasks (such as emergency egress immediately postflight). The goal of this project is to develop a neural network model of postural equilibrium control that can map sensory input signals from the visual, and proprioceptive systems into appropriate motor control strategies for maintenance of stable, upright posture.

### 3.2. *Adaptive diagnosis systems*

The goal of this work is to develop a methodology for the design of diagnostic tools for NASA Shuttle systems. While keeping this in mind, the main focus is on the diagnostic capability for the Space Shuttle Main Engine (SSME) system. Several sensor measurements taken at a time represent a performance vector which can then be classified into good performance, abnormal performance or beginning of abnormal performance. Thus, neural networks can be trained to predict the start of an abnormal performance period. Since it is difficult to generate proper training examples solely based on sensor measurements, neural networks should be provided with some additional information.

A functional modeling approach is being implemented to provide this additional information rather than relying wholly on the use of sensor measurements and conventional neural network paradigms. Functional modeling is necessary because it captures fundamental interactions between different components of a system. Also, every SSME is constrained to operate in the same way and functional modeling should therefore be applicable to all engines. Using AutoClass (a Bayesian classification algorithm developed at NASA Ames Research Center) an analytic description of the SSMEs nominal operating range has been generated. In addition, the GSplines algorithm is also being used to generate an analytic description of SSME nominal operations.

The planned diagnosis approach will use sensor measurements normally down-linked to shuttle flight controllers (JSC personnel) during ascent. The functional models of the SSME operations will also generate similar data during the ascent phase. Neural networks are planned to correlate the data which are the output of these functional models and compare them with the down-linked data to identify normal behavior, start of abnormal behavior, abnormal behavior and faults. Signal processing of the various sensors will be accomplished with use of Space Time Neural Networks (described in the next section). During the course of developing a diagnostic system for the SSME, the process of designing diagnosis tools in general will be investigated.

### 3.3. *Space–time neural networks*

A spin-off of a project to monitor space shuttle main engine performance with neural networks resulted in the development of an innovative neural network architecture especially useful at sequence discovery and non-linear filtering. This neural network borrows concepts from associative memory and filtering techniques. In its present form, the Space–Time Neural Network (STNN) dynamics (Fig. 22) adapt and learn sequences of finite lengths. Currently, efforts are underway to extend the STNN to sequences of arbitrary lengths.

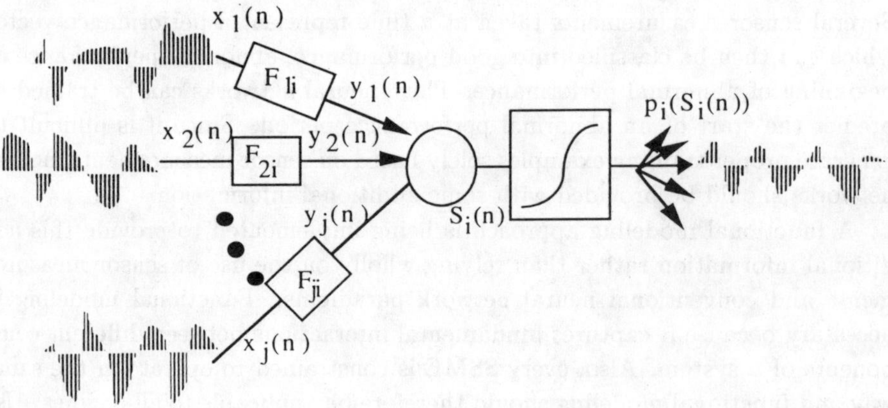


Fig. 22. A Space–Time Neural Network processing element.

The Space–Time Neural Network is basically an extension to a standard backpropagation network in which the single interconnection weight between two processing elements is replaced with a number of Finite Impulse Response (FIR) filters. The use of adaptable, adjustable filters as interconnection weights provides a distributed temporal memory that facilitates the recognition of temporal sequences inherent in a complex dynamic system such as the TSS. Neural network technology has impressively demonstrated the capability of automatically modeling spatial information. On the other hand, the application of parallel distributed models to the processing of temporal data has been severely restricted. The STNN architecture[36] is a novel technique which adds the dimension of time to the strong spatial modeling capabilities found in neural networks. Another dimension can be added to the standard processing element used in conventional neural networks by replacing the synaptic weights between two processing elements with an adaptable-adjustable filter. As shown in Fig. 22, the inputs are processed through the filters before they are summed at the summing junction.

Instead of a single synaptic weight, there are now several weights representing not only spatial association, but also temporal dependencies. In this case, the synaptic weights are the coefficients to adaptable digital filters:

$$y(n) = \sum_{k=0}^{N} b_k x(n-k) + \sum_{m=1}^{M} a_m y(n-m)\,. \tag{10}$$

Here the $x$ and $y$ sequences are the input and output of the filter and the $a_m$'s and $b_k$'s are the coefficients of the filter.

A Space–Time Neural Network (as shown in Fig. 23) includes at least two layers of filter elements fully interconnected and buffered by sigmoid transfer nodes at the intermediate and output layers. A sigmoid transfer function is not used at the input.

Forward propagation involves presenting a separate sequence-dependent vector to each input, and propagating those signals throughout the intermediate layers until the signal reaches the output processing elements. In adjusting the weighting structure to minimize the error for static networks, such as the standard backpropagation, the solution is straightforward. However, adjusting the weighting structure in a recurrent network is more complex because not only must present contributions be accounted for but contributions from past history must also be considered. Therefore, the problem is that of specifying the appropriate error signal at each time and thereby the appropriate weight adjustment of each coefficient governing past histories to influence the present set of responses.

Since STNNs are expected to learn complex dynamical behavior through mapping and filtering, time series of complex input/output data is required.

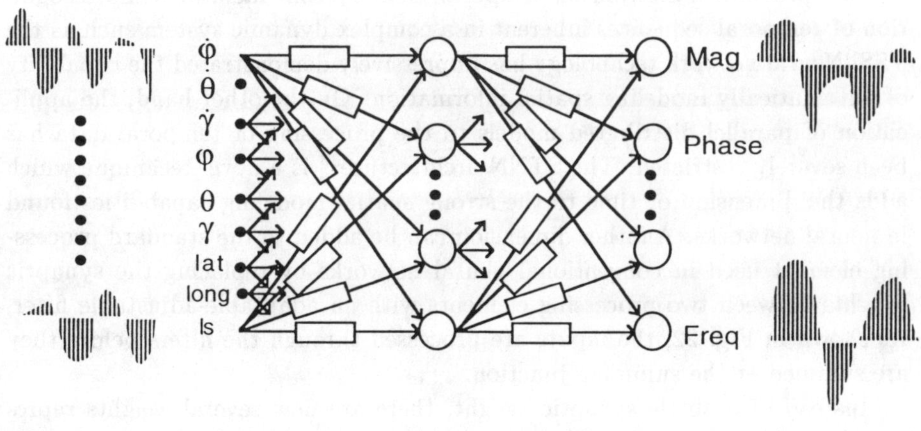


Fig. 23. A depiction of a STNN architecture showing the distribution of complex signals in the input space.

It should be emphasized that the data must be a long time series because the starting part of the series is used to initialize the filter coefficients. If large filters are used in order to learn very complex dynamics, too many data pairs are used up to set up the learning environment. Thus, a large series is expected for the training segment. STNNs are being applied in two applications at JSC: (1) monitoring and diagnostic of SSMEs, and (2) Skip rope Identification during Tether Operations in Space. Both of these applications are expected to be implemented in the Mission Control Center.

### 3.4. *Skip rope identification using STNN*

The Tethered Satellite System (TSS) mission[20] involves the upward deployment of the Italian satellite on a 20 km conducting tether (see Sec. 2.2). Electric current will be induced as the tether is dragged through the magnetic field of the Earth. Because of interaction between the orbital environment and tether dynamics, there are several oscillations in the tethered system. Particularly of interest are the "skip rope" oscillations resulting from the interaction of the Earth's magnetic field with the current pulsing through the tether. Identification and control of the skip rope oscillations (Fig. 24) are very important. Mission success is dependent on how successfully these oscillations can be damped during satellite retrieval. Because of this situation, the mission profile has a very special Onstation-2 phase where the tether is not retrieved unless the skip rope magnitude is smaller than a critical value. Since the skip rope oscillations excite very peculiar attitude oscillations for the satellite, the magnitude and phase of the skip rope can be detected using the satellite rates and angle data. The coupling between the attitude oscillations and the skip rope behavior is complex and very non-linear.

The Space–Time Neural Network (STNN) developed at JSC provides the capability to learn the time variations as well as spatial variations using digital filters before the data is provided to the hidden as well as the output layer. The approach is to apply the STNN to learn the coupling between the satellite attitude oscillations and the skip rope behavior.[37] Inputs to the STNN are satellite attitude rates, tether length, tension and other relevant parameters. Outputs are the skip rope magnitude and phase. Using a very high fidelity bead model in the Orbital Operations Simulation, training as well as test data is generated for the 2.4 km Onstation-2 test case. A set of 3500 examples was created to train as well as test the STNN for tether skip rope identification. In order to achieve a reasonable generalization, the network must learn the training data very well.

Several STNN scenarios[38,39] were configured to identify the skip rope magnitude and phase. These scenarios include 20 and 30 hidden nodes in two different layers, and have 10, 20 and 40 zeros for the digital filters between the input and hidden layers. Preliminary test results suggest that the STNNs require long learning times. This is consistent with the understanding that the coupling between the skip rope and the satellite oscillation is very complex and non-linear. Analysis of several test cases was performed to learn the effects of learning rate and momentum parameters. It was observed that the skip rope identification problem requires a high momentum (in the vicinity of 0.9) and a low learning rate (around 0.05 or less) in the weight update formula of the back propagation. The root mean square error is typically reduced to less than 4% while maximum error in the prediction remains very high. Thus, it is implied that a lot more training is required. It is also possible that a functional modeling approach with a comparison on the measurement data is more appropriate in this case. For test case #4 with momentum $\alpha = 0.9$ and a learning rate value $\eta = 0.05$, the root mean square error reduces to 0.04 in less than 100 cycles (Fig. 25). The maximum error is only 0.43, that is to say that the magnitude and phase estimated by STNN at some point in time were off by this amount. This estimate is very good, because the estimate based on correlation is off by a much larger amount.

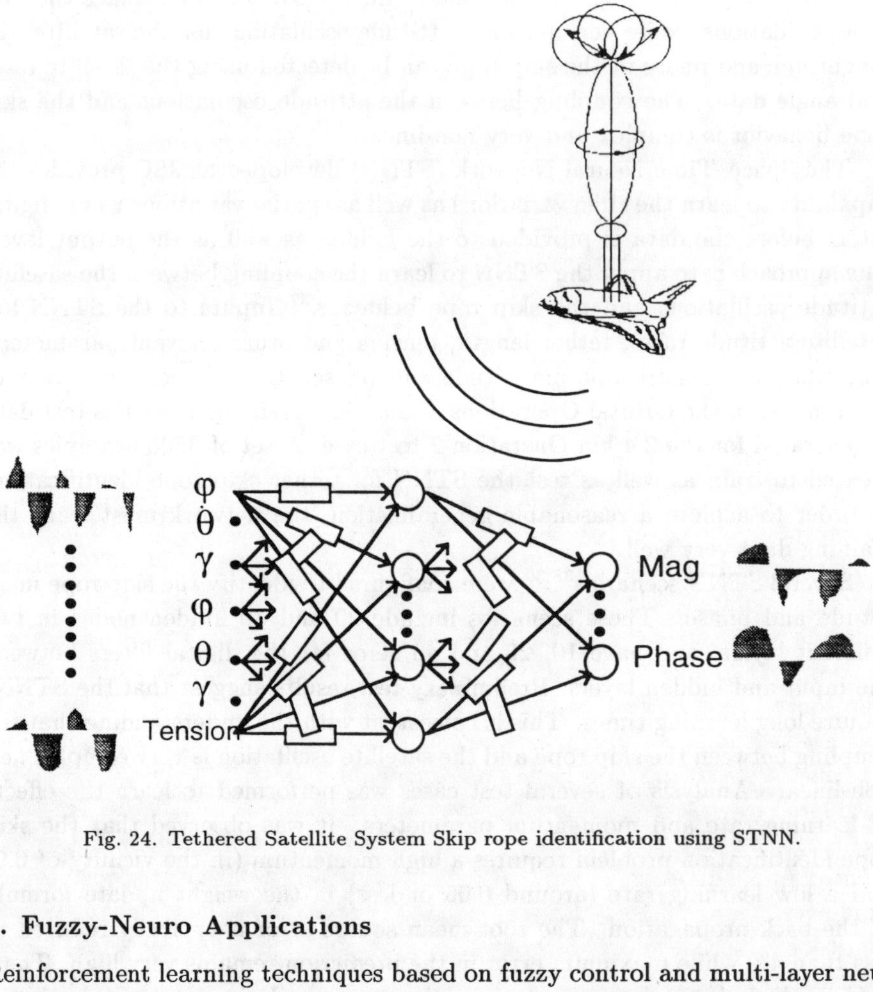


Fig. 24. Tethered Satellite System Skip rope identification using STNN.

## 4. Fuzzy-Neuro Applications

Reinforcement learning techniques based on fuzzy control and multi-layer neural networks have been successfully demonstrated by Hamid Berenji and his group[40,41] at Ames Research Center (ARC) using the inverted pendulum as an example. As a joint project between JSC and ARC, a concept was developed for applying this technique to spacecraft docking operations.[42] Two architectures — Approximate Reasoning-based Intelligent Control (ARIC) and Generalized ARIC (GARIC) — have been developed and applied to the space operations. The general description of these architectures are provided and then the space applications and results are discussed.

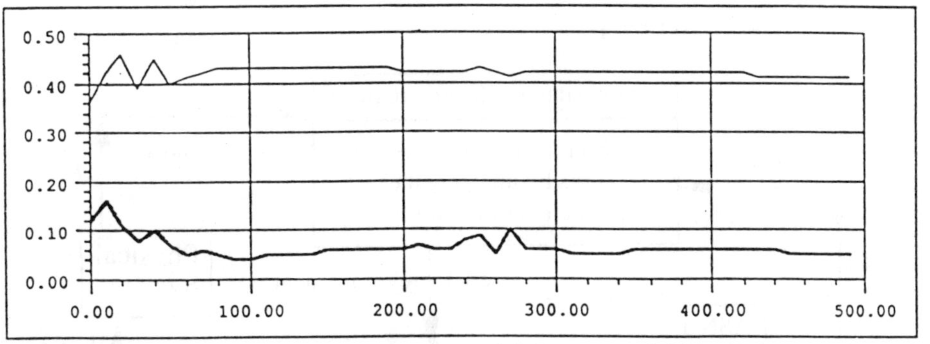


Fig. 25. Maximum and RMS errors for the test case no. 4 for STNN based tether skip rope identification experiment.

### 4.1. *Architecture of ARIC*

Typical Back Propagation Neural Networks or Space–Time Neural Networks undergo the training processes during which a supervisor computes the error (or deviation) in the expected output and updates the weights. Desired outputs are assumed available from the supervisor in the form of supervised learning. In reinforcement learning, there is no supervisor to critically judge the control action chosen as the output by the neural network. The learning system is told indirectly about the effects of its chosen control action. First, the effect of the control action is observed and the amount of reinforcement is evaluated by applying a temporal difference method. Then, a reward or blame is distributed to the individual elements contributing to that performance. For control problems, such as cart-pole balancing, the state space is partitioned into non-overlapping smaller regions and then the credit assignment is performed on a local basis. This is most appropriate when a state space-based control is used. For a rule-based control, partitions should correspond to the regions pointed to by rules, and individual rules engaged in solving the problem must be assigned proper credit. In fuzzy reinforcement learning, these partitions can overlap leading to the use of fuzzy partitions in the antecedents and consequents of fuzzy rules. The reinforcements coming from the environment are then used to refine the definition of the fuzzy labels in the rules.

The ARIC architecture as shown in Fig. 26 extends Anderson's method[43] by including the prior control knowledge of expert operators in terms of fuzzy control rules. The two main elements in this hybrid architecture are: the

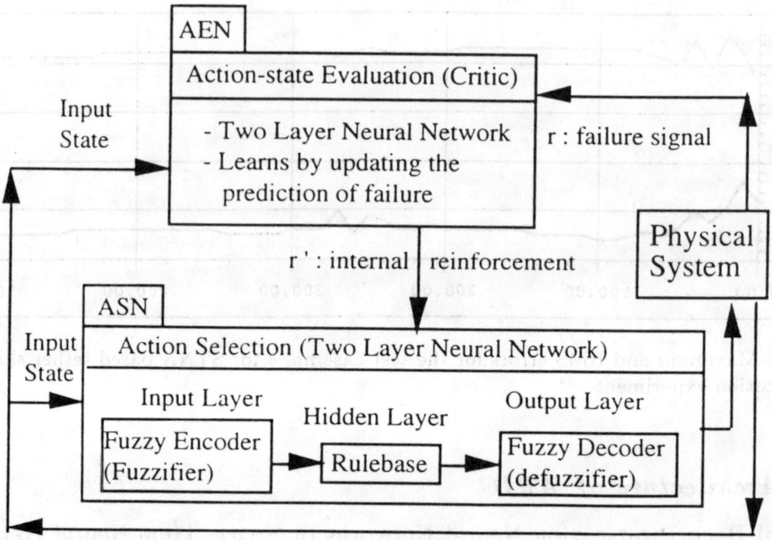


Fig. 26. The ARIC architecture with fuzzy controller.

Action Selection Network (ASN) which includes a fuzzy controller based on the operators knowledge, and the Action-state Evaluation Network (AEN), which acts as a supervisor (or a critic) and provides advice to the main controller. Both, the ASN and AEN, are multi-layer neural networks with the output layer employing reinforcement learning and the hidden layer using a modified error backpropagation scheme.

The fuzzy controller implemented in the ASN consists of a fuzzifier, a rulebase and decision making logic, and a defuzzifier. The design of the rulebase is based on a hierarchical process and considers the interaction of multiple goals. The multi-layer neural network consists of the input layer, hidden layer and an output layer. The input layer has a fuzzifier whose task is to match the values of the input variables against the labels used in the preconditions of fuzzy control rules. The hidden layer in the ASN corresponds to the rules used in the controller and includes the decision making logic. Knowledge from an operator or an expert is implemented in this hidden layer and typically connects the input to output via label references. The output layer includes the decoding or defuzzification process. It combines the conclusion of the individual rules by using the center of area method.

Let $w(i)$ represent the degree that rule '$i$' is satisfied by the input state variables in $X$,

$$w(i) = \text{Min}\left\{d_{i1}\mu_{i1}(x_1), \ldots, d_{\text{in}}\mu_{\text{in}}(x_n)\right\} \tag{11}$$

where $\mu_{ij}(x_j)$ represents the degree of membership of the input $x_j$ in a fuzzy set representing the label used in the first precondition of the rule $i$, $d_{ij}$ represents the connection weights on the input from node $j$ to a hidden layer node $i$, and $n$ is the number of inputs. Please refer to Fig. 27 to observe the relationship between the input layer and the hidden layer via connection weights $d_{ij}$ presented as matrix D. Then, $m(i)$, which represents the results of applying $w(i)$ on the conclusion of rule $i$, is calculated from

$$w(i) = \mu_{C_i}[m(i)] \tag{12}$$

where $\mu_{C_i}$ represents the monotonic membership function of the label used in the conclusion of rule $i$. The amount of the control action, $u(t)$, for the combined set of control rules is then calculated from

$$u(t) = \text{Sum\_numerator}/\text{Sum\_denominator}\,, \tag{13}$$

$$\text{Sum\_numerator} = \sum_{i=1}^{h} f_i * m(i) * w(i) \tag{13a}$$

$$\text{Sum\_denominator} = \sum_{i=1}^{h} w(i) * f_i \tag{13b}$$

where $h$ is the number of nodes in the hidden layer which is equivalent to the number of rules used in the model. The variable $f_i$ represents the connection weight on the link from node $i$ of the hidden layer to the output node. Again, please refer to Fig. 27 to see the relationship between the output layer and the hidden layer via connection weights $f_i$ presented as matrix F. The inputs and outputs are also connected by a standard feed forward neural network with connection weights $e_i$ presented as matrix E in Fig. 27.

In the back-propagation scheme, weights for each link are updated using the error which is the difference between the actual output and the expected output. Again, it should be noted that the expected output is provided in the so-called "training" data. Since there is no supervisor (or critic) to provide a measure of error or deviation, the internal reinforcement $r'$ (see Eq. 17) predicted by the AEN is used as a measure of error to update the weights of the links. Since another contributing factor to this error is the difference between the selected control action and its expected value given the current weights and the current state, the product of internal reinforcement and this difference is used as an error in back propagating and updating the weights in the network. The learning algorithm in ARIC architecture is described in detail in Ref. 44.

The Action-state Evaluation Network (AEN) in ARIC basically apportions the blame for the failure among states and actions in the sequence leading to a failure. The only information received by this network is the state of the physical system (as shown in Figs. 26 and 27) and a signal, $r$, indicating whether or not a failure has occurred. The state of the physical system is provided in terms of state variables $(x_1, x_2, \ldots x_n)$. The AEN structure consists of $m_h$ hidden units and $n$ inputs from the environment. Each hidden unit receives $n+1$ inputs and has $n+1$ weights, while each output unit receives $n+1+m_h$ inputs and has $n+1+m_h$ weights. Here, the AEN plays the role of an adaptive critic element and constantly tries to predict reinforcements associated with different input states. The equations for calculating the output as well as updating the weights are as follows:

If A, B, and C are the matrices of connection weights as shown in Fig. 27, then,

$$y_i[t_1, t_2] = g\left(\sum_{j=1}^{n} a_{ij}[t_1]x_j[t_2]\right) \tag{14}$$

where, $y_i$'s are the output of the nodes in the hidden layer and

$$g(s) = 1/(1+\exp(-s)) \tag{15}$$

The output of the evaluation network is

$$v[t_1, t_2] = \sum_{i=1}^{n} b_i[t_1]x_i[t_2] + \sum_{i=1}^{m_h} c_i[t_1]y_i[t_1, t_2] \tag{16}$$

where $t_1$ and $t_2$ are time points and double time dependencies are used to avoid instabilities in the updating of weights. Please note that the weights in the above equation at time $t_1$ are multiplied by the $x_i$'s at time $t_2$. If the same time index is used, one cannot detect whether the change in $v$ was caused by the change in the weights (i.e., $b_i$ and $c_i$) or was it caused by the change in the state of the system (i.e., $x_i$). Writing the equation in the above manner with different time steps allows one to compare different $v$'s over time and notice whether the system has moved to a better state (i.e., higher reinforcement) or to a worse one (i.e., lower reinforcement).

The AEN evaluates the action recommended by the action selection network as a function of the failure signal and the change in state evaluation based on the state of the system at time $[t+1]$:

$$\begin{aligned} r' &= 0 && \text{start}\,; \\ &= r[t+1] - v[t,t] && \text{failure}\,; \\ &= r[t+1] + \gamma v[t,t+1] - v[t,t] && \text{otherwise} \end{aligned} \tag{17}$$

where $r'[t+1]$ is called the heuristic reinforcement and plays the role of an error in the AEN's weight modifications. The discount rate $\gamma$ is between 0 and 1, and $v$ is the state evaluation in terms of predicted reinforcement.

### 4.2. *ARIC and 6 DOF space operations*

Translational and rotational fuzzy controllers developed at JSC have been implemented in the ARIC architecture (Figs. 26 and 27) and several test cases have been performed using high fidelity simulation of the shuttle.[45,46] The rotational or attitude controller developed for the shuttle has been implemented in the ARIC architecture (Fig. 27) and test cases for attitude hold have been performed.

The space shuttle attitude controller is expected to perform four basic operations: 1) attitude hold or maintaining the desired attitude within a small region of the desired value, typically known as a deadband; 2) attitude maneuver or going from one attitude to another attitude; 3) rate hold or maintaining a desired rate on a given axis; and 4) rate maneuver or going from one rate value to another rate value for a given axis.

Typical controllers based on the phase plane concept have angle errors and rate errors as input values. The controller output is a command for generating a correcting torque. For the space shuttle, the rotational corrective torques are generated by thrusters. The fuzzy logic attitude control rulebase described in Sec. 2.4 with seven fuzzy labels has been defined (Fig. 10a) and implemented in the ARIC architecture. The rulebase has 31 rules (Table IIIa) based on the attitude error and rate error. Figure 27 illustrates an ARIC model for

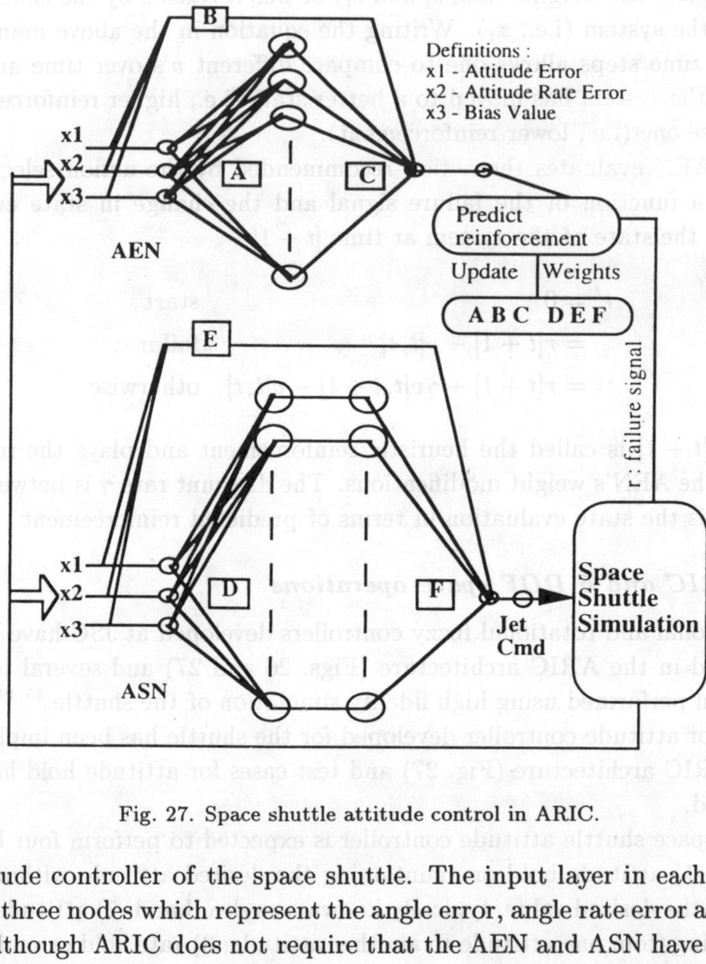


Fig. 27. Space shuttle attitude control in ARIC.

the attitude controller of the space shuttle. The input layer in each network includes three nodes which represent the angle error, angle rate error and a bias node. Although ARIC does not require that the AEN and ASN have an equal number of hidden layer nodes, both have been modeled with 31 nodes in their hidden layers. Also, both networks have a single node at their output layer. It should be noted that the AEN requires a single node at the output layer since a single value for internal reinforcement is calculated there. However, the ASN places no restriction on the number of nodes to be placed at the output layer.

Several experiments, testing ARIC's performance using a high fidelity simulator called the Orbital Operations Simulator at JSC were run. For the AEN, the only reinforcement comes in the form of failure signals from the physical system. Therefore, a failure criteria was developed which uses the angular excursion as a parameter and is consistent with the control design goals. Results show that the fuzzy reinforcement learning technique has no problem in controlling the angles or angular rates.[47,48] It performs attitude control properly (Fig. 28) meeting the control design goals and provides adequate learning ability during these operations.

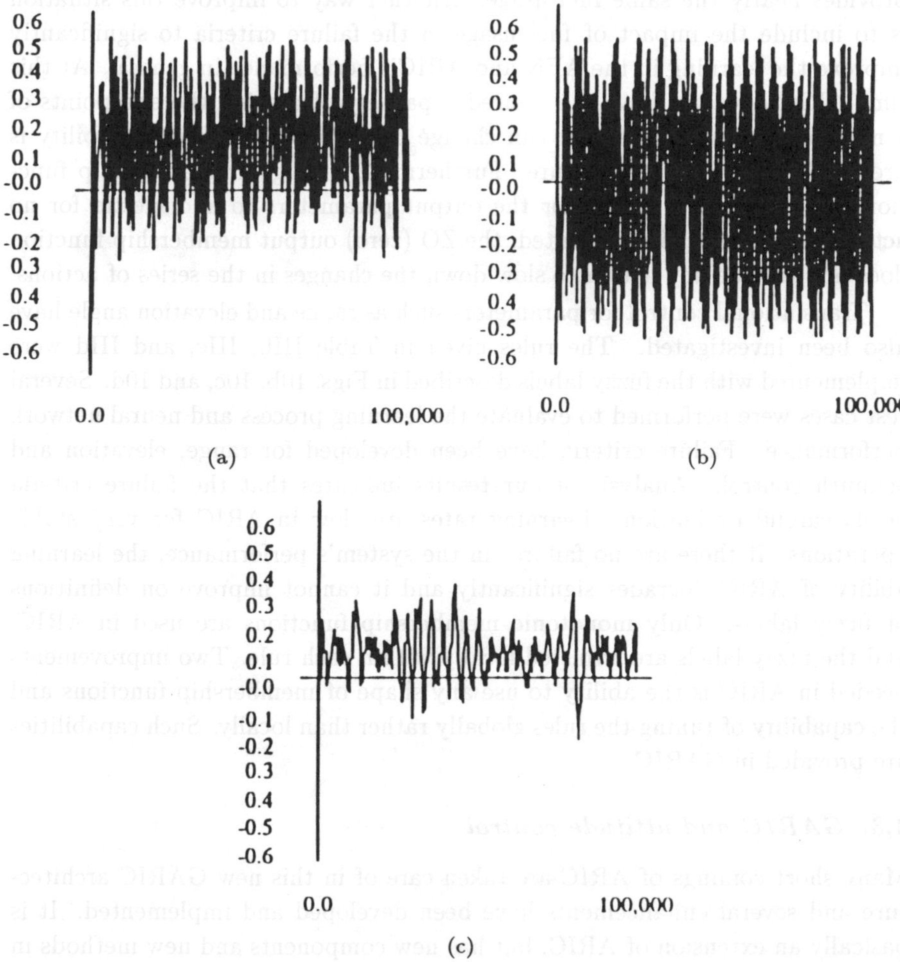


**Fig. 28.** ARIC's performance during attitude control (deadband = 0.5 deg).

Since the impact of fuel usage was not considered in the failure criteria, the fuel usage did not compare well with the non-learning fuzzy control system which is designed to achieve efficient fuel usage. Our detailed analysis of the defuzzification method and the definitions of PS and NS membership functions shows that a small hysteresis exists during rate reversal. A small overlap of 0.01 deg between these two membership functions removes the hysteresis and provides nearly the same fuel usage. Another way to improve this situation is to include the impact of fuel usage in the failure criteria to significantly improve the learning in the AEN and ARIC's performance in general. At this time ARIC does not have the desired capability of moving the end points of a membership function and it can change only the slopes. This capability is provided in GARIC architecture. Furthermore, triangular membership functions are required in ARIC for the output parameters to add inertia for no action. As currently implemented, the ZO (zero) output membership function does not provide any action to slow down the changes in the series of actions.

Translational control for parameters such as range and elevation angle have also been investigated. The rules given in Table IIIb, IIIc, and IIId were implemented with the fuzzy labels described in Figs. 10b, 10c, and 10d. Several test cases were performed to evaluate the learning process and neural network performance. Failure criteria have been developed for range, elevation and azimuth control. Analysis of our results indicates that the failure criteria needs careful evaluation. Learning rates are slow in ARIC for very stable operations. If there are no failures in the system's performance, the learning ability of ARIC degrades significantly and it cannot improve on definitions of fuzzy labels. Only monotonic membership functions are used in ARIC, and the fuzzy labels are adjusted locally within each rule. Two improvements needed in ARIC is the ability to use any shape of membership functions and the capability of tuning the rules globally rather than locally. Such capabilities are provided in GARIC.

### 4.3. *GARIC and attitude control*

Many short comings of ARIC are taken care of in this new GARIC architecture and several enhancements have been developed and implemented. It is basically an extension of ARIC, but has new components and new methods in implementing the fuzzy controller to achieve global learning. The architecture of GARIC is schematically shown in Fig. 29. It has three main components: 1) the Action Selection Network (ASN) maps a state vector into a recommended action $F$ using fuzzy inference; 2) the Action Evaluation Network (AEN) maps a state vector and a failure signal into a scalar score which indicates state goodness, and generates internal reinforcement $r'$, and 3) the Stochastic Action Modifier using both $F$ and $r'$ produces an action $F'$ which is applied to the plant.

The ensuing state is fed back into the controller, along with a Boolean failure signal. Learning occurs by fine tuning the free parameters in the two networks: in the AEN, the weights are adjusted; in the ASN, the parameters describing the fuzzy membership functions change.

The AEN in the GARIC architecture is similar to that used in ARIC and no major modifications or enhancement have been implemented. It produces a prediction of future reinforcement for a given state. This prediction is used to update the weights and modify the fuzzy labels in the Action Selection Network. Changes in this prediction are provided to the Stochastic Action Modifier to guide the selection of actions.

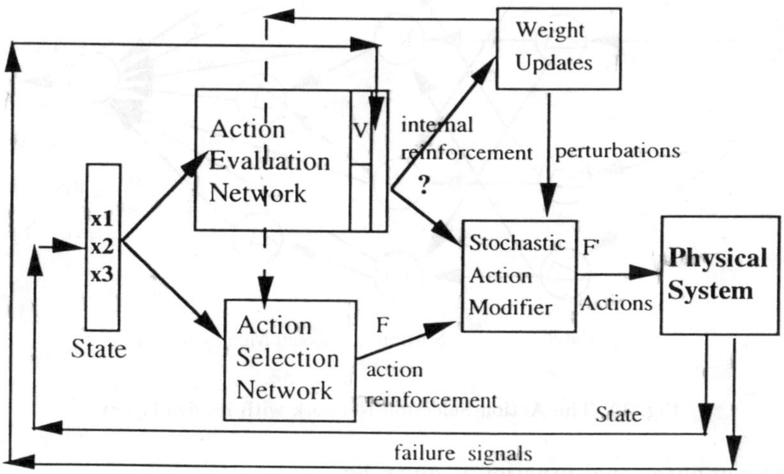


**Fig. 29.** The architecture of GARIC.

The Action Selection Network implements the fuzzy control rules in the form of five layer nodes, each layer performing one stage of the fuzzy inference process (Fig. 30). The connections are fed forward with each node performing a local computation. However, this computation may be different from the conventional weighted sum of inputs and sigmoid function. Layer 1 consists of real-valued input variables and does not perform any computations. These variables are linguistic variables of interest whose values are defined in terms of labels. Layer 2 consists of nodes which correspond to one possible value of the linguistic variables in layer 1. For example, if HIGH is one of the values that an input $x$ can take, a node computing $\mu_{\text{HIGH}}(x)$ belongs to layer 2. This node will have exactly one input, and its output $\mu_{\text{HIGH}}(x)$ is connected to all nodes in layer 3 where the clause 'if $x$ is HIGH' is used in the "if" part of the rule.

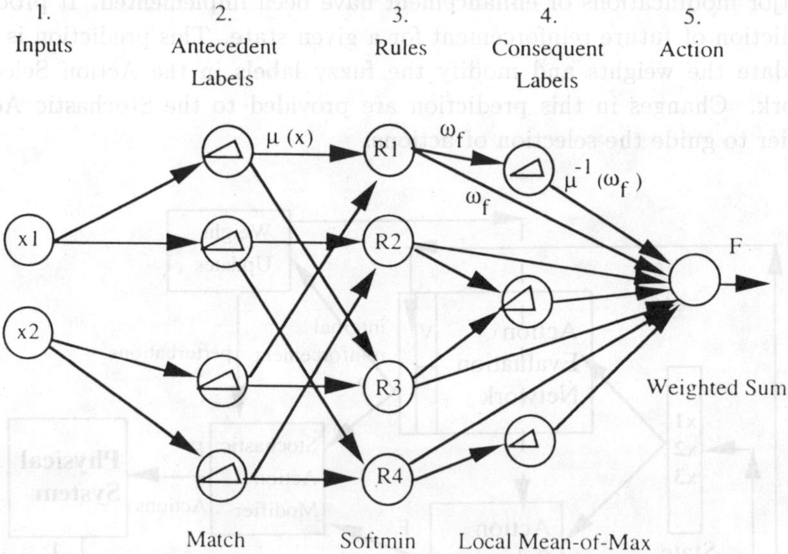


Fig. 30. The Action Selection Network with its five layers.

The computational function is given by

$$\mu = f(x, c_{\text{V}}, s_{\text{VL}}, s_{\text{VR}}) \tag{18}$$

where, V indicates a linguistic value (e.g. HIGH), and $c_{\text{V}}$, $s_{\text{VL}}$, and $s_{\text{VR}}$ correspond to the center, left and right spread of the fuzzy membership function of label V as shown in Fig. 31. The value $c_V$ serves as a reference point for the fuzzy label V and the spread on both sides permit asymmetry from this center point. More parameters may be included if desired, and their use could be defined in the function in (13).

An instance of a smooth membership function is

$$\mu(x) = \frac{1}{1 + \left|\frac{x - c}{s}\right|^b} \tag{19}$$

where $s = s_{\mathrm{VL}}$ or $s_{\mathrm{VR}}$ accordingly as $x < c$ or $x \geq c$ and $b$ controls the curvature. For triangular shapes, this function is given by

$$m_{C,s_{\mathrm{L}},s_{\mathrm{R}}}(x) = \begin{array}{ll} 1 - |x - c|/s_{\mathrm{R}}\,, & c \leq x \leq c + s_{\mathrm{R}} \\ 1 - |x - c|/s_{\mathrm{L}}\,, & c - s_{\mathrm{L}} \leq x \leq c \\ 0 & \text{otherwise}\,. \end{array} \tag{20}$$

Triangular membership functions are easy to use, simple and have been proven as sufficient in many applications. The center and spreads may be considered as weights on the input links, analogous to the approach taken in neural networks with radial-basis-function units.

Layer 3 implements the conjunction of all the antecedent conditions in a rule. Each node in layer 3 corresponds to a rule in the rulebase. Layer 2 nodes which participate in the 'if' part of the rule supplies the inputs to the node in layer 3. The node itself in layer 3 performs the *softmin* operation which is a continuous and differentiable operation as follows.

$$\mathrm{O}_{\mathrm{R3}} = \omega_r = [\Sigma_i \mu_i \exp(-k\mu_i)] / [\Sigma_i \exp(-k\mu_i)]\,. \tag{21}$$

Here, $\mu_i$ is the degree of match between a fuzzy label occurring as one of the antecedents of rule $r$ and the corresponding input variable. The *softmin* operation computes the degree of applicability, $\omega_r$, for a given rule $r$, and the parameter $k$ controls the hardness of the *softmin* operation. As $k$ tends to infinity, the usual minimum operator of the fuzzy logic is recovered. However, for $k$ finite, the *softmin* operation results in a differentiable function of the inputs making it convenient for calculating gradients during the learning process. Experiments have shown that the choice of $k$ is not critical.

Layer 4 nodes correspond to consequent labels. Each node receives its inputs from all rules which use this particular consequent label in the "then" part of the rule. Based on the input value $\omega_r$, this node computes the corresponding output action as suggested by rule $r$ using the inverse mapping written as

$$\mu^{-1}_{c_{\mathrm{V}},s_{\mathrm{VL}},s_{\mathrm{VR}}}(w_r)\,, \tag{22}$$

where V is the specific fuzzy consequent label, and $c$, $s_{\mathrm{L}}$ and $s_{\mathrm{R}}$ are membership parameters as defined earlier.

Inverse mapping implies a suitable defuzzification procedure applicable to an individual rule. In general, the mathematical inverse of $\mu$ may not exist if the function is not strictly monotonic. A simple procedure very similar to the mean-of-maximum method is proposed to determine this inverse.

If $w_r$ is the degree to which rule $r$ is satisfied, then, $\mu_{\mathrm{V}}^{-1}(w_r)$, is the $x$-coordinate of the centroid of the set $\{x : \mu_{\mathrm{V}}(x) \geq w_r\}$.

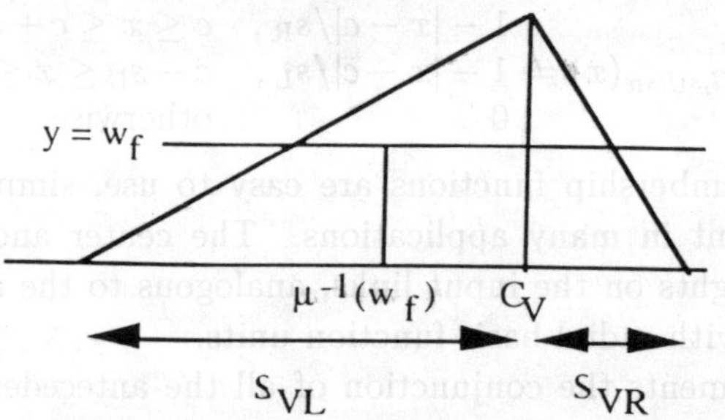


Fig. 31. Triangular Membership Function.

The difference between the proposed method and the mean-of-maximum method is when it is applied. The mean-of-maximum method is applied after all rule consequents have been combined, whereas the proposed method is applied locally to each rule before the consequents are combined. Therefore, the method is referred as the Local Mean-Of-Maximum (LMOM) method. In Fig. 31, the set limited to the line $y = w_f$ is used to determine the centroid value for a rule whose output is $w_f$. For triangular functions, LMOM gives,

$$\mu^{-1} c_{\mathrm{V}}, s_{\mathrm{VL}}, s_{\mathrm{VR}}(w_r) = c_{\mathrm{V}} + 0.5^*(s_{\mathrm{VR}} - s_{\mathrm{VL}})^*(1 - w_r) \tag{23}$$

If the rule output $w_r = 0$, the limiting value of the inverse $\mu^{-1}(w_r \rightarrow 0^+)$ is used, which is $c_{\mathrm{V}} + 0.5^*(s_{\mathrm{VR}} - s_{\mathrm{VL}})$ for triangular functions. If the membership function is monotonic, then, $\mu^{-1}(w_r)$ is just the standard mathematical inverse with proper limiting values. Since sharing of consequent labels is allowed, a node in this layer may have multiple outputs carrying different values. For each rule feeding it a degree, the node should produce a corresponding output action for the next layer. This unusual feature, even though it is useful, has not been tried and experimented with yet in applications. In fact, for many classes of membership function, it can be eliminated, and therefore it was eliminated for triangular shaped functions by generating the output as

$$O_{V4} = (c_V + 0.5^*(s_{VR} - s_{VL}))^*(\Sigma_r w_r)$$
$$- (0.5^*(s_{VR} - s_{VL}))^*(\Sigma_r w_r^2). \quad (24)$$

Whenever $\mu^{-1}(x)$ is a polynomial in $x$, then one output is sufficient regardless of the number of inputs.

Layer 5 has as many nodes as there are output action variables. Each output node uses the rule strength as a weight and combines the recommended (defuzzified) values from each rule in the rule base. The weighted sum formula is very simple.

$$\mathrm{F} = (\Sigma_r w_r \mu^{-1}(w_r))/(\Sigma_r w_r) \quad (25)$$

or

$$\mathrm{F} = \Sigma_V \mathrm{O}_{V4}/\Sigma_R \mathrm{O}_{R3} \quad (26)$$

where the inputs come from layer 4 and layer 3 and we have used formulas (19) and (16) in reducing the equation. The node simply sums up each set of inputs and forms a quotient. In this way, a continuous value of the output variable is delivered as an action selected by the ASN. When the input space is completely covered by the antecedent label functions, the output F is always defined. It should be noted that the weights on input links in layer 2 and 4 are modifiable and all others are fixed at unity. Therefore, the gradient descent procedure will work effectively on only two layers.

The Stochastic Action Modifier generates an action $F'$ based on the internal reinforcement $r'$ from the previous time step and an action $F$ recommended by the ASN. The action $F'$, actually applied to the plant, is a stochastically generated value with mean F and standard deviation $\sigma(r'(t-1))$. This $\sigma(\ )$ function is non-negative and monotonically decreasing, e.g., $\exp(-r')$. This stochastic perturbation in the action leads to a better exploration of the state space and better generalization ability. The difference between $F$ and $F'$ is large when the internal reinforcement $r'$ is low, and small when it is high. A large random step away from the ASN value may occur when the last action performed is bad, but the controller will remain consistent with the fuzzy rules when the previous action selected is a good one. The actual form of $\sigma(\ )$, especially its scale and rate of decrease, should take the units and range of variation of the output into account. The perturbation at each time step is simply the normalized deviation from the ASN-recommended action and contributes as a learning factor in the ASN.

Weight updating in the AEN is based on the reward/punishment scheme. If positive internal reinforcements are received, the values of the weights are rewarded by being changed in the direction that increases its contribution to the total sum. If reinforcements are negative, then weights are adjusted to reduce its contribution to the total sum. Learning in GARIC and its application to the cart-pole balancing problem is discussed in detail in Ref. 49.

The Space Shuttle attitude control membership functions and rules described in Sec. 2.4, Fig. 10a and Table IIIa, were implemented in GARIC format within the OOS at Ames Research Center and several experiments were performed[48] to understand the learning in GARIC. For the ASN, there are two inputs, attitude error and rate error, each with 7 fuzzy labels (NB, NM, NS, ZE, PS, PM, PB). The single output of the ASN, jet firing commands, has five fuzzy labels (NM, NS, ZE, PS, PM). A total of 31 rules control the state space. Each rule has two antecedents, the first uses a fuzzy label for error and the second uses a fuzzy label for rate error, and one consequent that uses a fuzzy label for the jet firing command. Hence, the network has 2, 14, 31, 5, and 1 neurons in its five layers. For the AEN, a biased unit was also included as an input. Thus, there are three inputs, error, rate error and a bias, and one output. Since there were 31 rules in the rulebase, the hidden layer has 31 nodes. Hence the network has 3, 31, and 1 neurons in its three layers.

The goal for the attitude control is to keep the angle and rate deviations within a certain deadband. In a learning experiment, a failure occurs when the value of this deviation exceeds the allowed deadband. Every time a failure occurred in a GARIC execution, the control was shifted to a supervisory control routine to bring the state of the system back to within the deadband. This is indeed to restart the learning process after a failure as required by these architectures. The original fuzzy controller with a small modification in the NS membership function was implemented as a supervisory controller. The center $c_V$ of the NS membership function was shifted to $-3$ deg from its original value of $-2$ deg. This modification was sufficient to bring back the error to within its deadband of 0.4 deg after a failure, but not sufficient to hold the error there. The original fuzzy controller holds the errors within 0.5 deg but not within 0.4 deg. After a small number of trials (less than 10) GARIC learns to do this task by revising the membership functions. A similar experiment was performed to train the new controller to hold the errors within 0.3 deg, and GARIC learned this new task within 5 trials. The fuel consumption for each test case is about 222 lb which is in the same range as the non-adaptive fuzzy controller and less than the conventional controller in OOS. Since the conventional controller has more constraints on its design (such as hardware concerns about the jet life which is improved by using longer but as few firings as possible), this fuel comparison is done just to show that the GARIC performance is within the acceptable range with the conventional controller.

Although GARIC is very similar in structure to that proposed by Anderson,[43] the action selection network is a synthesis of fuzzy logic and neural networks. The fuzzy controller in GARIC extends Anderson's approach to provide for continuous representation of output values and inclusion of the expert's control rules into the action selection network. A knowledge base of fuzzy control rules is implemented in the Action Selection Network as the starting point. Tuning via learning in a continuous manner is then achieved in the neural networks in GARIC. Thus, the goal is not restricted to strategy learning only.

The stochastic action modifier, even though it has similarities to Gullapalli's method, uses a completely different approach for defining internal reinforcement. Single layer neural networks require identification of trace functions for keeping track of the visited states and their evaluations. Further, the single layer networks typically consider only the generation of output values, and leave the preconditioning of fuzzy rules as is. The approach suggested in GARIC is different in both respect. First, it does not use the trace functions. The new representation of fuzzy rules allows faster development and faster learning. Second, based on the reinforcement received from the environment, both the preconditions and the conclusions of the rules can be modified.

In summary, a new method of designing and tuning fuzzy logic controllers has been proposed in ARIC and GARIC architectures. Both architectures have been fully developed with applications to cart pole balancing as well as the space shuttle attitude control. The knowledge acquired through experience and used by an operator in controlling a process can now be modeled and implemented using linguistic terms. The rules can later be refined through the process of learning from experience. Neural networks are configured in a particular fashion so that the reinforcement signals can be used to update the weights for improvement in the fuzzy controller's performance. The architecture of GARIC provides a well balanced method for combining the experiential knowledge in the form of fuzzy rules and the learning power of artificial neural networks.

## 5. Software Tools

The Software Technology Branch at the Johnson Space Center had an active fuzzy logic and neural network research program. What follows is an overview of the various research and application tools developed within the Software Technology Laboratory.

### 5.1. *Network execution and training system*

Network Execution and Training System (NETS) is a portable C-based simulation of the popular back-propagation neural network which can run on most computers. NETS is intended as a tool for anyone interested in exploring the use of neural networks. It is easy to use, but sufficiently rich in features and it has been used extensively in support of in-house research. NETS is also available through Cosmic. NETS is currently in use at 288 government sites.

### 5.2. *Neural network environment on a transputer system (NNETS)*

Due to the extreme computational demands which arise in training of ANS, a neural network workstation was designed and built to be hosted by a network of Transputers.[50] This workstation software was recently released to Cosmic. The workstation was designed to allow for the simulation of a wide range of neural network architectures. There is a sophisticated graphics interface to allow the user to visualize the progress of the network toward convergence. The network of 40 Transputers yields the performance of a super-computer at a fraction of the cost.

### 5.3. *Faster training algorithms for neural networks*

A major barrier to the use of artificial neural systems is the potentially large amount of processing power required to accomplish the training process. Therefore much research is directed toward the improvement of training methods. This research has resulted in two better algorithms: the Difference Optimized Training Scheme for the Neocognitron Neural Network (DOTS) and Accelerated Training Method for the Back-Propagation Neural Network.

The neocognitron is an artificial neural model which was intended to simulate some of the functions of the optic nerve and visual cortex. As originally conceived, the network would "learn" to recognize objects by repeated exposure, a process which often required hundreds of passes to achieve satisfactory results. The DOTS algorithm[51] partitions the training images into a large number of possible templates, and intelligently selects a small collection of templates which describe the training data. From these templates, the synaptic connection strengths may be easily constructed. The time requirements for DOTS are much less than those for the iterative algorithm.

The back-propagation network has become an extremely popular tool for modeling complex functions and pattern recognition. For large networks, the standard training algorithm may be extremely time-consuming if for no other reason, because the network may have thousands of weights to be determined. The accelerated training algorithm exploits the fact that the input weights for the network (often the vast majority) can be represented as a linear combination of the input vectors. This representation of the input weights can result in a big reduction in the number of optimization variables which must be considered, and thus in a significant savings of time. The degree of effectiveness of the algorithm is strongly dependent on the particular problem under consideration.

### 5.4. *FuzzyCLIPS*

Development of FuzzyCLIPS was funded as a Small Business Innovative Research grant to Togai InfraLogic, and authors have recently completed this project and have delivered the prototype of the tool to JSC.[52] The project objective was to develop a prototype of a tool that allows AI/Expert System technology users to build three types of expert systems: a) Conventional rule-based expert systems where symbolic reasoning is used in production type rules using CLIPS syntax; b) fuzzy rule-based expert systems where fuzzy sets, inference methods and defuzzification methods are used in the fuzzy 'if... then...' rules, and c) hybrid expert systems where pure symbolic as well as fuzzy reasoning are combined.

FuzzyCLIPS prototype has been created with a seamless non-invasive binding to the CLIPS v 5.1 environment (Fig. 32). Users can create fuzzy, conventional or hybrid expert systems within this environment easily. Examples of such expert systems are Life Insurance Risk Assessment (purely CLIPS, and purely FuzzyCLIPS) applications, and Verification and Validation Tool that is

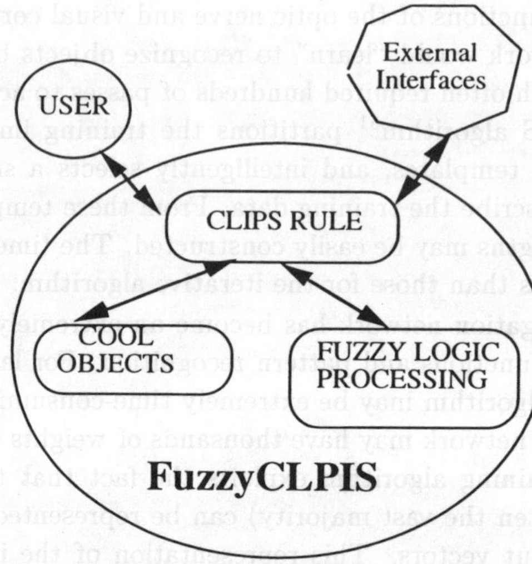


Fig. 32. FuzzyCLIPS environment.

also implemented in the FuzzyCLIPS environment. An architecture has been created such that FuzzyCLIPS can be extended for fuzzy relational databases, automatic rule generation and tuning, automatic membership function tuning using Genetic Algorithms, remote on line tuning and graphical user interface to provide ease of use to the users. Verification and validation capability has been implemented as a FuzzyCLIPS application to minimize the problems of size on PC platforms. A users manual has been prepared for the experienced users of the CLIPS environment.

As a tool, FuzzyCLIPS is extensible, flexible to incorporate new concepts and techniques, e.g. neural networks, genetic algorithms, etc. and has open architecture so that users can add their own capabilities. It is essentially a basic core around which other tools can be developed. The main platform for FuzzyCLIPS is a PC so that a large part of the user community can benefit from this tool. Currently it has no graphical user interface to provide maximum operational transition from CLIPS environment to FuzzyCLIPS environment. It is possible to upgrade the fuzzy logic processing capabilities to future versions of CLIPS. For maintaining portability, the source code is developed using ANSI C standards and no PC dependencies are implemented in the source code.

Within the FuzzyCLIPS environment, users can define membership functions for input/output parameters for the fuzzy expert systems and write rules using the fuzzy sets and operations. Membership functions can be piece-wise linear (triangle, trapezoidal), *s*-functions, pi-functions, arctan or any user defined functions. Rules are written following typical CLIPS syntax using input/output parameters, membership functions, and a connective of 'AND' type. The right hand side can have more than one consequent and the results of a rulebase are returned in a multi-field or array so that the user can utilize them properly. There is no limit on the number of input/output variables, rules and the size of the application except hardware limitations such as the memory of the user computer. Users have the capability to select one of three inference schemes (max-min, max-dot, and max-av) and one of three defuzzification schemes (centroid, height, and alpha) globally for all rulebases. Users also have the ability to set the certainty factor for a given rule as well as the firing threshold value. Thus, a rule's contribution is based on its certainty factor and it fires (meaning provides any contribution) only if its firing strength is above the threshold value set by the user. In this manner, users can create a Boolean logic-based expert system within the framework of FuzzyCLIPS.

## 6. Summary

Fuzzy logic is simple, easy to understand and reflects human-type thinking. Its architecture is very well suited for implementing heuristic knowledge or the knowledge gained through experience. Several applications have shown that the fuzzy logic based control is usually robust, non-linear and comparatively stable. It provides an ability to combine seemingly non- related parameters for higher order reasoning. Control of systems that are non-linear and difficult to model is easily achieved using fuzzy logic principles. In our experience, fuzzy controllers can be easily designed using heuristics and experiential knowledge. Tuning of membership functions and modifications to the rule base is also simplified to a point where rapid implementation and testing is possible. Maintenance of the algorithms is minimal. Since the algorithms are in a graphical form, the knowledge transfer from one generation to another generation is very easy.

Neural networks provide a framework to relate the input parameters with output parameters with the learning process. As the training samples are processed, neural networks learn the relationship in the form of modified weights. Thus, neural networks are trained and not programmed. Learning is as good as the samples (known as input and output data in supervisory learning), and therefore many times neural networks generalize very well when the new situations require interpolation. In cases where extrapolation is necessary, neural networks may fail. Therefore, an application developer must be very careful in selecting the training data as well as in determining the training time. Typically, the root mean square errors are used to indicate that a neural network has learned. But, it is not guaranteed that the performance of the neural network is what is desired by the developer. There is no concise method to indicate that the training provided is sufficient and will provide the desired performance. In complex applications, neural networks require too much training time. For example, the STNN training time for the tether skip rope effect is in terms of weeks rather than days. This limits their practical applications in many areas. Neural networks are often called opaque in terms of their ability to provide any explanation of their weights. They are purely numerical algorithms and cannot do any symbolic processing, much less any reasoning.

From a logic and reasoning point of view, fuzzy logic provides symbolic as well as numerical processing ability. Rules are written using the logic form but are processed using precise numerical computations. This characteristic of fuzzy logic has provided a path for merging fuzzy logic with the CLIPS environment so developers can build hybrid expert systems. Since fuzzy logic provides a framework to perform numerical as well as symbolic reasoning, it fills the gap between the expert systems and neural network processing. Fuzzy logic cannot map complex relationships unless the developer already has the knowledge to create rules. Fuzzy systems typically do not learn as part of its processing.

When combined with neural systems, a fuzzy-neuro system can handle the control and decision making aspect as well as learn the changing environment, user requirements, or new demands for performance. Thus, fuzzy-neuro systems can adapt to changes and can provide the flexibility required in many applications. As humans continue to deal with more and more complex systems in future, the fuzzy-neuro based expert systems will suitably perform in varied environments and will meet the demands placed on these complex systems.

## 7. Acknowledgments

A large part of this chapter is based on the paper submitted to SPIE technical conference 2061, September 1993. The authors very much appreciate the encouragement provided by Chris Culbert, Bob Savely and Robert Shelton of Software Technology Branch at the Johnson Space Center and Ken Kristie and Steve Imsen of Hitachi America Limited. The assistance provided by Hamid Berenji and Anil Malkani at Ames Research Center, Jeff Hoblit at LinCom Corporation, Luc P. Tran and Jonathan Weiss at McDonnell Douglas Space Systems Company, Michael Lembeck at Space Industries, Jack Aldridge at Advantex Inc., and Edgar Dohmann at Ortech Engineering Inc. is also very much appreciated.

## References

1. L. A. Zadeh, "Fuzzy sets," *Information and Control* **8**, 338–353 (1965).
2. L. A. Zadeh, "Fuzzy algorithms," *Information and Control* **12**, 94–104 (1968).
3. L. A. Zadeh, "Outline of a new approach to the analysis of complex systems and decision processes," *IEEE Trans. Syst., Man and Cyberns.* **SMC-3**, 28–44 (1973).
4. D. Dubois and H. Prade, *Fuzzy Sets and Systems — Theory and Applications* (Academic Press, New York, 1980).
5. H. J. Zimmermann, *Fuzzy Set Theory and Its Applications* (Kluwar-Nijhoff, 1985).
6. G. J. Klir and T. A. Folger, *Fuzzy Sets, Uncertainty, and Information* (Prentice Hall, New Jersey, 1988).
7. B. Kosko, *Neural Networks and Fuzzy Systems* (Prentice Hall, Englewood Cliffs, New Jersey, 1991).
8. S. Yasunobu and S. Miyamoto, "Automatic train operation system by predictive control," *Industrial Applications of Fuzzy Control*, ed. M. Sugeno (North-Holland, Amsterdam, 1985), pp. 1–18.
9. T. Shingu and E. Nishimori, "Fuzzy based automatic focusing system for compact camera," *Proceeding of IFSA89* (1989), pp. 436–439.
10. T. Tobi, T. Hanafusa, S. Itoh, and N. Kashiwagi, "Application of fuzzy control system to coke oven gas cooling plant," *Proceedings of IFSA89* (1989), pp. 16–22.
11. First Industrial Conference on Fuzzy Logic Systems sponsored by MCC, the consortium of Corporations, Austin, June 1990.
12. IEEE International Conference on Fuzzy Systems FUZZ-IEEE '92 and '93, sponsored by the IEEE Neural Network Council, San Diego and San Francisco, CA, March 1992 and 1993.
13. First and Second International Workshop on Industrial Applications of Fuzzy Control and Intelligent Systems sponsored by Texas A&M University, College Station, Texas, November 1991, and December 1992.
14. R. N. Lea and Y. Jani, "Applications of fuzzy logic to control and decision making," in *Proceedings of Technology 2000 Conference, NASA Conference Publication 3109* **2** (1990).

15. R. N. Lea and Y. Jani, "Fuzzy logic in autonomous orbital operations," *International Journal of Approximate Reasoning* **6** no. 2, 151 (1992).
16. J. A. Mica and R. N. Lea, "Space shuttle orbiter remote manipulator system simulator (RMSS) software program description document (PDD), Goddard RMSS," *NASA GSFC Document* (May 29, 1994).
17. R. N. Lea, Y. Jani, and J. Hoblit, "Fuzzy logic based robotic arm control," in *Proceedings of the Second IEEE International Conference on Fuzzy Systems*, vol. 1 (1993), pp. 128–133.
18. Togai InfraLogic, Inc., TilShell User Manual.
19. J. Mica, R. N. Lea, and Y. Jani, "Fuzzy logic control of a six degree of freedom remote manipulator system," in *Proceedings of the WNN/FNN '94 conference*, (1994).
20. S. Coledan, "Tethered satellite advances," *Space News* **2** no. 15, 8 (1991).
21. R. N. Lea, J. Villarreal, C. Copeland, and Y. Jani, "Fuzzy logic based tether control," North American Fuzzy Information Processing Society – NAFIPS '91 (1991).
22. R. N. Lea, J. Villarreal, C. Copeland, and Y. Jani, "Tether operations using fuzzy logic based length control," in *Proceedings of FUZZ-IEEE '92* (1992), p. 1335.
23. R. Kahl and S. Bailey, "Mars rover sample return: Project description and mission operations review," NASA JSC (New Initiatives Office), Houston, TX., presented at MRSR Phase A Midterm Review, Nov. 1989.
24. R. N. Lea, L. Walters, and Y. Jani, "A fuzzy logic approach to mars rover trajectory planning and control," in *Proceedings of the 1st ISMCR*, Session D.3 (1990), pp. D.3.1.1.
25. R. Lea, "Fuzzy logic approach to mars rover guidance," presented to The International Conference on Fuzzy Logic and Neural Networks IIZUKA '90, Iizuka, Japan, July 1990.
26. R. N. Lea, M. Murphy, and L. Walters, "Fuzzy logic control for autonomous collision avoidance," *NAFIPS '91 Workshop Proceedings*, University of Missouri-Columbia (1991).
27. R. N. Lea, J. Hoblit, and Y. Jani, "Performance comparison of a fuzzy logic based attitude controller with the shuttle on-orbit digital auto pilot," *NAFIPS '91 Workshop Proceedings* (1991), pp. 291–295.
28. R. N. Lea, J. Hoblit, and Y. Jani, "A fuzzy logic based spacecraft controller for six degree of freedom control and performance results," in *Proceedings of AIAA Guidance, Navigation and Control Conference*, vol. 3 (1991), pp. 1680.
29. R. N. Lea, I. Chowdhary, Y. K. Jani, and H. Shehadeh, "Design and performance of the fuzzy tracking controller in software simulation," in *Proceedings of the FUZZ-IEEE '92* (1992).
30. R. N. Lea and Y. Jani, "Design and performance comparison of fuzzy logic based tracking controllers," presented at AIAA Space Program and Technologies Conference, Huntsville, Alabama, March 1992.

31. L. P. Tran and J. Weiss, "Fuzzy control system for a remote focusing microscope," in *Proceedings of the Fifth Annual Workshop on Space Operations Automation and Robotics* (1991).
32. M. F. Lembeck, "TES/CRIM fuzzy logic thermal control system," presentation to Software Technology Branch/JSC, March 1992.
33. J. A. Freeman and D. Skapura, *Neural Networks Algorithms, Applications and Programming Techniques* (Addison-Wesley, Reading, MA, 1991).
34. P. D. Wasserman, *Neural Computing Theory and Practice* (Van Nostrand Reinhold, New York, 1989).
35. J. C. Bezdek and S. K. Pal, *Fuzzy Models For Pattern Recognition, Methods That Search for Structures in Data* (IEEE Press, 1992).
36. J. A. Villarreal and R. O. Shelton, "A space–time neural network," *International Journal of Approximate Reasoning* **6** no. 2, 133 (1992).
37. R. Lea, J. Villarreal, C. Copeland, and Y. Jani, "Application of space time neural networks to detect tether skiprope phenomenon in space operations," in *Proceedings of AIAA Guidance, Navigation and Control Conference*, vol. 1 (1992), p. 428.
38. R. N. Lea, J. Villarreal, C. Copeland, and Y. Jani, "Learning characteristics of a space–time neural network as a tether skiprope observer," in *Proceedings NAFIPS '92, NASA Conference Publication 10112*, vol. 1 (1992), p. 154.
39. R. N. Lea, J. Villarreal, C. Copeland, and Y. Jani, "Space time neural networks for tether operations in space," in *Proceedings of the Third International Workshop on Neural Networks and Fuzzy Logic, NASA Conference Publication 10111*, vol. 1 (1992), p. 127.
40. C. C. Lee and H. R. Berenji, "An intelligent controller based on approximate reasoning and reinforcement learning," in *Proceedings of IEEE International Symposium on Intelligent Control* (1989).
41. H. R. Berenji, "Strategy learning in fuzzy logic control," in *Proceeding of North American Fuzzy Information Processing Society Workshop* (1991).
42. H. R. Berenji, R. N. Lea, and Y. Jani, "Fuzzy logic controller with reinforcement learning for proximity operations and docking," in *Proc. 5th IEEE Int. Symposium on Intelligent Control*, vol. 2 (1990), p. 903.
43. C. W. Anderson, "Learning and problem solving with multilayer connectionist systems," Ph. D. thesis, University of Massachusetts, 1986.
44. H. R. Berenji, "An architecture for designing fuzzy controllers using neural networks," *International Journal of Approximate Reasoning* **6(2)** 267 (1992).
45. Y. Jani, "Application of fuzzy logic — Neural network based reinforcement learning to proximity and docking operations: Attitude control results," report no. 2 for the Research Activity no. AR.06 under Cooperative Agreement NCC 9-16, The Research Institute for Computing and Information Systems (RICIS), University of Houston at Clear Lake, August 1992.
46. Y. Jani, "Application of fuzzy logic — Neural network based reinforcement learning to proximity and docking operations: Translational controller results," report no. 3 for the Research Activity No. AR.06 under Cooperative Agreement NCC

9-16, The Research Institute for Computing and Information Systems (RICIS), University of Houston at Clear Lake, November 1992.
47. H. R. Berenji, R. N. Lea, and Y. Jani, "Approximate reasoning based learning and control for proximity operations and docking in space," in *Proceedings of AIAA Guidance, Navigation and Control Conference*, vol. 3 (1991), p. 1707.
48. H. R. Berenji, R. N. Lea, Y. Jani, A. Malkani, P. Khedkar, and J. Hoblit, "Space shuttle attitude control by reinforcement learning and fuzzy logic," in *Proceedings of FUZZ-IEEE '93 Conference*, vol. 2 (1993), p. 1396.
49. H. R. Berenji and P. Khedkar, "Learning and tuning fuzzy logic controllers through reinforcements," *IEEE Transactions on Neural Networks* **3** no. 5, (1992).
50. G. McIntire and J. Villarreal, "Design of a neural network simulator on a transputer array," in *First Annual Workshop on Space Operations Automation and Robotics (SOAR 87)*, NASA CP-2491, pp. 111–119 (1987).
51. R. O. Shelton, "A difference optimized training scheme for the neocognitron neural network," in *Proceedings of Instrument Society of America Conference* (1988).
52. Y. Jani, R. Steele, L. Carreno, and T. Bogot, "FuzzyCLIPS final report," in *NASA Small Business Innovative Research Report* (1993).

# REACTIVE FUZZY CONTROL OF AUTONOMOUS ROBOTS

ENRIQUE H. RUSPINI
*Artificial Intelligence Center, SRI International*
*Palo Alto, California 94025, USA*
*E-mail: ruspini@ai.sri.com*

The control of complex systems in uncertain environments requires the development of regulators that, beyond adapting to changing operational circumstances, are capable of determining the attainability of various, possibly conflicting, goals and of specifying actions for their achievement. The task of controlling these systems is further complicated by additional problems caused by difficulties in the observation of their state and that of their environment. These complexities demand the utilization of intelligent approaches that are not only capable of reasoning about the state of the world and about the ability of the system to attain its goals but also of doing these deliberations on the basis of imprecise, vague, and uncertain information.

The ability of the concepts and tools of fuzzy logic to represent and manipulate imprecise and vague models and observations, combined with their capabilities to handle elastic system constraints and goals makes the methodology attractive as an effective approach for the treatment of these problems. In this paper we present such an approach, which is based on the notion of *behaviors* — i.e., sequences of system actions intended to attain both explicit goals, or to react, in a robust manner, to unforeseen circumstances. The approach is discussed in the context of an application to the control of an autonomous mobile robot operating in an indoor office environment.

Both purposive and reactive behaviors are controlled by computational constructs called *control structures*, which utilize sets of if-then fuzzy rules to compute *desirability measures*. Desirability measures compute the utility of each potential control action from the viewpoint of multiple, possibly conflicting, goals. The actions being regulated include a variety of motion and perception actions that promote attainment of these objectives. From such a viewpoint, the controller may be regarded as a tool that allocates system resources to various behaviors on the basis of deliberations about the current operational context. Desirability measures corresponding to various objectives are combined by means of explicit, context-dependent, metarules in order to determine the actual control that is sent to the mobile device.

Results of application of these ideas to the development of an actual controller are illustrated with examples from experiments with SRI's Artificial Intelligence robot, Flakey.

## 1. Introduction

The effective regulation of a wide variety of complex systems requires the synthesis of control strategies that, beyond attaining specific performance goals, must also respond to unexpected changes in their operational environment. The conditions of imprecision and vagueness under which these systems operate and must be observed require the production of robust, flexible, controllers that are capable of responding to unforeseen circumstances while still seeking to attain explicit, designer-specified, objectives.

In other words, any controller capable of effectively governing the behavior of this type of system must rely on procedures that seek to attain both *purposive* objectives (explicitly specified by designers) and *reactive* goals (implicitly required by the need to maintain continued, effective, performance).

Furthermore, the controller must be capable of smoothly activating, deactivating, and transferring between the diverse behaviors required to attain either purposive or reactive goals. The control strategy must also be capable of *blending* on the basis of the relative importance, in the context of the existing situation, of the various associated objectives. In addition, many applications require the consideration of potentially conflicting objectives attainable to diverse degrees. The controller must, therefore, be capable of deliberating, in real time, about the attainability of various objectives and about the actions that may be taken to reach them.

This paper presents a fuzzy-logic-based approach for the reactive control of systems operating in uncertain environments. The presentation is made in the context of the problem of planning and controlling the perception and motion actions of an autonomous mobile robot operating in an indoor-office environment. The controllers developed by means of this approach are capable of governing a variety of purposive and reactive behaviors seeking to blend their characteristics so as to attain — to the best possible degree — each of the multiple objectives.

In applications such as those exemplified by autonomous robot control, multiple behaviors may be active at any time, each being activated to a degree determined both by the importance of the goal and by the current operational context. Unlike approaches based on the optimization of global performance measures, deliberations about the importance of these decision factors are based on *metarules* that are explicitly specified by the control-system designer. Reliance on rule-based approaches both to specify single behaviors and to describe their context-dependent importance facilitates, in addition, the explanation of the rationale leading to specific control recommendations.

Fuzzy logic provides an attractive framework for the development of reactive controllers because of its ability to express *elastic* restrictions on system and control behavior[1] while being also capable of representing imprecise and vague system knowledge. Furthermore, interpretations of fuzzy logic as a logic of utility and similarity[11,12] simplifies either the explicit description of preferences between goals or that of the procedures that mediate between different objectives.

The fuzzy-logic techniques employed by our approach to blend multiple behaviors on the basis of explicit, user-specified, trade-off considerations about objective importance are based on the concept of *control structure.* Control structures are special constructs that produce possibility distributions measuring the desirability of each possible control action as a function of system state.

We discuss briefly the semantic bases of our approach in Sec. 2, which is presented in Sec. 3. Results of the application of this approach to the design of a mobile robot controller are presented in Sec. 4.[a]

## 2. Semantic Bases of the Approach

Our approach is based on semantic models that interpret fuzzy logic as a formalism concerned with notions of utility and similarity.[12,13] In particular, our utilitarian view of fuzzy logic was strongly influenced by seminal ideas of Bellman and Zadeh[1] and by the "logics of preference" of Rescher.[7] In these views, the truth value (measured by a number between 0 and 1) is regarded as a measure of the desirability of a proposition $p$ coming about, or, in other words, the degree by which $p$ is a "good thing."

Our formulation extends the original ideas of Bellman and Zadeh proposing specific procedures for the representation of multiple objectives and for the blending of the behaviors that seek them. A significant aspect of the methodology is the ability to explicit represent the context-dependent importance of

[a]Application of this fuzzy logic approach to the development of a robot motion controller is the result of the collaboration, at the Artificial Intelligence Center of SRI International, of N. Helft, K. Konolige, D. Musto, K. Myers, A. Saffiotti, L. Wesley, and the author.

goals or to declare the rules that must be employed to mediate between multiple objectives.

Our formulation is also different from that proposed by Rescher in that each of the logical expressions utilized to model system characteristics or behavior in our scheme is associated with a measure that constrains the value of the utility of each solution as measured from the perspective of that constraint. Rescher's approach, on the other hand, relies on global measures of adequacy based on averages of individual desirability values — an assumption that unnecessarily forces the conclusion that those measures have the properties of probability distributions.

Our approach also integrates epistemic aspects by introduction of modalities that allow the representation of ignorance about the utility of control/decision choices or about the consequences of certain assumptions. Finally, our approach permits the context-dependence of the importance of goals[10] without demanding specification of numerical weights or numerical measures of relative importance.

### 2.1. *Desirability measures*

In our representation scheme, based on the notion of elastic constraint of Bellman and Zadeh, a number between 0 and 1 is assigned to every pair composed of a potential solution and a system/control constraint. These numbers, measuring the desirability of each solution from the perspective of the constraint, model either the desirability of a particular state of affairs, or the adequacy of representations of approximate knowledge about the system. In the case of the latter type of knowledge, these values bound the cost of the potential errors that may be incurred if the system does not satisfy the assumption stated by that particular model statement.

Alternatively, we may think of each constraint (or model statement) as a function that assigns to each possible solution, or *possible world* in our semantic-model terminology, a number that measures its degree of compliance with the restriction. Solutions that fully meet the constraint are assigned a desirability value of 1 while those that must not be considered from such a viewpoint receive a measure equal to 0. Intermediate values between 0 and 1 measure partial compliance.

Formally, the *desirability measures* that express each constraint are functions that map the set $\mathcal{U}$ of all potential problem solutions (or *universe of discourse*) into the unit interval [0, 1] of the real line, i.e., they are possibility distributions on $\mathcal{U}$. The number $\mathbf{D}(w)$ associated with the generalized constraint $\mathbf{D}$ and the solution $w$ may be also thought of as the truth value of the proposition "The solution $w$ is satisfactory from the viewpoint of $\mathbf{D}$."

### 2.2. *Combination of desirability measures*

In order to specify complex constraints by means of a logical-representation scheme with adequate computational properties, it is necessary to provide a formalism that allows the derivation of the truth value of complex statements from those of their component elements.

In system-analysis problems, for example, it is often required that the system satisfy several generalized constraints $\mathbf{D}_1, \mathbf{D}_2, \ldots, \mathbf{D}_n$, or, equivalently that it meets their conjunction

$$\mathbf{D}_1 \wedge \mathbf{D}_2 \wedge \ldots \wedge \mathbf{D}_n .$$

In fuzzy logic, the truth value of the conjunction $\mathbf{D} \wedge \mathbf{D}'$ of the desirability functions $\mathbf{D}$ and $\mathbf{D}'$ is computed by means of functions $\circledast$, called *triangular norms*,[5] that model the desirability of the conjunction of two desirability measures:

$$(\mathbf{D} \wedge \mathbf{D}')(w) = \mathbf{D}(w) \circledast \mathbf{D}'(w) , \quad w \text{ in } \mathcal{U} .$$

Correspondingly, the disjunction of of two desirabilities $\mathbf{D}$ and $\mathbf{D}'$ is expressed by

$$(\mathbf{D} \vee \mathbf{D}')(w) = \mathbf{D}(w) \oplus \mathbf{D}'(w) , \quad w \text{ in } \mathcal{U} ,$$

where $\oplus$ is a function known as *triangular conorm.*

*Complementation* functions $\sim$ from $[0,1]$ into $[0,1]$, which are nonincreasing and satisfy the constraints $\sim 0 = 1$ and $\sim 1 = 0$, generalize the logical notion of negation.

In our approach, we represent the logical implication operator $\rightarrow$ by the *pseudoinverse* or *residuation* operator $\oslash$, i.e.,

$$(\mathbf{D} \rightarrow \mathbf{D}')(w) = \mathbf{D}'(w) \oslash \mathbf{D}(w) ,$$

where the pseudoinverse $\oslash$ of a T-norm $\circledast$ is defined by the expression

$$a \oslash b = \sup \{ c : \ b \circledast c \leq a \} .$$

The importance of the residuation operation and its role in the tautology

$$(\mathbf{D} \circledast (\mathbf{D} \rightarrow \mathbf{D}')) \rightarrow \mathbf{D}' ,$$

which generalizes the classical modus ponens, have been pointed out by Trillas and Valverde.[16] We have also chosen this particular form of modeling the implication operator over other alternatives because of its ability of properly gauging the extent by which the truth value of a rule consequent exceeds that of its antecedent.

Implication operators are of central importance in any fuzzy-logic-based inferential formalism because of their function as propagators of the degree of matching between the current state $S$ of the system and the antecedent $A$ of the rule $R \equiv A \rightarrow B$ into an estimate of the truth value of $B$.

## 3. Control Structures

The concept of *control structure*[9] was originally developed in the context of autonomous mobile agent applications as an extension of the intuitive notion of landmark or place.[6] Informally, control structures are the computational agents that quantify, within the vicinity of perceivable environmental features, the desirability of potential control actions as promoters of specific goals. When trying to "reach the end of a corridor," for example, motions that keep the robot parallel and at a safe distance from a wall are preferred to those that would place the robot in danger of collision with the wall or that do not result in significant advances along the desired direction.

While most of the control structures implemented in our early robotic studies were concerned with assessing the benefits of *motion* actions, these constructs were conceived, however, as general mechanisms to allocate resources so as to attain various goals, including those associated with perception and communication. For example, control structures may activate, in certain situations, a variety of sensors and motion actions to better determine the location of the mobile agent.

The overall architecture of our reactive controllers is hierarchical[3] with layers of regulating functions controlling simple behaviors at lower levels and mediating their interactions. Our approach — in addition to considering constraints, observations, and system knowledge as *elastic* constraints — also regards the various behaviors capable of being induced by control structures as conditions capable of being achieved to various degrees. Motion behaviors, for example, may be executed at different levels of performance quality ranging continuously from ideal to completely unsatisfactory.

Initially, knowledge about the state of the system, its laws of behavior, and its operational goals are used to produce a *plan* to attain such objectives.

This plan is not, however, a specification of a sequence of actions but, rather, a collection of interrelated control structures that act as instructions for a fuzzy inferential machine. This fuzzy machine computes the various desirability measures that the controller combines to determine the actual actions sent to the plant.

In our robotics application we have relied upon traditional AI techniques to develop such plans. These techniques may also be used to generate several plans, which may be ranked numerically by means of fuzzy-logic-based techniques that measure the extent by which the plan promotes the desired goals. This plan-generation approach may be also applied in a hierarchical fashion to produce lower-level plans having simpler control structures as components.

Control structures, which are input — together with the estimates of the current state — to the fuzzy inferencer, are usually specified by means of if-then fuzzy rules. Two major types of rules are specified as part of a control structure:

1. *Control rules* of the form "If the state is $A$, then the control is $B$," where $A$ and $B$ are fuzzy sets (of state and controls, respectively). These rules induce behaviors that seek attainment of a single goal or subgoal.

2. *Contextual metarules* of the form "If the state is $S$, then activate (or deactivate) the control structure $S$," which permit measurement of the importance of activating a rule in a given context. These rules, which characterize the applicability of the goal associated with the control structure in a given context, are utilized to mediate between conflicting objectives. Contextual metarules are exemplified by commands such as "Do not start following wall until you are in the vicinity of point $P$," or "If the load is balanced, then proceed along wall." This aspect of our approach was inspired by the techniques employed by Berenji[2] to deal with multiple conflicting goals in control problems.

Rules in the first class are utilized to compute the desirability of each potential control action whenever the operational context is such that the control structure is fully activated while rules in the second class determine the level of activation of the control structure as a function of that context. Multiple control structures may be active at any point in time, each seeking a specific goal such as "following a wall" or "avoiding an obstacle."

The degree of activation of each control structure is determined by metarules in the second class, which effectively *blend* each of the active behaviors and choose the course of action that best satisfies them. The degree of satisfaction is measured by the extent by which the fuzzy metarule antecedents — defining the context of applicability — match the system state.

Control structures may be activated to promote both *purposive* behaviors leading to goals such as "reach the end of wall A," or *reactive* behaviors, such as "avoid an obstacle" or "confirm your location." The inferential engine determines, on the basis of the contextual metarules, the applicability of each control structure blending its control recommendations on the basis of its assessment of their relative importance. In the immediate vicinity of an obstacle, for example, reactive control structures intended to prevent collisions prevail over those seeking purposeful behaviors. When that danger is reduced, the reactive structure is gradually deactivated while other active structures are given more weight.

The fuzzy inferencer implements both knowledge propagation (*generalized modus ponens*) and summarization (*defuzzification*) procedures to determine desirability distributions and to choose recommended control actions, respectively. A detailed discussion of the functions utilized in our mobile agent experiments is given in several papers by Saffiotti *et al.*[14,15]

## 4. Experiments with a Mobile Robot

The problems associated with the planning and control of an autonomous mobile agent exemplify the issues leading to the development of the methodology discussed in the previous sections of this paper. In this section, we present results of its application to the control of the actions of Flakey, the autonomous mobile robot platform of the Artificial Intelligence Center (AIC) of SRI International.[14,15] The controller developed in this effort was evaluated in in-house experiments and at the First International Robotics competition of the American Association for Artificial Intelligence.[4]

The AIC mobile robot is a custom-built platform, with a height of 1 meter and a diameter of 0.6 meters. The robot moves and turns using two independently-powered wheels. The maximum linear velocity of the robot is about .5 meters/sec. The mobile agent sensors include sonars, wheel encoders, and a video camera. Flakey also has several on-board computers dedicated to the support of various motion, reasoning, perception, and communication tasks.

A two-layered architecture based on the previously discussed ideas was implemented to develop a planner for robot activities that decomposed user-specified goals into a set of rule-based control structures. A fuzzy inference engine that activated and deactivated these structures was then employed to control the execution of multiple reactive behaviors. Plans and control rules were both produced on the basis of knowledge of the robot characteristics and of approximate maps of its operational environment.

Control structures, associated with operational landmarks, were implemented by means of fuzzy if-then rules of the form

$$\text{IF } A_i \text{ THEN } C_i\,,$$

where $A_i$ is a fuzzy predicate defined over state variables and where $C_i$ is a fuzzy set of control values. The result of application of these rules is an approximation of a desirability measure associated with a specific behavior.

Collision-avoidance behaviors, intended to keep the robot safely away from occupied areas may include, for example, the following rule:

```
IF obstacle-close-on-right AND NOT obstacle-on-left,
THEN turn slow-left
```

When a rule set $\mathcal{R} = \{R_1, \ldots, R_n\}$ is active in state $s$, the controller, computes the desirability distribution

$$Des_{\mathcal{R}}(s,c) = (A_1(s) \otimes C_1(c)) \oplus \ldots \oplus (A_n(s) \otimes C_1(c))\,,$$

utilizing the generalized modus ponens. The centroid defuzzification method is then used to compute the actual control sent to the plant.

In our robotics experiments, control structures are defined by triples

$$S = \langle \mathrm{A}, Des, \mathrm{C} \rangle\ ,$$

where $A$ is an object in the robot's workspace, *Des* is a function that specifies the desirability of potential control actions as a function of the relations between the state of the robot and the characteristics of that object, and $C$ is a fuzzy predicate describing the *context* where the structure is applied. Typically, desirability measures are computed using four to eight rules based on four to twenty predicates.

The control structure

$$\langle \mathrm{CP}, \text{Go-To-CP}, \textit{near}(\mathrm{CP}) \rangle$$

where CP is a "control-point", i.e. a fuzzy location in state space, the associated measure GO-TO-CP gauges the adequacy of potential control actions in terms of their ability to guide the robot from its present state to CP, and the context *near*(CP) specifies where this control structure should be activated, was utilized, for example, to specify a behavior steering the robot to CP.

Several control structures may be active at any time, each inducing its corresponding associated behavior. The actions recommended by each active structure are then blended to generate a context-dependent combined desirability measure. In our robotics application, when $n$ control structures $\{S_1, \ldots, S_n\}$, with corresponding desirability measures $\{Des_1, \ldots, Des_n\}$ and contexts $\{C_1, \ldots, C_n\}$ were active, they were combined using the expression

$$Des(s,c) = (Des_1(s,c) \otimes C_1(s)) \oplus \ldots \oplus (Des_n(s,c) \otimes C_n(s)) .$$

The actual control value was then chosen using centroid defuzzification.

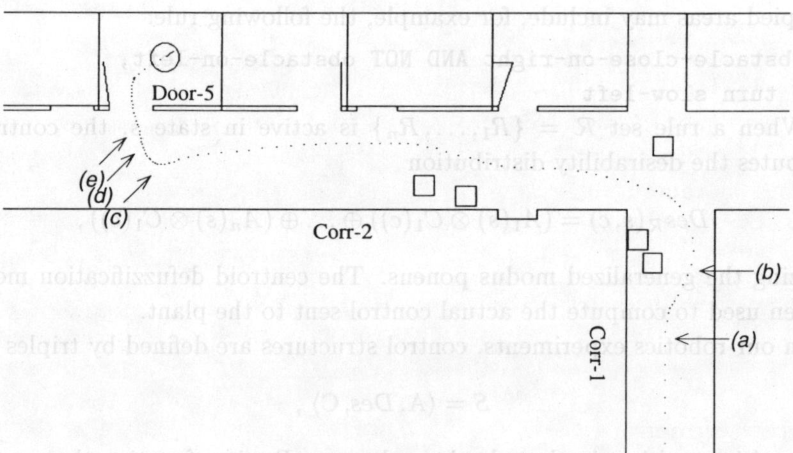


Fig. 1. Plan execution.

The actual execution of a plan to guide the robot to a specified office is illustrated in Fig. 1. The following four control structures were the most important components of that plan:

$$S1 = \langle \text{Obstacle}, \text{KEEP-OFF}, \text{near(Obstacle)} \rangle$$
$$S2 = \langle \text{Corr2}, \text{FOLLOW}, \neg\text{near(Obstacle)} \wedge \text{at(Corr2)} \wedge \neg\text{near(Corr1)} \rangle$$
$$S3 = \langle \text{Corr1}, \text{FOLLOW}, \neg\text{near(Obstacle)} \wedge \text{at(Corr1)} \wedge \neg\text{near(Door5)} \rangle$$
$$S4 = \langle \text{Door1}, \text{CROSS}, \neg\text{near(Obstacle)} \wedge \text{near(Door5)} \rangle$$

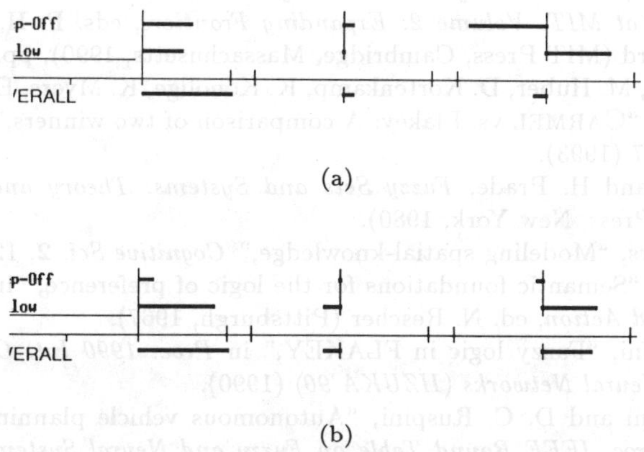


Fig. 2. Behavior blending.

While following the first hallway, for example, a purposeful behavior, FOLLOW, was combined with the KEEP-OFF behavior resulting in a trajectory that follows the wall while avoiding collisions with obstacles. The obstacle detected at point (a) caused, for example, the KEEP-OFF behavior to dominate thus leading the robot to deviate from the straight trajectory preferred by FOLLOW. As the danger of collision decreased, FOLLOW regained its importance guiding the robot from (b). Figure 2 show, for each behavior, the level of activation at points (a) and (b) for each of these two behaviors (solid lines). The combined controls are shown at the bottom of each figure.

## Acknowledgments

This work was partially supported by the U.S. Air Force Office of Scientific Research under Contract No. F49620–91–C–0060. Additional support was provided by SRI International.

## References

1. R. E. Bellman and L. A. Zadeh, "Decision-making in a fuzzy environment," *Management Science* **17**, B141–B164 (1980).
2. H. Berenji, Y.-Y. Chen, C.-C. Lee, J.-S. Jang, and S. Murugesan, "A hierarchical approach to approximate reasoning-based controllers for dynamic physical systems," in *Proc. 6th Conf. on Uncertainty in Artificial Intelligence*, eds. P. P. Bonissone and M. Henrion (Morgan-Kaufmann Publishers, 1990), pp. 362–369.

3. R. A. Brooks, "A robust-layered control system for a mobile robot," in *Artificial Intelligence at MIT. Volume 2: Expanding Frontiers*, eds. P. H. Winston and S. A. Shellard (MIT Press, Cambridge, Massachusetts, 1990), pp. 2–27.
4. C. Congdon, M. Huber, D. Kortenkamp, K. Konolige, K. Myers, E. Ruspini, and A. Saffiotti, "CARMEL vs. Flakey: A comparison of two winners," *AI Magazine* **14(1)**, 49–57 (1993).
5. D. Dubois and H. Prade, *Fuzzy Sets and Systems: Theory and Applications* (Academic Press, New York, 1980).
6. B. J. Kuipers, "Modeling spatial-knowledge," *Cognitive Sci.* **2**, 129–153 (1978).
7. N. Rescher, "Semantic foundations for the logic of preference," in *The Logic of Decision and Action*, ed. N. Rescher (Pittsburgh, 1967).
8. E. H. Ruspini, "Fuzzy logic in FLAKEY," in *Proc. 1990 Int. Conf. on Fuzzy Logic and Neural Networks* (*IIZUKA'90*) (1990).
9. E. H Ruspini and D. C. Ruspini, "Autonomous vehicle planning using fuzzy logic," in *Proc. IEEE Round Table on Fuzzy and Neural Systems and Vehicle Applications* (1991).
10. E. H. Ruspini, "On truth, utility, and similarity," *Proc. 1st Int. Conf. on Fuzzy Engineering* (1991).
11. E. H. Ruspini, "Approximate reasoning: Past, present, future," *Information Sciences* **57–58**, 297–317 (1991).
12. E. H. Ruspini, "On the semantics of fuzzy logic," *Int. J. Approximate Reasoning* **5**, 45–88 (1991).
13. E. H. Ruspini, "On truth and utility," in *Symbolic and Quantitative Approaches to Uncertainty*, eds. R. Kruse and P. Siegel (Springer, 1991).
14. E. Ruspini, A. Saffiotti, and K. Konolige, "Progress in research on autonomous vehicle motion planning," in *Industrial Applications of Fuzzy Logic*, eds. J. Yen and R. Langari (1995).
15. A. Saffiotti, K. Konolige, and E. H. Ruspini, "A multivalued logic approach to integrating planning and control," *Artificial Intelligence* (1995).
16. E. Trillas and L. Valverde, "On mode and implication in approximate reasoning," in *Approximate Reasoning and Expert Systems*, eds. M. M. Gupta, A. Kandel, W. Bandler, J. B. Kiszka (North Holland, Amsterdam, 1985), pp. 157–166.